Contributions to Environmental Sciences & Innovative Business Technology

Contributions to Environmental Sciences & Innovative Business Technology (CESIBT) is an interdisciplinary series of peer-reviewed books dedicated to addressing emerging research trends relevant to the interplay between Environmental Sciences, Innovation, and Business Technology in their broadest sense. This series constitutes a comprehensive up-to-date interdisciplinary reference that develops integrated concepts for sustainability and discusses the emerging trends and practices that will define the future of these disciplines.

This series publishes the latest developments and research in the various areas of Environmental Sciences, Innovation, and Business Technology, combined with scientific quality and timeliness. It encompasses the theoretical, practical, and methodological aspects of all branches of these scientific disciplines embedded in the fields of Environmental Sciences, Innovation, and Business Technology.

The series also draws on the best research papers from EuroMid Academy of Business and Technology (EMABT) and other international conferences to foster the creation and development of sustainable solutions for local and international organizations worldwide. By including interdisciplinary contributions, this series introduces innovative tools that can best support and shape both the economical and sustainability agenda for the welfare of all countries, through better use of data, a more effective organization, and global, local, and individual work. The series can also present new case studies in real-world settings offering solid examples of recent innovations and business technology with special consideration for resolving environmental issues in different regions of the world.

The series can be beneficial to researchers, instructors, practitioners, consultants, and industrial experts, in addition to governments from around the world. Published in collaboration with EMABT, the Springer CESIBT series will bring together the latest research that addresses key challenges and issues in the domain of Environmental Sciences & Innovative Business Technology for sustainable development.

Ahmad Rafiki · Léo-Paul Dana ·
Muhammad Dharma Tuah Putra Nasution
Editors

Open Innovation in Small Business

Creating Values for Sustainability

Springer

Editors
Ahmad Rafiki
Universitas Medan Area
Medan, Indonesia

Léo-Paul Dana
ICD Business School of Paris, Groupe IGS
Halifax, NS, Canada

Muhammad Dharma Tuah Putra Nasution
Universitas Pembangunan Panca Budi
Medan, Indonesia

ISSN 2731-8303 ISSN 2731-8311 (electronic)
Contributions to Environmental Sciences & Innovative Business Technology
ISBN 978-981-99-5141-3 ISBN 978-981-99-5142-0 (eBook)
https://doi.org/10.1007/978-981-99-5142-0

This Springer imprint is published by the registered company Springer Nature Singapore Pte Ltd.
The registered company address is: 152 Beach Road, #21-01/04 Gateway East, Singapore 189721, Singapore

Paper in this product is recyclable.

Contents

The Role of Business Model Innovation (BMI) in Social Enterprises During Pandemic COVID-19 in Indonesia: A Case of Islamic Boarding Schools' Business Units

1

Yudi Ahmad Faisal, Runita Arum Kanti, Amelia Rizki Alamanda, and Ardi Apriliadi

Abstract

The business units of Islamic boarding schools known as *pesantren* in local terminology as part of the engines of the local economy in Indonesia are among the social enterprises hit hardest by the COVID-19 pandemic. Theoretical and practical observations in the Indonesian social enterprise sectors indicate that business model innovation (BMI) proves to overcome and recover the business during the pandemic whereby they have successfully transformed their business models. This study analyzes why and how changes in this BMI are conducted by those social enterprises by conducting case studies of three Islamic boarding schools' business units. To explore the questions, this study conducted in-depth interviews with administrators, especially religious leaders (*kyai*) and business managers for each case, which we triangulated with secondary data for analysis. The results show that BMI has been employed during the crisis to create new sources of income and ensure sufficient levels of liquidity, with the key role of business unit managers and staff with the support of religious leaders. This research contributes to the scarcity of innovation-based model in the context of social business in developing countries like Indonesia.

Y. A. Faisal (✉) · A. Apriliadi
Management and Business Study Center (LMFEB), Faculty of Economics and Business, Universitas Padjadjaran, Bandung, Indonesia
e-mail: yudi.ahmad@unpad.ac.id

R. A. Kanti · A. R. Alamanda
Department of Accountancy, Faculty of Economics and Business, Universitas Padjadjaran, Bandung, Indonesia

A. Rafiki et al. (eds.), *Open Innovation in Small Business*, Contributions to Environmental Sciences & Innovative Business Technology,
https://doi.org/10.1007/978-981-99-5142-0_1

Keywords

Islamic boarding schools (*pesantren*) • Social enterprises • COVID-19 • Business model innovation

1.1 Introduction

The COVID-19 pandemic is a challenge for all stakeholders in the economy and business world including governments, communities, and companies (Clark et al. 2020). This pandemic has spawned crises such as strategic challenges, and associated disruptions and production capacity, increasing financial uncertainty and instability, thereby impacting reassessment, and restructuring of operations for both public and private organizations (Cortez and Johnston 2020). This challenge is also experienced in Indonesian business, a country dominated by Muslims that reaches 87.18% from a population of 232.5 million people (Global Islamic Economy Report 2018).

The pandemic has significant impacts on Indonesian social enterprises particularly business units of Islamic boarding schools known as *pesantren*. These institutions existed in Indonesia long before the modern formation of the country. Historically, there is no exact year of their formation. However, there is historical data that it existed since the seventeenth century, particularly in East Java. *Pesantren* traditionally has three functions, namely, education, Islamic preaching, and community empowerment including economic and business activities. Its business represents micro, small, and medium enterprises, particularly under a cooperative scheme. According to the data of the Indonesian Ministry of Religious Affairs, in 2018, the archipelago has 28,961 *pesantren* in which 44.2% of them have economic potentials.[1] The pandemic has had an impact on all components and value chains of their business units. These include declining sales, difficulties in accessing capital, hampered distribution, and difficulties in obtaining raw materials. On a large scale, the decline of their business sectors, which is one of the pillars supporting the local economy, could impact the sustainability of the institution itself.

This study comprehends why and how successful recovery efforts through Business Model Innovation (BMI) by conducting case studies in three *pesantren*'s business units in West Java Province whereby 31.8% of their population existed. As stated by Kraus et al. (2020), BMI promises a strong response to the COVID-19 crisis for business sectors including micro, small, and medium, and social enterprises. From this perspective, the paper sets questions as follows.

Can BMI be used to overcome the COVID-19 crisis in the social business including Islamic boarding schools' business units?

[1] Ministry of Religious Affairs, the Republic of Indonesia, 2016, and Department of Sharia Economics and Finance (DEKS), Central Bank of Indonesia, 2017.

What is the driving force behind BMI in the Islamic boarding schools' business units during the COVID-19 crisis, and what is the roles of the manager of those institutions?.

This study employs data triangulation with findings from interviews of each Islamic boarding school element (religious leaders, teachers, business managers, and staff) and available secondary data. The results empirically prove whether BMI plays an active role as a strategic response of this institution in dealing with the pandemic crisis. The findings from this study contribute to social research, especially themes related to social businesses and how they transform their business models to survive during the crisis. In addition, this research can also fill out the research gap, especially on factors that increase and hinder the implementation of BMI. This result shows how BMI practices in developing countries like Indonesia.

1.2 Literature Review

1.2.1 Pesantren: Its Term and Functions

Pesantren refers to traditional Islamic educational schools or institutions and existed for decades before the establishment of modern Indonesia (Yasmani 2002). There are four main components of this institution. These include *kyai* (an Islamic religious leader), *santri* (students), an educational system, and *pemondokan/ asrama* (dormitory). These components create a unique social capital derived from physical characteristics, perspectives on life, ideology, and traditional education but prioritizing scientific traditions, and distinctiveness in conveying the learning process (Goeztav 2012). This institution plays a role in teaching the values of Islamic morality to students and society in general through various Islamic preaching media. These roles are noble traditions that are timeless (Syam 2009). In general, these noble values are derived from Islamic teachings, namely, the Qur'an and Sunnah (traditions of the Prophet Muhammad) (Mcshane and Von Glinow 2003). In its traditional form, *kyai* is a central figure who has the authority to regulate institutional life and is a role model for students and the local community (Marhumah 2011; Dhofier 2011). A *kyai* is the founder, or descendant of the founders of it. Therefore, it is only natural that its role is closely related to the culture of the institution. In addition, this religious leader and their students have a strong emotional connection. This connection creates social and commercial links. They, for instance, set up policy of not receiving financial assistance from central and local governments. This means that they must earn own income. They will use internal sources including the number and the ability of students in creating commercial products for the consumption of the *pesantren* or for external markets.

Considering the pivotal function of this school or institution in the country, the government enacted the *Pesantren* Law No. 18 Year 2019 (Republic of Indonesia 2019). According to this law, this institution must have three basic functions, namely, education, Islamic preaching, and community empowerment. The first and second functions are closely related to the primary functions that have existed since the beginning of its establishment. Meanwhile, the third function is also related to economic and business functions to generate revenue either to support their main activities or to generate social programs for community empowerment.

1.2.2 Pesantren Business Units as Social Enterprises

To enhance community empowerment, Islamic boarding schools establish a business unit. Portions of the profits earned from the business will support the main activities of them. It is also common for these schools to open a shop selling daily staple products for its students and the teachers' families. Students may also be involved in business unit management. They also provide daily consumed goods to their surrounding community. The spirit of their business is commonly set out in accordance with the principles of brotherhood, cooperation, and partnership, as well as service of members and society that is embedded in the institution. Moreover, cooperative becomes a considerable option as a form of business organization and entrepreneurship developed by these schools. Figure 1.1 depicts their business model.

Islamic boarding schools have started to introduce economic aspects, the positive side of this development is the birth of entrepreneurship so that they could independently support their main functions. This becomes an alternative driver of social change and development that is centered on the community itself, as well as a center for value-oriented development (Karni 2009).

The Islamic boarding schools' business unit is a form of social enterprise/ business (Yunus et al. 2010). Companies with a social mission are designed like ordinary businesses, but not all the generated profits are for re-investment or dividends of its owners but are also directed to various activities that have implications for social life. In addition, the terms social enterprise and social business are often interchanged.

The term social business refers to organizational entities whose main goal is to solve social and environmental issues. This entity uses a business approach so that it can survive and grow in realizing this social mission (Peerally et al. 2019). This

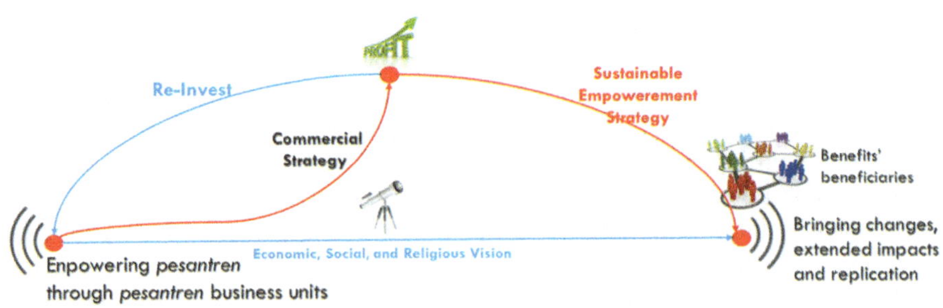

Fig. 1.1 A *pesantren* business model. *Source* Authors' proposed model (2022)

concept is a new hope for people who have been living in misery and poverty so that they can benefit from this social mission, including increasing their standard and life expectancy (Hasan 2017). To gain insight into the spectrum of social enterprise, here are variants of its definition (Table 1.1).

From the above definition of social enterprises, *pesantren* business units as part of social enterprises have the following characteristics (Table 1.2).

Table 1.1 Literature reviews on social enterprises or business

Authors (year)	Definition
OECD (1999)	Any private activity conducted in the public interest is conducted with an entrepreneurial strategy, but the aim is not to maximize profits but to achieve certain economic and social goals
Nicholls (2005)	An entity that is run in a professional, innovative, and sustainable manner aimed at resolving the crisis caused by the malfunctioning of the free market system. Social entrepreneurs develop various business models, both for-profit and non-profit, and provide a wider social impact
Haugh (2005)	A collective term for various organizations that trade for social causes. They adopt different legal formats but share the same principle of pursuing business-based solutions to achieve social goals, and reinvesting surpluses for the benefit of society
Mair and Marti (2006)	A unit that creates innovative social values that are conducted in commercial, social businesses, as well as government organs
Sullivan Mort et al. (2003)	Part of Non-Government Organizations, and not a pure business driven by a social mission
Dart (2004)	A modern form of non-profit organization by combining commercial elements to achieve radical social goals
Simonov et al. (2014)	Reflects an innovative model of non-profit sector

Table 1.2 *Pesantren* business units as social enterprises

No	Dimensions
1	The emphasize on revenue generation and sustainable society empowerment strategies
2	The common organization types are non-profits such as foundation, and profits such as cooperatives, limited liabilities companies, or *Commanditaire Vennootschap*
3	Focus on partially profitable and non-profit activities
4	Beneficiaries are mostly involved
5	Strategies are developed with the help of internal and external actors
6	Its main target is to achieve religious, social, and economic missions
7	The context is on Islamic economy and legal frameworks are being developed

Source Authors (2022)

1.2.2.1 Business Model Innovation (BMI)

In the context of a crisis, Business Model Innovation (BMI) is important for the company's business processes. Theoretically, BMI consists of two crucial elements, namely, (i) value proposition: target segment, product or service offering, and revenue model; and (ii) operating model: value chain, cost model, and organization (Lindgardt et al. 2012). During and after the crisis, the concept of BMI became a reference in designing new sources of income for companies so that they could survive and sustain (Breier et al. 2021). This model offers a new perspective on how a company adds new activities, or integrates new ways into old activities, and changes who does activities within a company (Shahrezaei et al. 2018).

BMI must survive and succeed in a business environment that is full of challenges and constantly changing (Giesen et al. 2010). This offers customers a business model that modifies and presents new concepts in offered product or service (Mitchell and Bruckner Coles 2004). BMI refers to a company's search for new logic and designing original activity systems to create and capture value for its stakeholders (Zott et al. 2011; Foss and Saebi 2017). Furthermore, BMI is a new model business design or a modification of the existing one (Massa and Tucci 2013).

1.2.3 BMI in Social Enterprises

Literatures show that business processes change and evolve over time in different contexts. Social-based companies use business models in parallel to survive in this ever-changing environment (Gebauer et al. 2017). The concept of a business model describes the management of a company, its resources, and income (Hockerts 2015). Furthermore, business model innovation is increasingly important in the context of a comparison between one company and another where the ways of using technology, operations, ideas, and income may differ from one to another (To et al. 2019). Business model innovation is currently receiving widespread attention in the business literature, with a focus on creating revenue streams that have implications for corporate social mission plans (Saebi et al. 2019).

Value creation and propositions are important so that they can provide added value to both the company and customers through innovative processes and are driven by internal and external factors (Gebauer et al. 2017). Social enterprises are creative in selecting what business model to do in a different context (Tykkyläinen and Ritala 2021). In addition, innovation in social entrepreneurship is conducted with the aim of fulfilling the main social mission to create social value and to remain competitive.

Table 1.3 Research sampling

Case	Name of schools	Source of data	Generation (ancestors of schools' founders)	Seasonal business	Establishment
A	*Darul Amal Jampangkulon*	Two interviews	Second generation	No	1997
B	*Al Isytirok*	Two interviews	Second generation	No	2009
C	*Miftah Nurul Huda*	Two interviews	Second generation	No	2017

Source Authors (2022)

1.3 Methodology

1.3.1 Research Design

To analyze how Islamic boarding schools' business units innovate in response to the COVID-19 crisis and why these innovation efforts can be increased or hindered, this study adopted a multiple case study method. This method is very appropriate for understanding complex realities where theoretical knowledge needs to be proven empirically. From the observation of the literatures, there is still no research on BMI that is applied to social enterprises, especially these business units in Indonesia, and the context of COVID-19 has a new quality compared to other crises. However, this research on BMI is not specific to certain business sectors. Therefore, our qualitative research approach aims to extend existing theory by bridging contexts.

Using purposive or theoretical sampling, this study selected three business units of *pesantren* in West Java (see Table 1.3) that were severely affected by the pandemic but are showing signs of recovery and are therefore likely to exhibit BMI. As mentioned previously, West Java is a province in Indonesia that has the largest population of Islamic boarding schools.

1.3.2 Data Collection and Analysis

For each business unit in our sample, two interviews were conducted: one with the religious leader/teacher/students, and one with manager. The semi-structured interview format (Appendix 1) allows this study to adapt the questions to the respondent's condition. With the consent of the interviewee, the interview was recorded. Data was triangulated with publicly available information from *pesantren* profiles, business unit websites (if any), and other secondary data.

After conducting the interviews, the results of the interviews conducted by the enumerators on the respondents were copied, analyzed, coded per case or business unit, and analyzed as a whole. The next process is coding, where this research contextualizes the theory with existing data. A cross-case analysis of a sample of those business units this research compares cases and looks for common themes that can be verified in an interactive loop, to gain an in-depth understanding and find potentially generalizable insights. The reliability and validity of our findings are ensured by multiple cases, independent coding, and consolidation of the data together.

1.4 Discussion and Findings

1.4.1 Case Description

Case A

Islamic boarding school of *Darul Amal Jampangkulon* was established in 1992. Five years later, the school opened a business unit with the aim of assisting the operational activities within the school. The business includes mini markets, agriculture, animal farming, and printing services managed by nine middle managers. They have internal segment markets including students, teachers, teachers' family, and the founders' family, as well as external segment markets such as the surrounding. In 2015, they opened ten branches; seven branches in West Java, two branches in Central Java, and one branch in Yogyakarta Special Region, where the main activities of these are engaged in education, Islamic preaching, and social activities.

Case B

Islamic boarding school of *Al Isytirok* was established in 1978. Two decades later, their second generation continued as the successor and introduced a business unit in 2009 to support education and Islamic preaching activities. This business helped existing activities to be independently financed. Livestock and agriculture are the main focuses of their business with ten business unit managers. Meanwhile, the main partners (business partners/consumers) of them include restaurants and distributors (Table 1.4).

Case C

Islamic boarding school of *Miftah Nurul Huda* is the youngest among the observed, which was founded in 2005. Previously, this school required external funds from donation to meet operational needs; however, in 2017, bean sprout cultivation was established to finance their operational. In addition, the prominent level of demand for vegetables and the location of the school which is not too far from the market are the opportunities for them to grow their business unit. Currently, they have twelve employees. Suppliers and the community are the main partners for their business.

Table 1.4 Identification of business model

	Case A	Case B	Case C
Segment markets	Students, internal markets, external markets	Restaurant, distributors	Distributors, individual external markets, institutional markets
Value proposition	Modern retailers, agriculture, stockbreeding, Printing services	Stockbreeding, agriculture	Agriculture
Competitive advantage/ value creation	• Islamically permissible goods (*halal*) and hygiene (*toyyib*) • Produced by experienced and skillful students (*santri*)	• Quality control in all processes of catfish breading from seed to harvest • Processed by students who already have good skills in catfish cultivation	• Basic Ingredients (peanuts) are of excellent quality • Processed by students who already have skills in bean sprout cultivation
Revenue model/value capture	• Offline (mini market) • Online • Consignee	Offline (restaurant and collectors)	• Offline (collectors and Community) • Online (using social media through WhatsApp groups for Family, Class, Alumni, and Friends)

Source Internal Analysis (2022)

1.4.2 Key Findings

The following analysis provides insight into the above cases. Table 1.5 provides an overview of the main components of the case.

Case A

During the COVID-19 pandemic, the *pesantren* business unit did not produce usual products and services for usual customers such as students/internal community, and its surrounding area. This business unit changed its products/services according to the needs of the community during a pandemic, such as making basic food packages for those who are isolating in their own homes and strengthening the agricultural sector. At the time of the initial pandemic, this business unit relied on the revenue generation on the consignment system, because they were temporarily closed. In addition, the business unit also cooperates with agencies by providing staple food packages and lockdown or isolation packages.

Interestingly, during the pandemic that business unit did not utilize external source of funds including banks. Furthermore, they did not receive any government assistance funds. They preferred to obtain funding from the foundation and investors totaling fourteen people to finance the opening of their new business unit.

Table 1.5 Components of main cases

Factors	Case A	Case B	Case C
Business model innovation			
Changing of segment markets	No	No	No
Changing of products and services	Yes	No	No
Changing of source of income	Yes	No	No
Supporting partners			
Fund supports	No	No	No
Another supports	No	No	No
Government supports			
Subsidize, loans, and financial support	No	Yes	No
Owners/foundation supports			
Fund Supports	No	No	No

Source Internal Analysis (2022)

For trade, agriculture and animal husbandry lands are sourced from the foundation and from *waqf* (Islamic endowment) land.

The owners only provided financial assistance at the beginning of the business unit establishment, but during the pandemic, there was no financial assistance. The foundation also provided capital assistance at the beginning of the opening of the business unit, which consists of fourteen investors. They also provided building assistance and other facilities at the initial business unit operation. The development of the business unit is handed over to the Daily Management Board (BPH) of the business unit.

Case B

The *pesantren* business unit did not change its target customers during the COVID-19 pandemic, namely restaurants, and collectors. The business unit continued to produce chicken, beef, catfish, chili, cucumber, and rice. Its main selling commodity is catfish. There is no change in the method of how the business unit earns income because public consumption of agricultural and livestock products tends to be stable.

The funding is obtained purely from the foundation and grants from the local government through the *One Pesantren One Product* (OPOP) policy, a program to help developing business units of Islamic boarding schools. The business unit has won that government program. Hence, it has no obligations to other parties including banking for funding. Agricultural land and livestock are sourced from the foundation and from Islamic endowment land, in addition to the Central Bank of Indonesia's financial assistance to build a greenhouse.

The provision of capital assistance provided by the founders or foundation is only given at the beginning of the opening of the business unit. Meanwhile, other assistance, such as buildings and other facilities, is provided in the same way as capital assistance, which is given at the start of the business unit's operation. So, for the development of such a business, the responsibility is fully given to the manager.

Case C

The *pesantren* business unit has target customers such as individual buyers, companies/institutions such as restaurants and education, with a size of more than hundred students. During the COVID-19 pandemic, there is no change in target customers. The business unit continued to produce bean sprouts as their main selling product. Online marketing was strengthened during the pandemic using social media through WhatsApp groups for Family, Alumni, and Friends.

Funding for business units comes from the foundation and has never collaborated with other parties, including banks. Business land for agriculture is sourced from the foundation and from *waqf* (Islamic endowment) land. This school also received grants from the OPOP program. Regarding the aspect of funding its operations, the business unit does it independently because of sales profits.

The foundation provides capital assistance at the beginning of the opening of the business unit. In the first year of 2017, the business unit received an initial capital of five million rupiahs. The foundation provides building assistance and other facilities at the beginning of the business unit operation. Development of the business unit is handed over to the business unit itself.

1.5 Key Findings and Conclusion

1.5.1 Key Findings

Business Model Innovation (BMI) is an alternative strategy for social enterprises including *pesantren* business units, to overcome the pandemic crisis. This study found that BMI can be a game changer to give a greater impact on small businesses (see also, Foss and Saebi 2017). This study also noted strategic factors for a BMI in *pesantren* business units, namely, the existing business model of the business unit, support from religious leaders and managers, and pressure due to emergency environments. The existing revenue model of the business unit enhances the implementation of BMI when these revenue models are negatively impacted by the crisis. In cases where demand for the goods produced by the business units remains stable, BMI was not considered as a necessary strategy. In addition, business units that have a focus on specialized products, namely, business units that only sell one or two specific goods, experienced minor changes in the demands for those goods. In addition to the impact of the crisis on the revenue stream, overall pressure has also emerged as an essential criterion. The business units that have received financial support (from banks or the governments), are not threatened by burdens of financial pressure, and therefore react less proactively than others in their consideration to implement BMI. Religious leaders and managers provide

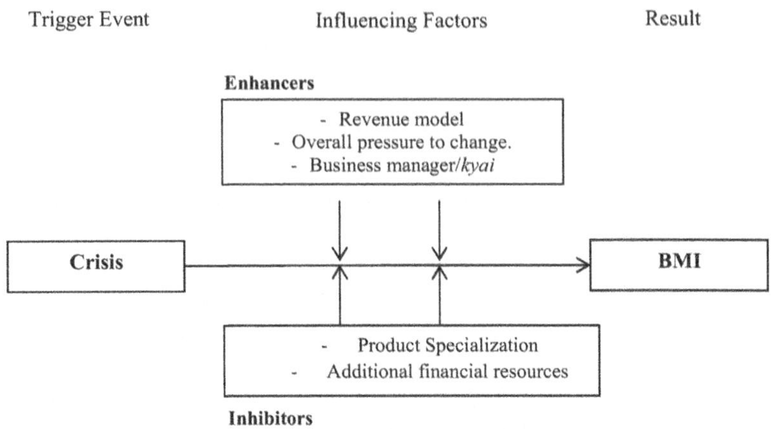

Fig. 1.2 BMI business model during crisis. *Source* Adopted and modified by the authors from Breier et al. (2021)

moral support which then encourages business units to innovate their BMI. Meanwhile, this study finds that the foundation (*yayasan*) does not give support in the form of emergency capital in the time of pandemic. From the above discussions, it can be suggested that the BMI business model during the crisis for those business units is as follows (Fig. 1.2).

Our results are in-line with the previous studies that crisis can be a driving force to change business model (Sigala 2020; Breier et al. 2021) started to initiate BMI in the social enterprises. *Pesantren* business units which were experiencing a decline in their performance can now modify their business model by implementing BMI. As we might be aware that during the crisis, the decline of market demand impacted the decrease in revenue of these businesses. This situation creates pressure on these business units to change the way they manage the business for sustainability and innovation. Enhancing factors such as business managers' and religious leaders' psychological support, revenue model, and financial pressure attract changes and give rationale to implement BMI quickly. However, during a crisis, certain business units did not suffer a shortage in product demands. These business units also received government grants which cause the business units not to adjust their BM if liquidity is already secured.

1.5.2 Theoretical Contributions

One of the contributions of this research is the importance of crisis management for *pesantren* business units during the crisis. Some researchers have initially introduced research in the context of terrorism, natural disasters, and financial crises (Anson 1999; Butler and Baum 1999). Recently, literatures showed that crisis management is also for the disease crisis. Interestingly, it can be argued that BMI can address issues of micro, small, medium, and social enterprises sectors (Kraus et al. 2020; Breier et al. 2021). This paper supports the importance of BMI in time of crisis and suggests that BMI can take a strategic role in the pandemic for social enterprise sectors. From our perspective, it shows that innovation arises from pressures and responsibility to help others. When certain pressures arise, the business unit managers become innovators as they want to improve the business performance of the *pesantren* and could execute their ideas on the backup of their religious leaders. Unless there are certain pressures for BMI, the role of religious leaders/business managers to initiate BMI is minimal.

Surprisingly, the business units of the *pesantren* which have a focus on one specific or main selling product fared better during the pandemic compared to those which have a diversified output. This may suggest that a specialization strategy may be beneficial for social enterprises in a time of crisis. Furthermore, the business unit also needs to understand of their customer needs. The current business environment imposes business players to concern on customer-based products as a recipe to develop their business model innovation as suggested by Balan-Vnuk and Balan (2015).

1.5.3 Implication for Practices

Our research focuses on recommendations for social enterprises in the form of *pesantren* business units, especially on how BMI plays a significant role for micro, small, and medium enterprises, and cooperative sectors for sustainability and survival during the pandemic crisis. The pandemic impacted negatively to business performance of most of those business units due to social distancing and lockdown policies and blocking down innovative efforts. This study also suggests that BMI can help social enterprises to modify their revenue streams during the crisis and help them to design future business strategies. The managers of social business units should actively adopt the modification of their business model toward innovative strategies. One of the important skills is digitalization which provides solutions to address the pandemic crisis through various digital channels (Clark et al. 2020). In the future, digitalization will be a driving force for those business units to increase their business performance and gain a greater external market potential. Kraus et al. (2020) explained on how modification business model addresses a crisis and sets plans for the future. Recommendations regarding product specialization can also be inferred from this study. In the case of this study, a business unit that creates a specific output has a competitive advantage in the form of a stable demand from consumers (Oftedal 2013). In our research, it is found that innovation is not only related to products and services but also business model. This business model can also be utilized through collaboration with external parties such as professional trainers. These initiatives can connect the *pesantren* with business ecosystem that improves ideas and innovations (Beritelli 2011; Kallmuenzer 2018) that can help to develop stronger resilience and recovery potential from future crises.

1.5.4 Limitation and Future Research Opportunities

The study has limitations particularly its methodology and the crisis. However, we can argue that this study is a pioneer study on the relationship of BMI in the time of crisis particularly for *pesantren* business unit sectors. The limitation of this study includes the samples that we choose through purposive sampling techniques. We assume by including a greater number of samples that generalization will be more precise. Our findings limit to social business units managed by *pesantren* in the West Java Province. We predict that BMI will be one of the important themes in discussing social enterprises in the country. This paper proposes a foundation for the next studies. Furthermore, it suggests using mixed methods with quantitative approaches. Future research should extend both the scope of types of social enterprises and the geographical area of sampling to further explore the phenomenon and expand the validity of findings.

Acknowledgements Centre of Business and Management Study (LMFEB) is a unit under the Faculty of Economics and Business, Universitas Padjadjaran. This research is funded by Internal Grant of Universitas Padjadjaran year 2021/2022.

References

Anson C (1999) Planning for peace: the role of tourism in the aftermath of violence. J Travel Res 38(1):57–61. https://doi.org/10.1177/004728759903800112

Balan-Vnuk E, Balan P (2015) Business model innovation in nonprofit social enterprises. University of Adelaide Press. https://www.ptonline.com/articles/how-to-get-better-mfi-results

Beritelli P (2011) Cooperation among prominent actors in a tourist destination. Ann Tour Res 38(2):607–629. https://doi.org/10.1016/j.annals.2010.11.015

Breier M, Kallmuenzer A, Clauss T, Gast J, Kraus S, Tiberius V (2021) The role of business model innovation in the hospitality industry during the COVID-19 crisis. Int J Hosp Manag 92:102723. https://doi.org/10.1016/j.ijhm.2020.102723

Butler RW, Baum T (1999) The tourism potential of the peace dividend. J Travel Res 38(1):24–29. https://doi.org/10.1177/004728759903800106

Clark C, Davila A, Regis M, Kraus S (2020) Predictors of COVID-19 voluntary compliance behaviors: an international investigation. Glob Transitions 2:76–82. https://doi.org/10.1016/j.glt.2020.06.003

Cortez RM, Johnston WJ (2020) The Coronavirus crisis in B2B settings: crisis uniqueness and managerial implications based on social exchange theory. Ind Mark Manage 88:125–135. https://doi.org/10.1016/j.indmarman.2020.05.004

Dart R (2004) The ligitmacy of social enterprise. Nonprofit Manag Leadersh 14(4):411–424

Dhofier Z (2011) Tradisi pesantren: studi tentang pandangan hidup Kyai dan visinya mengenai masa depan Indonesia. LP3ES

Foss NJ, Saebi T (2017) Fifteen years of research on business model innovation: how far have we come, and where should we go? J Manag 43(1):200–227. https://doi.org/10.1177/0149206316675927

Gebauer H, Saul C, Halidmann M, Kramer S (2017) When one business model is not enough for a social business. Strateg Dir 33(1):10–12. https://doi.org/10.1108/SD-06-2016-0096

Giesen E, Riddleberger E, Christner R, Bell R (2010) When and how to innovate your business model. Strategy Leadersh 38(4):17–26. https://doi.org/10.1108/10878571011059700

Goeztav D (2012) Menggagas Pesantren sebagai Pusat Peradaban di Indonesia. http://thejasspro.blogspot.com/2012/04/menggagas-pesantren-sebagai-pusat.html

Hasan K (2017) Social marketing and social business. Strateg Market Manage Asia 475–509. https://doi.org/10.1108/9781786357458
Haugh H (2005) A research agenda for social entrepreneurship. Soc Enterp J. https://doi.org/10.4337/9781789907964
Hockerts K (2015) How hybrid organizations turn antagonistic assets into complementarities. Calif Manage Rev 57(3):83–106. https://doi.org/10.1525/cmr.2015.57.3.83
Kallmuenzer A (2018) Exploring drivers of innovation in hospitality family firms. Int J Contemp Hosp Manag 30(3):1978–1995. https://doi.org/10.1108/IJCHM-04-2017-0242
Karni AS (2009) Etos studi kaum santri Wajah baru pendidikan islam. PT. Mizan Pustaka
Kraus S, Clauss T, Breier M, Gast J, Zardini A, Tiberius V (2020) The economics of COVID-19: initial empirical evidence on how family firms in five European countries cope with the corona crisis. Int J Entrep Behav Res 26(5):1067–1092. https://doi.org/10.1108/IJEBR-04-2020-0214
Lindgardt Z, Reeves M, Stalk G, Deimler M (2012) Business model innovation: when the game gets tough, change the game. In: Own the future: 50 ways to win from the boston consulting group, pp 291–298
Mair J, Marti I (2006) Social entrepreneurships research: a source of explanation, prediction, and delight. Int J World Bus 41(1):36–44. https://doi.org/10.1039/c5tc02806j
Marhumah E (2011) Konstruksi Sosial Gender di Pesantren: Studi Kasus Kiai atas Wacana Perempuan. LKiS.
Massa L, Tucci CL (2013) Business model innovation. Oxford Handb Innov Manage. https://doi.org/10.4135/9781483399836.n2
Mcshane S, Von Glinow MA (2003) Organizational behaviour (Internatio). McGraw-Hill Education
Mitchell DW, Bruckner Coles C (2004) Business model innovation breakthrough moves. J Bus Strateg 25(1):16–26. https://doi.org/10.1108/02756660410515976
Nicholls A (2005) Measuring impact in social entrepreneurship: new accountabilities to stakeholders and investors. Seminar on Social Enterprises
OECD (1999) Social enterprises. OECD
Oftedal EM (2013) Business model innovation. Encycl Corp Soc Responsib
Peerally JA, De Fuentes C, Figueiredo PN (2019) Inclusive innovation and the role of technological capability-building: the social business Grameen Danone foods limited in Bangladesh. Long Range Plan 52(6):1. https://doi.org/10.1016/j.lrp.2018.04.005
Global Islamic Economy Report (2018) State of the global islamic economy report 2018/2019. In: Thomson reuters. https://ded.ae/DED_Files/StudiesAndResearch/SGIE-Report-2018-19_Eng_1540649428.pdf
Republic of Indonesia (2019) Undang-Undang Republik Indonesia Nomor 18 Tahun 2019 tentang Pesantren. https://peraturan.bpk.go.id/Home/Details/122743/uu-no-18-tahun-2019
Saebi T, Foss NJ, Linder S (2019) Social entrepreneurship research: past achievements and future promises. J Manag 45(1):70–95. https://doi.org/10.1177/0149206318793196
Shahrezaei MAS, Goharpey F, Yeganeh JK (2018) Effect of selective localization of cellulose nanowhiskers on viscoelastic phase separation. Polym Eng Sci 58(6):928–942. https://doi.org/10.1002/pen
Sigala M (2020) Tourism and COVID-19: impacts and implications for advancing and resetting industry and research. J Bus Res 117:312–321. https://doi.org/10.1016/j.jbusres.2020.06.015
Simonov SG, Khamatkhanova MA, Peshkova NN (2014) Development of social entrepreneurship on a given territory Sergey. Life Sci J 11(7):314–317
Sullivan Mort G, Weerawardena J, Carnegie K (2003) Social entrepreneurship: towards conceptualisation. Int J Nonprofit Voluntary Sector Market 8(1):76–88. https://doi.org/10.1002/nvsm.202
Syam N (2009) Kepemimpinan dalam Pengembangan Pondok Pesantren: Manajemen Pesantren. LKiS Printing Cemerlang
To CKM, Au JSC, Kan CW (2019) Uncovering business model innovation contexts: a comparative analysis by fsQCA methods. J Bus Res 101:783–796. https://doi.org/10.1016/j.jbusres.2018.12.042

Tykkyläinen S, Ritala P (2021) Business model innovation in social enterprises: an activity system perspective. J Bus Res 125:684–697. https://doi.org/10.1016/j.jbusres.2020.01.045

Yasmani (2002) Modernisasi Pesantren: Kritikan Nurcholis Madjid terhadap Pendidikan Islam Tradisional. Ciputat Press

Yunus M, Moingeon B, Lehmann-Ortega L (2010) Building social business models: lessons from the Grameen experience. Long Range Plan 43(2–3):308–325. https://doi.org/10.1016/j.lrp.2009.12.005

Zott C, Amit R, Massa L (2011) The business model: recent developments and future research. J Manag 37(4):1019–1042. https://doi.org/10.1177/0149206311406265

2 Co-creation Strategy in Empowering Indigenous Women's Innovation in Indonesia: Empirical Evidence of Weaver Community Entrepreneurs in West Nusa Tenggara

Retno Kusumastuti, Asra Virgianita, and Vishnu Juwono

Abstract

Innovation is the main key to the sustainability of all industries. The fashion industry with traditional woven fabrics as the basic material is one of the complex industries due to its association with the social environment in terms of cultural value, market taste, chain value, life-cycle products, and even government policy. This study analyzes the problems and strategies in empowering women entrepreneurs in the context of indigenous peoples in Indonesia by boosting innovation based on the indigenous knowledge conversion. Focusing on the indigenous innovation created by the weaving community in West Nusa Tenggara, this study employs several concepts, namely, resource-based view and indigenous knowledge, knowledge conversion, and social innovation. Studies of indigenous knowledge in Indonesia are commonly drawn from the anthropological perspective and merely a small number of them use perspectives from business administration. This study applies mixed methods and collects data through questionnaires, observations, and in-depth interviews. The primary and secondary data are collected during several months of field study with a

R. Kusumastuti (✉)
Department of Business Administration, Faculty of Administrative Sciences, Universitas Indonesia, Depok, Indonesia
e-mail: r.kusumastuti@ui.ac.id

A. Virgianita
Department of International Relations, Faculty of Social and Political Sciences, Universitas Indonesia, Depok, Indonesia
e-mail: asrahui@ui.ac.id

V. Juwono
Departement of Public Administration, Faculty of Administrative Sciences, Universitas Indonesia, Depok, Indonesia
e-mail: vjuwono@ui.ac.id

A. Rafiki et al. (eds.), *Open Innovation in Small Business*, Contributions to Environmental Sciences & Innovative Business Technology,
https://doi.org/10.1007/978-981-99-5142-0_2

local partner as a gatekeeper. The questionnaire is derived from the concepts of knowledge conversion practice. Observation is also conducted to obtain a deeper understanding of the process of making traditional fabrics. In-depth interviews are conducted to explain the results of the survey. The informants for this study represent all kinds of stakeholder, namely, women entrepreneurs, associations, clusters of women weavers, financial intermediaries, local governments, small business employees, the local community, etc. According to the resource-based view, it is possible for the indigenous knowledge to be a source of competitive advantage because it can create the indigenous innovation to produce inimitable unique traditional fabrics. Each local area in Indonesia has its own distinctive characteristics. Indigenous knowledge is a tacit knowledge originating from the ancestors and is usually passed through generations. Indigenous knowledge is a knowledge-based local wisdom that is difficult to imitate. Innovative activities are done by the indigenous peoples as a part of human activities relatively rare. The results of this study indicate that empowering small businesses that create traditional fabrics and belong to local people, particularly women weavers, faces several problems. These problems include inadequate support from the local government, the limited capacity of local human resources, difficult access to capital, difficulties in marketing, limited supply of raw materials, and debates on local cultural values vs economic values. Therefore, this study proposes co-creation activities with outside stakeholders. The involvement of different relevant stakeholders is important to overcome the existing problems. The traditional woven co-creation in design requires the involvement of social actors in the social learning process. The indigenous co-creation of traditional woven fabrics is a social innovation learning activity that requires context and the adequate involvement of actors. This collaboration model emphasizes the important of an open innovation activity as a keyword for empowering the indigenous communities and boosting indigenous innovation. This policy recommendation is significantly valuable for decision-makers at the local government level.

Keywords

Co-creation strategy • Indigenous innovation • Knowledge conversion

2.1 Introduction

The problems of development and its impacts to indigenous peoples are global issues that start in the late 1990s, in addition to various criticisms of development models and theories that have thus far been applied. Sen (1999) in *Development as Freedom* criticizes the dominance of the economic growth rather than the human aspects (human life and human freedom) as the focus of the development models. Sen personally considers that development is a process of expanding real freedoms that people can enjoy, not simply increasing basic incomes. This opinion emphasizes the importance of the concept of development based on the local wisdom as an effort to provide access to marginal communities, including indigenous peoples, to develop their potential and define their own needs. It is a concept as well as a criticism of the traditional model of development that favors social modernization.

There are numerous definitions of indigenous peoples, one of which is defined by the World Bank: "Indigenous peoples are culturally distinct societies and communities. The land on which they live and the natural resources on which they depend are inextricably linked to their identities, cultures, livelihoods, as well as their physical and spiritual well-being." It is estimated that the indigenous peoples around the world amount to approximately 370 million in total (or approximately 5% of the global population), spread over 90 dominant countries in Asia, with 15% of which are extremely poor (World Bank 2022).

The number of indigenous peoples in Indonesia is estimated to reach 50–70 million or nearly 20% of the entire population in the country (International Work Group for Indigenous Affairs 2020). This amount is far larger than the estimated number of indigenous peoples in Asia and the world. However, it is not easy to apply the traditional development strategies to these communities due to different reasons, such as the subordinated structure of the communities to the power of the state and the reluctance of local people to innovate and actively participate in the development process. Nevertheless, the potential of these indigenous communities is fairly large as an actor of change to drive the local and global economy (IWGIA 2020).

Related to the concept of development based on the local wisdom, this study shall focus on the problems in empowering indigenous peoples in Indonesia through indigenous knowledge and innovation. Various community empowerment programs in Indonesia have targeted different indigenous groups. In reality, the empowerment of indigenous peoples is an effort to develop the independence and welfare of the communities by increasing knowledge, attitudes, skills, behaviors, abilities, and awareness, as well as utilizing resources through the establishment of policies, programs, activities, and assistance according to the prioritized needs of local communities. Phillips and Pittman (2009) explain that the community empowerment is a learning process of how people collectively seek solutions to a problem. Good community empowerment is not merely oriented to the development of social welfare and independence, but also the preservation of the aspects of harmony with the environment, the emphasis on regional culture, and the strengthening of national identity. The literature on community development generally refers empowerment as social capital or social capacity, which signifies the ability of the community to organize and mobilize their resources for the accomplishment of consensually defined goals.

The results of the development are expected to be sustainable since it is ingrained in the rural communities themselves. Interestingly, if we observe the internal perspective of these indigenous groups, the marginalized community learning process effectively transpires. However, this effort requires innovation to be able to provide the proper response to the occurring transformation. Historically, these indigenous groups have also learned how to preserve and live in harmony with their environment. Therefore, a variety of indigenous knowledge is created as principles in the management of the surrounding environment.

The study of innovation activities in corporations is easier to carry out due to its more mature structure and strong legal framework (Duncan 1976). Different from corporations, indigenous peoples are extremely complex. Community learning is an activity in the community, in which innovation learning is included. According to Johansens, community learning refers to diverse learning processes in various contexts, in which uncertainty and transformation are problematic (Johanssen and Nyberg 2013). This grand theory also refers to the concept by Bandura (1977) in Johansens that the social learning process is carried out by individuals through observations as well as social interactions within groups. These learning activities eventually will transform into social innovation. Moreover, individual learning in reality is part of social learning (Moulaert et al. 2013).

Social innovation is normally influenced by individuals with creative learning abilities (Jansen et al. 2005). Interestingly, all innovation activities in the community are generally temporal, particularly for marginalized groups. It is caused by the lack of coordination and support from the local government (Loban et al. 2013).Studies of social innovation in the community are limited, and studies of indigenous communities and indigenous innovation as the forerunner of indigenous enterprises are even rarer. Countries with moderately advanced studies of indigenous innovation are Canada, Britain, China, Norway, Japan, and New Zealand.

Studies on indigenous knowledge in Indonesia are commonly conducted from the perspective of anthropology while merely a small number of them use the perspective of the administrative science (Kusumastuti et al. 2022a, b). Based on the resource-based view, the management of indigenous knowledge sourced from local wisdom shall be a source for indigenous communities to innovate and create a competitive advantage. The innovation produced by indigenous peoples is called an indigenous innovation.

The learning process carried out by these community groups is generally communal, allowing this phenomenon to be explained as part of a social innovation (Kusumastuti et al. 2022a, b). Mostly, social innovation carried out by indigenous communities is an effort to maintain the sustainability of these communities and even a response to the unavailability of what they need (Lee 2014). The bottom-of-the-pyramid approach interestingly observes how to encourage and create innovation in relatively disadvantaged community groups. Indeed, the innovation carried out by these community groups shall provide an impact on social transformation, reducing poverty (Prahalad 2010). In line with this concept, it is deemed necessary to identify the existing indigenous knowledge and innovation since they are regional wealth and assets to be protected and provided with legal protection according to the Law No. 19 of 2002 on Copyright. Unfortunately, the study of the Indigenous Peoples' Alliance of the Archipelago (AMAN) reveals that a great number of indigenous knowledge and innovation have not been properly institutionalized. Forensics regarding indigenous innovation is not merely expected to maintain the learning process of innovation, but also to increase its leverage in assisting indigenous communities in increasing their economic independence.

This study analyzes the problems and strategies for empowering indigenous peoples through a case study in West Nusa Tenggara, particularly the weaving community. This province is selected because it is one of the provinces that have more than 200 women weaver entrepreneurs who strongly implement local culture. Furthermore, weavers are selected as the object of the study by considering that one of the main focuses of the activities of indigenous peoples in West Nusa Tenggara is the woven fabric industry. This industry, with its variety of motifs and traditional processes, is one of the industries with high cultural values and selling points. However, the woven fabric industry in West Nusa Tenggara has not been able to economically provide a significant contribution to both the weaver community and the environment in general. Utilizing a resource-based view and indigenous knowledge and innovation as the main concepts, this study aims to identify the problems in empowering indigenous peoples and formulate a strategy to overcome these problems.

This study offers a novelty related to the mapping of indigenous knowledge and innovation with a bottom-up approach through an analysis concerning not only the aspects of administrative management but also socio-economic aspects. So far, the existing community empowerment programs have been approached using a top-down approach, which needs to incorporate the social context using anthropological, political, legal, and economic perspectives. This study argues that in empowering indigenous peoples, particularly in developing indigenous businesses, it is important to prioritize a bottom-up approach by emphasizing the collaboration between actors in all activities. Thus, this study offers the concept of the co-creation of indigenous enterprises.

2.2 Research Methods

This study applies qualitative method by using single case study. The data is collected through direct observation/site visit to the location of the weaving community, and in-depth interviews with key persons from the government, representatives of the weavers, representatives of NGOs working with the weavers, and local community leaders. By using single case, this study will explore the phenomenon through variety of data sources to ensure the exploration is conducted through variety of lens.

Regarding the objectives of the study, several concepts are employed, specifically the resource-based view, indigenous knowledge and social innovation, and knowledge-creating process by Nonaka et al. (2008). Resources-based view (Barney and Clark 2007) states that the internal wealth owned by an organization has the potential to be utilized to become an organizational excellence if it has the following characteristics; Value (V): It is valuable in the sense that the resource has the potential to open opportunities for the organization or even neutralize threats to the organization; Rarity (R): It is rare in the sense that the resource is unusual compared to that of other organizations; Imitability (I): It is particularly difficult to imitate by others since it comes from a unique and different context and history; Organization (O): Related to whether the management within the organization exploits the resource.

The second concept, indigenous knowledge, can be interpreted as a concept referring to the tacit knowledge established over time by a group of people occupying a particular area. Indigenous knowledge in general includes a set of classification systems, a set of knowledge as empirical evidence based on observations related to the surrounding environment, a management system for natural resources, as well as a tool to pass the knowledge from one generation to the next generation (Johnson 1992; Rahman 2000). Meanwhile, social innovation mentioned in this study refers to the definition by Moulaert et al. (2013) and Michellini (2012), that is "*social innovation is about the satisfaction in social relations within processes; it is about people and organization who are affected by deprivation or lack of quality in daily life and services, who are disempowered by lack of rights or authoritative decision making and who are involved in agencies and movement favoring social innovation.*" It emphasizes social innovation as a means to contribute to urban and community development. This perspective adds social innovation to the category of innovative learning activities by focusing on human-centered community development.

The knowledge creation model by Nonaka, in this regard, is relevant and essential to complete the analysis in this study since it recognizes that the knowledge gained for innovation activities is the result of creation among individuals within the organization. The organizational context is a highly influential factor in encouraging the innovation activities of traditional fabric producers/weavers. The role of these individuals in turn is highly important to encourage the organizational learning process. Using the SECI Model (Socialization, Externalization, Combination, and Internalization), the learning process shall be visible through the local wisdom-based co-creation. This model illustrates how each organization shows a different focus of learning, for example, a strong focus at the level of individuals, groups, and even the entire organization. Having said that, the organizational learning process through the process of knowledge creation is the essence of creativity and innovation (Nonaka et al. 2008).

Based on SECI Model by Nonaka et al. (2008), there are four types of conversion of tacit and explicit knowledge with the following explanation:

1. Socialization is the process of converting tacit into another tacit knowledge. Tacit knowledge is a type of unwritten knowledge. It is formed as naturally as social interaction. The knowledge remains in the minds of the people rather than written in books.
2. Externalization is the process of converting tacit into explicit knowledge. Unlike the unwritten tacit knowledge, explicit knowledge is written. An example of this knowledge conversion is the activity of writing a book to share the knowledge to other people, to ensure that the knowledge can be preserved.

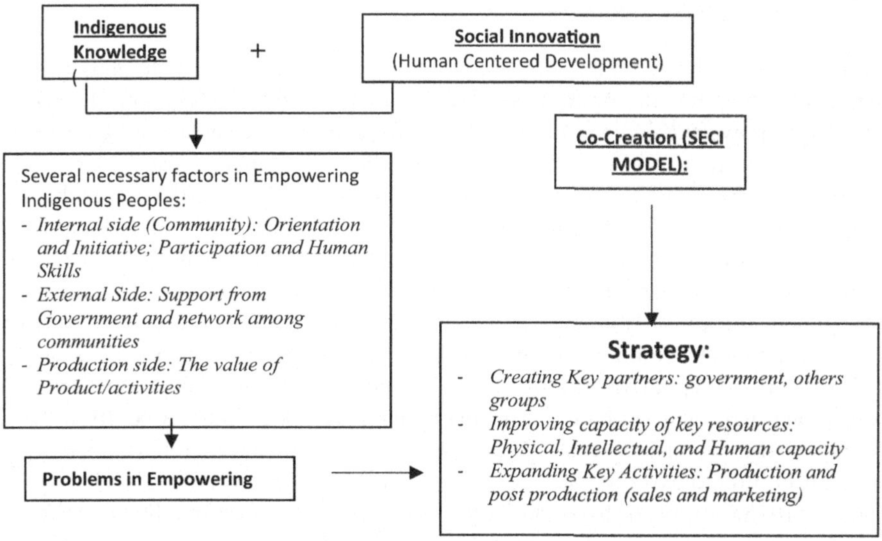

Fig. 2.1 Analytical model. *Source* Author's proposed analytical model (2022)

3. Combination is the process of converting explicit into another explicit knowledge. This process can be interpreted as combining explicit knowledge from numerous other sources and eventually creating a new form of explicit knowledge. An example of this activity is compiling financial reports.
4. Internalization is the process of converting explicit to tacit knowledge. An example of this process is the formulation of written rules in corporations. Over time, the knowledge will spread and become tacit knowledge.

The concept of co-creation is widely applied in established, large-scale, and industrial technology-based companies. A study carried out by Hirano et al., (2013) using Fujitsu Company as a case study discloses that innovation can be generated through the process of co-creation. This process involves the exchange of social values and discoveries for the utilization of technology in a more optimal manner in Fujitsu (Fujitsu, 2020).[1]

In contrary to this dominant approach, this study applies the co-creation concept in unestablished organizations rarely utilizing high technology, while similarly applying the concept of innovation obtained through co-creation. The analytical model of this study is presented as follows (Fig. 2.1).

[1] https://www.fujitsu.com/global/about/resources/publications/technicalreview/topics/article002.html.

2.3 Research Results

Observing the collected data, this study obtained several problems in empowering indigenous peoples in West Nusa Tenggara, applicable to explain the problems in empowering indigenous peoples in Indonesia. The most crucial problem discovered through an in-depth interview with the government is a budgetary problem (Interview with Government Officer, May 12 and 13, 2018), namely, the lack of budget allocated to support the production process, the strengthening of human resources, and the promotion of woven fabrics by indigenous peoples. The strengthening of human resources, for instance, is carried out quarterly by the local government for approximately 2–3 days. Nonetheless, in order to produce quality products, adequate trainings and guidance are important to conduct (Fig. 2.2).

The lack of capital owned by fabric weavers to finance the production process is another vital problem. Raw materials have to be purchased from Java, adding to the production costs. Meanwhile, the access to capital from the government is limited and private sectors are reluctant to be involved considering the unclear prospect of profit from the production process. In addition, the inter-agency coordination is not optimal. In terms of production targets, community orientation is also important. Technological uptake or utilization remains low due to the expensive machinery and machine maintenance that require expertise. Nevertheless, it is necessary to emphasize that the woven products produced traditionally or without machines have higher economic and cultural values. The government

Fig. 2.2 Women weavers from the Gema Alam cluster. *Source* Author's Research Output (2022)

later appealed to change the size of traditional tools used for weaving to produce woven fabrics as materials for clothes or other derivative products. Unfortunately, innovation remains low as woven products focus on fabrics only.

Another problem is the production of woven fabrics is merely carried out as a side job rather than the primary one for the weavers. However, it is interesting to cite the opinion of local community leaders that weaving for the local community is not merely limited to the production process. It also has cultural values, for example as a habit of forming the identity of the village (Interview with Community leader, May 15, 2018). In addition, networks among weaver communities are poorly established. Despite the roles of NGOs such as Gema Alam who is assisted by international donors (Oxfam), the consultation and reporting documentation of such roles are non-existent. Therefore, the assistance is frequently inaccurate or poorly directed, for instance providing large equipment or tool despite an inadequate capacity of electricity in the location. Similarly, universities frequently proceed independently without coordination with the government.

Examined from the perspective of the weavers (Interview with Gema Alam Community, May 14, 2018), several identified problems in the development of their business are mainly related to poor skills. Regeneration is inefficient because the weavers are dominated by the elderly with limited knowledge of innovation. A similar problem is also conveyed by local community leaders (Interview with Community leader, May 15, 2018). To overcome this problem, the weavers have created several programs such as weaving schools for any interested parties, assisted by several local NGOs and funded by international donors. Additionally, the function of the weaving hall is no longer limited to the storage of materials, but also a place to study, for example learning to operate a computer. Other weavers who are relatively successful in developing woven products and good markets disclose one of the strategies to develop their products, namely, continually participating in exhibitions, not merely to sell their products, but also to survey fashion trends and products as well as motifs favored by visitors (Interview with Champion Weaver, May 15, 2018. In this regard, it is significant to note that the opportunity to participate in exhibitions is normally facilitated by the government with a limited budget.

Based on the results of the interviews, this study summarizes the following main problems:

1. From the government's point of view: inadequate budget and the lack of coordination with fellow government institutions as well as other actors, specifically NGOs and universities.
2. From the weaving community's point of view: low capacity of human resources, resulting in weak innovation and regeneration; a conflict between the purpose of the product as a product with economic value and cultural value; the perspective on the weaving business as a side business; and the lack of networking among communities, hindering the learning process between groups.

3. From NGOs' point of view: weak coordination with the government and the dependence on donors for a budget.

Direct observation during field study (Site Visit to Gema Alam Community, Pringasela Area, West Nusa Tenggara Province, May 12, 2018) further confirms several problems as well, including the reliance on traditional tools, the majority of elderly weavers in the weaving community, limited innovation, extremely limited types of diversified products made of woven fabrics, and non-established networks among communities. According to the survey results, several problems are discovered as follows:

1. The weavers do not have partners in the process of production, sales, and marketing. They distribute their products by directly selling to buyers (65.54%) and by partnering (15.12%), while the others (19.34%) utilize other distribution channels such as through salespeople, their shops, websites, exhibitions, and assistance provided by the local government.
2. The dominant resources owned by the weavers are looms. They generally obtain traditional looms independently for no more than IDR 10,000,000 modern looms are rarely used, particularly for woven fabrics from Pringgasela that emphasize classic motifs. The looms affect the fabrics produced. According to women weavers at Gema Alam Community (2018), the fabrics woven by traditional looms can last up to tens of years. Meanwhile, the intellectual resources possessed by weavers are particular abilities/skills to create motifs. A considerable number of weavers can create motifs (88.55%), while others can weave (7.78%). It shows the lack of ability to weave. In terms of human resources, most of the weavers merely rely on one person as a skilled worker. The remaining weavers are merely of a technical nature.
3. In terms of activity, approximately 90.9% of weavers admit that their main activity is to produce woven fabrics. There are not many weavers who can create more values to the woven fabrics. It is highly rare to find weavers selling other woven products such as bags, clothes, or wallets. Furthermore, the majority of weavers focus on the production process. Those trying to expand in the context of sales and marketing are rather rare. The pattern of sales and marketing is dominated by direct sales to buyers.

2.3.1 Strategies for Empowering Indigenous Peoples Based on Knowledge Conversion Activities in the SECI Model

Referring to various constraints discovered in the field as discussed above, strategies that can be formulated based on the knowledge conversion grouping of the SECI model are as follows (Table 2.1).

Table 2.1 Knowledge conversion and managerial instruments

Stage of SECI	Societal condition in the cluster	Managerial instruments
Socialization	– Individuals are willing to share indigenous knowledge – High commitment/loyalty – Cooperation among women weavers	– Design of experience sharing – Observation – Mentoring
Externalization	– High group commitment	– Open dialogue – Community of practices
Combination	– High employee commitment – No risk of informational abuse	– Consultative decision-making – Symmetric information in a group
Internalization	– Little fear of mistakes	– Several policies and programs that encourage learning by doing activities

Source Author's Research Output (2022)

1. **Socialization**

Generally, girls in weaving-producing areas are taught how to weave since they are 20 years old. However, currently, teenage girls who are interested in weaving are increasingly rare. Most of them choose other paths, such as going to school or university and leaving their villages. If this condition lasts, there will be a loss of transfer of indigenous knowledge of weaving from generation to generation.

In the context of knowledge conversion, the initial stage of knowledge formation is socialization. The socialization activities are usually carried out by the weavers when they gather to learn about weaving techniques, weaving designs and patterns, sizes generally used for certain types of fabrics, the dyeing process, the process of making natural dyes, the drying process, yarn dyeing, and other skills related to the production process of woven fabrics. General strategic managerial tools to strengthen these activities are:

a. Providing well-planned design-sharing experience programs from the government and outside weavers to help maintain the knowledge transfer process.
b. Providing space for those wishing to learn to weave by observing the weaving process through participatory observation.
c. Providing the opportunity to have trainings for young generation, with experienced weavers as mentors (Fig. 2.3).

2. **Externalization**

This process occurs when the weavers have started establishing associations or clusters of weavers with a clear division of tasks that is reflected in the organizational structure. Several areas have naturally-formed groups of weavers such as Fetung Bayan cluster. The construction of the Fetung Bayan Art Shop, for instance, is inseparable from the government support in terms of:

Fig. 2.3 Workshop and artshop of the Fetung Bayan cluster. *Source* Author's Research Output (2022)

(1) Provision of an adequate budget. The available funds are intended for land acquisition, the establishment of art shop buildings and showrooms for the woven products, the purchase of group equipment, the supply of raw materials such as threads and natural dyes, and non-physical development programs such as training needed by weavers. This strategy is quite successful in the West Lombok area and can be implemented by other regions.

(2) An open dialogue between various components of society. This strategy, though needs improvement, will facilitate various interested parties. The government needs to provide more informal communication channels to capture various aspirations of the weaving community, particularly those who are not members of a cluster or group of weavers.

(3) The establishment of a community of practices among weavers who share common concerns. This community of practices occurs naturally according to the needs of the weaving community. In this regard, the government can cooperate with NGOs or community leaders. This community of practice is usually initiated and led by a social entrepreneur respected by the community.

3. **Combination**

This activity occurs when a cluster represents the weaver groups. Weaver groups who are members of the cluster represent their interests to external parties, for example banks, donor agencies, or related training institutions. The working group in the form of a cluster indirectly facilitates the weavers. In turn, this working group will be developed into a cooperative whose function is to control fabric quality, monitor selling prices, and provide financial support through mutual savings and loans.

The strategy used should open the space for building **consultative decision-making** where every decision is made after being communicated and discussed in deliberation. In this regard, weavers can consult and discuss related problems to receive assistance in solving them. Important decisions for weavers cannot be made personally because they have to accommodate the interests of the cluster of weavers and not endanger the weavers. This condition renders the weavers in need of each other, leading to the creation of a high level of commitment by the weavers involved and joined in the cluster. In general, this cluster is managed and owned collectively by the weavers. The management is determined and elected by deliberation and consensus.

In addition, each member has a common understanding of all information, which is called symmetric information in a group. It will build trust and a strong bond among the weavers. They perceive that this forum is established and managed collectively and is beneficial for the common good. The possibility of unilateral misuse of information is low. All weavers who are members of the cluster have the same information and knowledge. No weavers will be left behind since they will help each other in terms of continuity of production, availability of basic materials for weaving, the certainty of sales, and even the direct establishment of networks with customers to assist the process of learning innovation related to updating motifs and patterns by the times (Fig. 2.4).

4. **Internalization**

The work of the weavers with traditional fabrics from naturally pigmented ingredients has encouraged them to take a leading role in their life. It also provides them space to learn to interact with the community. In this stage, the most important aspect is to build a role model for prospective young weavers. This stage is important in terms of character building to maintain the continuity of the increasingly rare weavers. This role will be paramount in the human learning process where young people will imitate the closest people in the family or influential figures around them. This process is relevant for transferring indigenous knowledge which is usually passed down from one generation to another. Furthermore, it will be easier for new weavers to learn by imitation.

The weaver cluster generally holds a regular monthly meeting among fellow weavers to facilitate the members to understand the values of weaving and traditional knowledge that becomes the backbone of the production of woven products.

Fig. 2.4 Discussion among the weavers. *Source* Author's Research Output (2022)

The values of local wisdom contained in woven fabrics represent values in the cycle of life from birth, growing up as teenagers, getting married, to death.

An important strategy in internalization is developing **various policies and programs that encourage learning by doing activities** for the members. These activities will be certainly instrumental and greatly influential provided that they are supported by adequate information technology applications. A great amount of traditional knowledge is not well-documented and thus at risk of extinction considering that this weaving practice is carried out by the elderly. It is necessary to document various skills in terms of creating and combining motifs and patterns, basic materials for natural dyes, and the meaning of each motif woven by the weavers (Figs. 2.5 and 2.6).

2.3.2 Co-creation of Indigenous Innovation with the Resource-Based View Approach, Social Innovation, and Knowledge Conversion

The learning process carried out by the women in the weaver groups is communal, thus this phenomenon can be explained as part of social innovation. Social innovation carried out by indigenous communities is mostly the efforts to maintain the sustainability of these communities and a response to the unavailability of what these community groups required (Lee 2014).

Fig. 2.5 The learning activities of weavers among generation. *Source* Author's Research Output (2022)

Fig. 2.6 The learning activities of weavers in their house as a part of their daily lives. *Source* Author's Research Output (2022)

In line with this concept, it is deemed necessary to identify existing indigenous knowledge and innovation as these are regional assets that should be protected by the government as stipulated in the Law No. 19 of 2002 concerning Copyright. In practice, violations frequently occur considering the weak institution authorized to manage copyright (Lodra 2012). The institution authorized to manage copyright is the Intellectual Property Rights Institution of each region, such as the District Office of Trade as the extension of the Director General of Intellectual Property Rights in Indonesia. Unfortunately, a study by The Alliance of Indigenous Peoples of the Archipelago (AMAN) reveals that a great amount of indigenous knowledge and innovation have not been properly institutionalized. In the future, it is expected that forensics on indigenous innovation will not only maintain the learning process of innovation but also increase its leverage in assisting indigenous communities in increasing economic independence.

Previous studies have discovered a business model formulated by the indigenous peoples in the form of the co-creation of indigenous social enterprises with a high content of local wisdom. Putri and Kusumastuti (2020) show that traditional woven co-creation in design needs the involvement of social actors in the social learning process. The indigenous co-creation of traditional woven fabrics is a social innovation learning activity that needs not only the adequate involvement of actors but also context. This research results are consistent with the empirical evidence showed by Michellini (2012) that social innovation can be an effective instrument for social change. In this case study, the groups of women weavers, both Gema Alam or Fetung Bayan, were established to empower women weavers through collaboration in value creation process (Fig. 2.7).

Fig. 2.7 Indigenous innovation. *Source* Author's Research Output (2022)

2.3.3 Co-creation Strategies with Outside Stakeholders

The following are nine aspects that motivate the co-creation activities (Frow et al. 2015).

1. Access to resources
2. Enhancing customer experience
3. Creating customer commitment
4. Enabling self-service
5. Creating a more competitive offer
6. Decreasing costs
7. Faster time to market
8. Emergent strategy
9. Building brand awareness.

The collaboration model emphasizes cooperation known as an open innovation activity, which is a model of a learning process for empowering the indigenous communities and boosting indigenous innovation through cooperation with other stakeholders outside the women entrepreneur cluster. The concept of "open" refers to the involvement of numerous stakeholders empirically existing in the indigenous enterprise activities in creating value propositions for customers. The comprehensive model developed from this study is presented in Fig. 2.8. This type of relationship is important to overcome the existing problems. Traditional weavers **involve customers in the process of creating new designs, new models, or even new diversified products and combinations** of woven fabrics with other materials. This specific co-creation model is called **co-design.** As a social innovation learning activity, the indigenous co-design of traditional woven fabrics needs context and the adequate involvement of actors. Another reason for working with external stakeholders is to improve **customer experience** by creating designs as a result of collective work. This activity produces product innovations needed and in demand by customers. In numerous cases in West Nusa Tenggara, the weavers frequently have limited knowledge because of their limited exposure to the development of the outside world. A collaboration with customers allows for new knowledge adoption by absorbing requests for consumer needs more easily and quickly.

The co-creation strategy widely used by weaving groups when collaborating with actors outside the cluster is usually based on the need for **easy access to resources**. This co-creation model is called **co-production**. The frequent issue is the provision of basic materials for weaving, namely yarn, from outside West Nusa Tenggara. In recent years, there has even been a scarcity of natural dyes due to the increasing difficulty in finding the plants used to make natural dyes. To anticipate this, a great number of clusters have started collaborating with suppliers from the island of Java regarding yarn delivery. The weavers have also cooperated with local farmers to initiate the planting of traditional plants as a source of natural dyes. In addition, the weavers have worked with customers in producing fabrics,

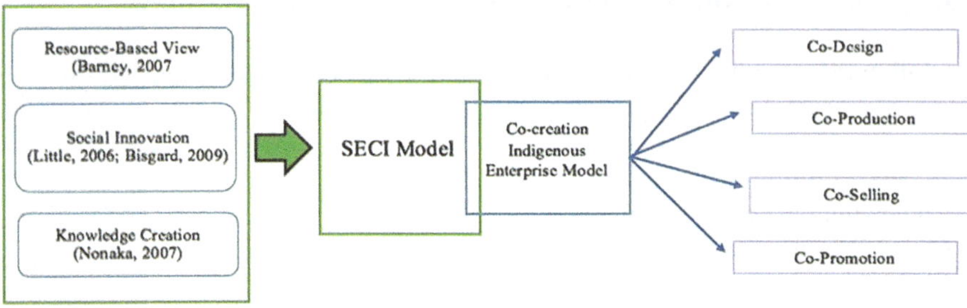

Fig. 2.8 Co-creation indigenous women entrepreneur business model. *Source* Author's Research Output (2022)

for example, combining woven fabrics with embroidered fabrics and producing woven fabrics with contemporary patterns, colors, and motifs.

To increase customer experience, the weavers provide a workshop storing non-machine looms or traditional looms for weaving that can be used by visitors in addition to equipment for dyeing woven fabrics. Moreover, they provide a showroom displaying ready-for-sale woven fabrics and their derivative products.

Other co-creation activities carried out by these weavers are co-promotion and co-selling, in which the women entrepreneurs cooperate to sell their products. These weavers exhibit their woven fabrics in the showroom in which the manager helps promote and sell them. During the Covid-19 pandemic, sales media were strengthened by information technology applications using Instagram, Facebook, Twitter, etc.

Other external stakeholders collaborating with the weaver cluster are designers, local government, non-governmental organizations, corporations, and financial intermediaries such as banking and insurance institutions, cooperatives, and associations. Designers generally come to the workshop to directly select low-priced, high-value, and high-quality woven fabrics and frequently hedge purchases for the next supply. The local government has a role in carrying out various assistance programs, such as technical assistance, mentoring, and various other development programs. Non-governmental organizations usually work with the community with the vision and mission of developing and empowering local communities, including the sustainability of highly competitive woven fabric production. The private sector has numerous roles, particularly when corporations provide support for business development and local community empowerment. Meanwhile, financial intermediaries play a role in ensuring that indigenous enterprises obtain various business capital schemes.

The cooperation network that needs to be built in co-creation activities of indigenous innovation has to have below functions:

1. Creating cooperation not only with the central government or local governments but also with other stakeholders.

2. Increasing the intellectual capacity and skills of existing human resources, both weavers and cluster managers.
3. Formulating policies that support these women entrepreneurs in accessing resources, both capital and raw materials.
4. Facilitating the expansion of distribution channels, marketing channels, and promotional media to access to sale channels to maintain the sustainability of woven fabric production.

2.4 Conclusion

The role of women weavers, particularly in West Nusa Tenggara, in utilizing natural resources has strengthened nationalism by preserving a culture with economic values, namely weaving. This weaving craft holds a meaning for women in NTB to establish a harmonious and good relationship among communities within the country, particularly indigenous communities in local areas.

According to the research results of empowering indigenous people, especially women weavers in West Nusa Tenggara, this study concludes the co-creation models, namely, co-design and co-production, generally used in the context of women entrepreneurs. Considering this point, this study offers several recommendations on strategies for empowering indigenous peoples based on the activities in the SECI model as follows. First, designing experience-sharing, observation, and mentoring activities for socialization activities. Second, establishing open dialogues and community of practices for externalization activities. Third, encouraging the spirit of consultative decision-making and building symmetric information in a group for combination activities. Fourth, formulating policies and programs that support the learning-by-doing activities.

References

Barney JB, Clark DN (2007) Resource based theory. Creating and sustaining competitive advantage. Oxford University Press, New York

Duncan RB (1976) The ambidextrous organization: Designing dual structures for innovation. In: Kilmann RH, Pondy LR, Slevin D (eds) The management of organization, vol 1. North-Holland, New York, pp 167–188

European Commission (2010) Empowering people, driving change: social innovation in the European Union, European Website on Integration. https://ec.europa.eu/migrant-integration/library-document/empowering-people-driving-change-social-innovation-european-union_en

Frow et al (2015) Managing co creation design: a strategic approach to innovation. Change Manag Strategy J

Hirano et al. (2013) https://www.fujitsu.com/global/about/resources/publications/technicalreview/topics/article002.html

IWGIA (2020) Indigenous world: Indonesia. https://www.iwgia.org/en/indonesia/3602-iw-2020-indonesia.html

Jansen JJP, Van den Bosch FAJ, Volberda W (2005) Exploratory innovation, exploitative innovation, and ambidexterity: the impact of environmental and organizational antecedents. Schmalenbach Bus Rev 57(4):351–363
Johanssen M, Nyberg L (2013) Using education and social learning in capacity building - The IntECR concept. Disaster Prev Manag 22(1):17–28. Emerald Group Publishing Limited 0965-3562. https://doi.org/10.1108/09653561311301943
Johnson M (1992) Lore: capturing traditional environmental knowledge. Denne Cult Inst, Ottawa, Canada
Kusumastuti R, Silalahi M, Sambodo M, Juwono V (2022a) Understanding rural context in the social innovation knowledge structure and its sector implementations. Manag Rev Q. https://doi.org/10.1007/s11301-022-00288-3
Kusumastuti R, Silalahi M, Asmara AY, Hadiyati R, Juwono V (2022b) Finding the context indigenous innovation in village enterprise knowledge structure: a topic modeling. J Innov Enterp 19(11). https://doi.org/10.1186/s13731-022-00220-9
Lee P (2014) Social innovation. Wahington Univ Law Rev 92(1):13–15
Loban H, Ciccotosto S, Boulot P (2013) Indigenous corporate governance and social enterprise. Indig Law Bull 8(8):22–24
Lodra IN (2012) URNA. Perlindungan pengetahuan tradisional dan praktek HaKI. J Seni Rupa 1(1):11–20
Michellini L (2012) Social innovation and new business model: creating shared value in low-income market. Springer, pp 1–8
Moulaert F, MacCallum D, Mehmood A, Hamdouch A (2013a) International handbook of social innovation: Collective action, social learning, and transdisciplinary research. Edward Elgar Publishing Limited, Cheltenham UK, Southampton USA
Nonaka I, Toyama R, Hirata T (2008) Managing flow: a process theory of knowledge-based firm. Palgrave MacMillan, p 14
Phillips R, Pittman RH (2009) An introduction to community development. Routledge, USA, pp 7–11
Prahalad CK (2010) The fortune at the bottom of the pyramid: Eradicating poverty through profits, 5th ed. Wharton School Publishing, Pearson Education, pp 6–18
Putri AM, Kusumastuti R (2020) Achieving sustainable future through indigenous people: the case of weaver community in South Pringgasela, West Nusa Tenggara Barat. Hasanuddin Econ Bus Rev 3(3):128–131
Rahman A (2000) Development of an integrated traditional and scientific knowledge base: a mechanism for accessing, benefit-sharing and documenting traditional knowledge for sustainable socio-economic development and poverty alleviation. Paper presented at United Nations Conference on trade and development (UNCTAD) expert meeting on and national experiences for protecting traditional knowledge, innovation and practices. Geneva, Switzerland, 30 October to 1 November
Sen A (1999) Development as freedom. In: The globalization and development reader: perspectives on development and global change, p 525
World Bank (2022) Indigenous peoples. https://www.worldbank.org/en/topic/indigenouspeoples#2
World Intellectual Property Organization. http://www.wipo.int/tk/en/tk/index.hml
https://sirusa.bps.go.id/index.php?r=indikator/view&id=18

Interview

Interview with Government Officer from the Office of Trade of West Nusa Tenggara Province on May 12, 2018
Interview with Government Officer from the Office of Industry and Trade of the District of West Lombok on May 13, 2018
Interview with Gema Alam Community, Pringgasela, West Nusa Tenggara Province, May 14, 2018

Interview with Community Leader, Pringgasela, West Nusa Tenggara Province, May 15, 2018
Interview with Weaver Champion, Pringgasela, West Nusa Tenggara Province, May 15, 2018

Underlying Factors of Green Innovation Adoption Among Indonesian Batik Enterprises

3

Gabriella Hanny Kusuma, Nurul Indarti, and Hardo Firmana Given Grace Manik

Abstract

The literature on innovation management indicates the rise of the green innovation topic recently, which emphasises how companies adopt and implement advanced friendly technology to achieve environmental sustainability as a part of sustainable development goals. Most of the studies were conducted in the context of large manufacturing companies. Little attention has been paid to the setting of small and medium-sized enterprises (SMEs) as the backbone of the economy in many countries (e.g. Takalo et al. 2021). In Indonesia, the contribution of SMEs is significant for absorbing 97% of the total workforce and generating 60.4% of total investment, according to the Coordinating Ministry for the Economy in 2021. Our study uses institutional theory as a point of reference. It aims to understand various factors that drive organisations to seek legitimacy or social fitness when they want to adopt or make a decision (Dimaggio and Powell, 1983; Oliver, 1991), which in this context is adopting green innovation. The study deployed qualitative multiple case studies by conducting in-depth interviews for 11 SMEs (14 participants) in batik sectors in Indonesia who were already implementing green innovation (from only green, a combination of brown and green and green until brown to green). SMEs in the batik

G. H. Kusuma
Faculty of Business and Economics, Universitas Atma Jaya Yogyakarta, Yogyakarta, Indonesia

University of New England, Armidale, NSW, Australia

N. Indarti (✉)
Faculty of Economics and Business, Universitas Gadjah Mada, Yogyakarta, Indonesia
e-mail: nurulindarti@ugm.ac.id

H. F. G. G. Manik
Faculty of Business, Universitas Kristen Duta Wacana, Yogyakarta, Indonesia

A. Rafiki et al. (eds.), *Open Innovation in Small Business*, Contributions to Environmental Sciences & Innovative Business Technology,
https://doi.org/10.1007/978-981-99-5142-0_3

sector were selected as a sample because batik is well known as traditional clothing from Java and is considered the original cultural heritage of Indonesia by UNESCO (Wang, 2019). Batik faces fierce competition due to mass and fast production from larger companies and the issue of chemical dyes and synthetic colours (Raya et al., 2021). Validity in this research was supported by the triangulation method combined with member checking. Data were analysed using a content analysis method referring to the procedure by Miles et al. (2014) to answer the research question. The current study shows the pressure of external forces, namely regulatory/coercive force, cognitive pressure and normative pressure, are the motives for organisational decisions and actions to adopt green innovation. Our findings substantiate and broaden the institutional theory (Bruton et al. 2010; Rui and Lu, 2020) by providing empirical evidence regarding SMEs in the batik sector in Indonesia. Limitations of the study are highlighted, which align with suggestions for future research avenues. Potential practical implications are also discussed.

Keywords

Green innovation adoption • Batik • Small and medium-sized enterprises (SMEs) • Qualitative study

3.1 Introduction

Research on green innovation has increased markedly in recent years (Takalo et al. 2021). Green innovation arises as a company's action to cope with the stakeholders' demand, especially the accord regulations of environmental protection, manifested in the goals of sustainable development (SDGs) and environmental consciousness of consumers (Albort-Morant et al. 2016; Ullah et al. 2021). In general, green innovation can be grouped into two types, green technology innovation and green management innovation (Abbas and Sağsan 2019). Through green technology innovation, a company adopts advanced eco-technology to save raw materials, energy and resources. Many companies can also design and implement a management system that puts environmental sustainability performance as one of the leading indicators of success (Journeault 2016).

SMEs face pressures in their R&D and green innovation efforts compared to large corporations because of their limited resources, diseconomies of scale, reputation, and capabilities (Arfi et al. 2018; Boohene et al. 2020). Therefore, SMEs need external backups, such as government subsidies and public support, to transform their business processes from brown to green (Capasso et al. 2019). SME owners or managers also need to strengthen their ecological entrepreneurial orientation by developing pro-environmental behaviours or always considering aspects of the planet, people and profitability simultaneously and proactively pursuing resource support for implementing green innovation initiatives (Elkington 1998; Widjojo and Gunawan 2020).

A recent literature review by Takalo et al. (2021) found that most studies on green innovation focused on the manufacturing, energy and food-processing industrial sectors among large companies. This study suggests exploring green innovation practices in other sectors. In addition, a few scholars (such as Ebrahimi and Mirbargkar 2017; Ullah et al. 2021) state that studies on green innovation practices among small and medium-sized companies (SMEs) as the backbone of the economy in many countries are still under-researched, therefore relevant to study.

Furthermore, previous studies highlight the importance of exploring the factors as drivers of green innovation adoption (Cuerva et al. 2014). Horbach et al. (2012) identified company strategy, regulation, market and technology influencing green innovation adoption in German companies, while Jun et al. (2021) found that external partnership, cooperation, and rules and regulatory stress had an insignificant effect on Pakistani SMEs. Further investigation is needed in other industrial or country contexts (Jun et al. 2021).

Based on the abovementioned argument, this current study investigates factors in adopting green innovation among SMEs in Indonesia's traditional clothing (batik) sector. Batik is a traditional clothing from Java, Indonesia, designated by UNESCO as the original cultural heritage of Indonesia (Wang 2019). Historically, batik was made from various natural materials of various plants (e.g. stems, wood, roots and leaves) and was positioned as a cultural heritage full of socio-cultural value, not just an economic commodity. However, amid the rapid growth of the traditional fashion industry in Indonesia, many SMEs in this sector chose to carry out mass and fast production so that high demand is met by using synthetic colours (Kusumawati et al. 2021; Raya et al. 2021). This choice is faced with several green innovation issues that must be solved. Batik SMEs in Java generally do not have an established chemical dye waste management system (Raya et al. 2021). They throw chemical dye waste into the rivers that cause agricultural irrigation to be polluted. In addition, the cost of chemical dyes is also increasingly expensive, which increases selling prices and can reduce the competitiveness of batik SMEs (Levi et al. 2021; Raya et al. 2021). Another aspect is unfair competition with big batik players who use printing technology for mass production, which quickly also becomes a separate issue that makes batik SMEs unable to compete (Raya et al. 2021).

Recently, some studies reported the awareness to return to nature for sustainability. Many SMEs have shifted the batik paradigm, and production method from brown (non-green production) to green or have continued to maintain the green method since the beginning (Capasso et al. 2019; Levi et al. 2021). Natural dyes are not only environmentally friendly but also more expensive than synthetic dyes and take a long time to produce because they are hand-made (Kusumawati et al. 2021). An example of the adoption of green innovation in batik SMEs is the use of natural dyes in the colouring process, such as green from mango (*Mangifera indica*), blue from Indigofera (*Indigofera tinctoria*), yellow from tegeran (*cugrania javanensis*) and creamy from rambutan (*Nephelium lappaceum*) (Firmansyah 2020; Raya et al. 2021).

As the study focuses on the decision to adopt green innovation among batik SMEs in Indonesia, which is influenced by external environmental pressures or the concept of legitimacy, we use the lens from institutional theory by Dimaggio and Powell (1983); and Kostova (1997) to understand reasons behind their decision to go green. Findings from this study are expected to provide empirical evidence in the adoption of specific innovation (i.e. green innovation), with particular reference to SMEs in Indonesia. In addition, the findings can be useful as a guideline for policymakers, such as the government, to set incentives and other support for SMEs in implementing green innovation, as suggested by Ullah et al. (2021). Based on this, the study is intended to answer the question, what factors affect *batik* SMEs in Indonesia in adopting green innovation?

The rest of the paper is organised as follows. Section 3.1 discusses the background and need of the study, and the definition of green innovation and its adoption literature review is elaborated in Sect. 3.2. The research methodology, findings, and discussion are described in Sects. 3.3 and 3.4, respectively. Section 3.5 ends with conclusions and further research agenda.

3.2 Literature Review

3.2.1 Institutional Theory as an Analytical Framework to Understand Adoption Behaviour

Institutional theory is gaining prominence in entrepreneurship and innovation research due to dissatisfaction with theories that focus on the individual's role as a 'superhero' developing a new product or service but downplaying the pressure of external forces as a motive for organisational decisions and actions (Bruton et al. 2010; Rui and Lu 2020). The underlying notion of this theory is that in addition to focusing on efficiency-seeking behaviour, organisations also seek legitimacy or social fitness when they want to adopt or make a decision (Bruton et al. 2010; Dimaggio and Powell 1983; Kostova 1997; Oliver 1991; Qi et al. 2021). In the context of the study, this is the adoption of green innovation within small and medium-sized enterprises.

Furthermore, to attain legitimacy or social fitness, many organisations do isomorphism (Dimaggio and Powell 1983; Oliver 1991). This behaviour reflects where organisational characteristics are modified according to environmental demands and pressures such as from the government, public community, business associations, etc. The literature defines three types of isomorphism that companies can experience in decision-making (Dimaggio and Powell 1983; Ruef and Scott 1998). The types of isomorphism are coercive, mimetic, and normative isomorphism. Coercive isomorphism occurs due to coercion of government regulations and public policies regarding applying new laws/regulations. For instance, the

Indonesian government set Law No. 32 on Environmental Protection and Management, which covers the general rules and specifics enforced to control pollution or set marketable emission permits.

Mimetic isomorphism is a phenomenon that occurs when there are environmental uncertainties due to new trends or disruptions that encourage organisations to model themselves to other organisations that are considered capable of dealing with these uncertainties. For example, Freeport Indonesia and Danone just received sustainable business awards in Indonesia in 2022 for the highly commended energy management categories, respectively (CNN Indonesia 2022). These show that both companies have significant achievements in water management, climate change and emissions. Various awards are available regionally and globally. Many mining and food beverage companies compete to get such recognition by learning to imitate, or benchmark from the companies considered role models.

Normative isomorphism arises from a company's tendency to follow established norms and values in the industry where it belongs, primarily through an industry association or employee organisation. For example, the Chamber of Commerce and Industry in Indonesia is a place for entrepreneurs from various industry sectors to exchange information and make mutual agreements on the norms of a profession in each industrial sector, such as educational standards and ways of collaborating with government and political parties.

Using institutional theory as an analytical framework, Qi et al. (2021) found that regulatory and imitative pressures are the main antecedents of green technology innovation of companies in China, with a sample focus on companies with environmental patents. Various companies in China are allegedly forced through regulations and peer pressure in the industry to use international environmental management system standards such as ISO 14002 and join green clubs whose members are companies that use these standards (Qi et al. 2021). Contrary to these findings, by examining 288 SMEs from various sectors in Pakistan, Jun et al. (2021) found that organisational factors, human resources, markets and customers, and government and technology support were significant determinants, while regulatory factors were not. They further explained that the situation behind this anomaly was due to the lack of enforcement of environmental regulations, lack of awareness regarding the existence of these regulations, light penalties for non-compliance, institutional corruption and alleged political affiliation of SME owners with the ruling parties in Pakistan.

As environmental damage becomes increasingly massive, having environmental ethics codified in a clear code of conduct could give the company a strong green entrepreneurial orientation so that stakeholders could give recognition (Rui and Lu 2020). Responsible leadership is the key to strengthening this green strategy orientation by consolidating the internal structure of a company to give full attention to the implementation of green innovation at all levels of the managerial hierarchy (Qi et al. 2021; Rui and Lu 2020). If not so, a company's commitment to cleaner production is doubted. It could eventually lead to various systemic negative impacts, ranging from rejection of products or services, cessation of operations by the government, public pressure through the media, etc. (Aguilera-Caracuel and Ortiz-de-Mandojana 2013; Li et al. 2017).

3.2.2 Definition of Green Innovation and Its Adoption

Literature on innovation management generally distinguishes between green innovation and innovation adoption. Green innovation can be defined as innovations aimed at creating and developing new green products (e.g. toxic materials), new green product designs, or new green manufacturing processes (e.g. energy saving) (Cheng 2020). In implementing green innovation initiatives, companies should have dynamic capabilities to reconfigure and regenerate their resources so that the green paradigm becomes the company's main DNA (Albort-Morant et al. 2016). Companies also need systemic collaboration and knowledge exchange with various external stakeholders (e.g. suppliers, universities, environmental sustainability activists, governments, etc.), especially if green innovation has not become their core competency (Arfi et al. 2018).

The adoption process is an information-processing process by a firm that will eventually lead to either the rejection or acceptance of the green innovation (Aboelmaged and Hashem 2019). Aboelmaged and Hashem (2019) further found that a high degree of trust in the information sources can contribute to transforming a positive attitude regarding green innovation into behaviour. The story of various companies' best practices in adopting green innovation could be disseminated to become role models and inspirations for potential adopters (Chen 2008; Zhang et al. 2020). With the increasing number of adopters of green innovations, green innovation networks are formed and create rules of the game or a system of legitimacy, including high barriers to entry for companies that do not adopt them (Qi et al. 2021). Some examples of green innovation actions companies could take regarding capacity building are setting measurable environmental goals and conducting environmental training programmes and environmental audits (Zhang et al. 2020). Meanwhile, in terms of product and process innovation, companies are encouraged to display environmental declarations related to the impact of products/services, lowering pollution and energy, producing recycled, reused and decomposed products, ensuring manufacturing processes reduce harmful substances and reducing water consumption, electricity, coal and oil (Zhang et al. 2020).

3.3 Research Methods

This study used a qualitative design with multiple-case studies to understand the phenomena that occur in real life in depth, namely, the decision to adopt green innovation (Creswell and Poth 2018; Yin 2018). Data were collected by involving 14 participants from 11 SMEs (see Table 3.1). Participants were selected by purposive sampling method with some criteria as follows: (1) The batik SMEs already implement the green innovation; (2) The SMEs had run for at least 2 years. Each

Table 3.1 Profiles of participants (SMEs)

No	Company name	Year of establishment	Number of employees	Type of adoption	Method of the production process
1	ECJ	2014	14	Only green	Handmade
2	JLW	2010	10	Only green	Handmade
3	MTE	2016	8	Only green	Handmade and printing
4	ARN	2012	7	Only green	Handmade and printing
5	DRB	2010	7	Brown and green	Printing
6	GLB	2008	35	Brown and green	Printing
7	SKB	2010	15	Brown and green	Handmade and printing
8	FLO	1994	10	From brown to green	Handmade
9	BRJ	2015	15	From brown to green	Handmade
10	BPL	1950	60	From brown to green	Handmade and printing
11	BIX	2000	7	From brown to green	Handmade

case was examined independently, and the results were compared based on the characteristics of the study (Yin 2018). The participants or SMEs were categorised as 'only green', 'both brown and green' and 'from brown to green'. Green SMEs use natural dyes such as leaves, tree branches, fruits, and flowers in colouring. On the other hand, brown SMEs are those who use synthetic and chemical dyes in the colouring process.

The data was collected through in-depth interviews and took around 90 min from October 2019 to January 2020. Each interview session was conducted in two stages. The first stage is by giving open questions without informing participants about the construction of the research objectives. The second stage is by structuring the questions related to the research objectives. The data collection terminated after saturation of data was achieved, i.e. when there was replication or repetition of information obtained from different informants (Creswell and Poth 2018).

Data analysis followed the procedure by Miles et al. (2014) which consists four steps. Meanwhile, the content analysis is used in this study to answer the research question which followed with interpretation of the contents of the data through a systematic classification process (Creswell and Poth 2018; Miles et al. 2014).

To ensure validity and credibility, this study uses the triangulation technique and member checking (Creswell and Poth 2018; Yin 2018). Triangulation includes source triangulation and method triangulation (observation and in-depth interview).

Member checking was conducted by sending back interview transcripts to the participants to ensure data congruity with the participants' perspectives. This process also was maintained to reduce bias in the study (Creswell and Poth 2018; Yin 2018).

3.4 Findings and Discussion

This section elaborates and discusses findings based on in-depth interviews with 11 Indonesian SMEs, with two main aspects: (a) the type of adoption of green innovation and (b) factors affecting its adoption.

(1) **The type of adoption of green innovation among SMEs in the batik sector**

The adoption of green innovation among 11 Indonesian SMEs in the batik sector is varied, which can classify into three conditions, namely 'only green', 'brown and green' and 'from brown to green'. The only green SMEs are those that produce green products or have used the green production process since the beginning. The brown and green SMEs are the SMEs that do both innovations. They produce green products and also non-green products. The brown to green SMEs are those who switch from non-green products or processes to green products or processes.

For SMEs that have chosen to produce green products from the start, their involvement in achieving one of the goals of sustainable development, namely sustainable production and consumption, is evident. According to Liu et al. (2021), the concept of responsible production is the management of a production-based process that maintains a green process throughout the production process, not just providing corporate social responsibility funds as a way to wash hands of the grown production process that has been carried out. In other words, responsible enterprises are committed to upholding environmental ethics or oriented towards moral obligations for the benefit of all stakeholders and the planet's sustainability (Rui and Lu 2020). Consistency in running green production can influence other SME owners in the batik industry to do the same. Such SME owners' principle is not short-termism. The JLW owner chose to use batik colouring from plants, minerals and various environmentally friendly organic sources since the enterprise establishment's beginning, as expressed: '*I chose to use environmentally safe material and produce natural dye batik from the beginning because I want to make a unique and responsible product*'.

Producing green products is not as easy as imagined. Indonesia's low market demand and quite expensive production prices are still the main challenges (Chekima et al. 2015). Raw materials for green products are claimed to be difficult to obtain in large quantities and the product engineering process takes a lot of time (Gleim et al. 2013). Facing this situation, some SME owners seem to be implementing an ambidexterity strategy by continuing to produce brown products to meet old markets and maintain short-term profits while exploring new markets by developing green products to meet community and government demands, also anticipating increasing future demand. The GLB owner explained his decision to develop the green products together with the existing products and stated,

> I see the opportunity in the green products, but the existing system already runs well, and the old product has the market and has cheaper production cost. Therefore, I keep the old one, but expand the product with green products.

The next type is from brown to green products. From the interview with SME owners, there are also SME owners who have had to move from brown to green purely due to government and community pressure. According to Liu et al. (2021), regulatory and societal pressure mechanisms are the main factors driving production process changes to become more sustainable in many enterprise contexts in various parts of the world. However, it is essential to note that there should be a balance between mandatory laws and regulations with incentives in the form of support funds and tax cuts for a smooth transition process (Rui and Lu 2020). This situation is supported by one SME (BPL). The BPL owner described the reason to eventually avoid the chemical dye in the batik production process.

> My company is required to change our production process because of the problem that occurs from the chemical. Government and community ask us to change if we still want to run the business.

(2) **Factors affecting the adoption of green innovation among SMEs in the batik sector**

This study used the lens of institutional theory which assumes that companies must compromise to fulfil personal values and certain norms such as from religious communities, society, industry and government to gain certain legitimacy or recognition is called isomorphism (Dimaggio and Powell 1983; Oliver 1991). In the context of our study, it is about legitimation or recognition regarding the implementation of green production and green innovation within a company in the batik sector. The isomorphism will occur when batik companies show similar behaviour in adopting green innovation, by using non-chemical materials, producing new green products, or new manufacturing processes (e.g. energy saving). This is considered a result of the institutionalisation process.

From the in-depth interview with 11 SME owners in the batik sector in Indonesia. We identified five factors that stimulate Indonesian SMEs to adopt green innovation in their business. Those factors arise from internal/personal and external pressures. The factors that occur from internal/personnel are personal religious values and personal environmental ethics, and organisational environmental values. On the other hand, the factors that occur from external pressure are national and international regulations, public community norms and trends and market demand. Each factor will be elaborated on in detail in the next subsection.

Internal/personal factors

a. *Personal religious values and personal environmental ethics*

Some batik SME owners implement green innovation because they are driven by the religious philosophy and code of environmental ethics taught by their respective religious figures. Local religious beliefs can be considered informal institutional forces that are not easily changed and the source of environmental ethics that at least adhere to by their followers, such as the eco-theology implementation of Eco-Islam in Saudi Arabian SMEs (Abdelzaher and Abdelzaher 2015) and Buddhism in China's polluting industries (Du et al. 2014). Environmental codes of ethics derived from religion are claimed to be stronger drivers of green innovation in the long run compared to mere regulatory coercion (Du et al. 2014). In other words, green innovation decisions based on environmental ethics and/or religious values consider the ideals of morality more than pragmatism seeking market or government validation. The ECJ, JLW and ARN owners support this argument as described by their motivations in the following quotations.

> My religion taught me to be responsible for the environment. If I use the dangerous chemical in the production process, I sin. - ECJ company

> God gives us so many options in nature. You can choose every colour that you want by using these natural materials. Hence, you do not have any reason to destroy nature by using dangerous chemical materials. - JLW company

> I am a Muslim. Hence, in business, the values of Islam will be my reference. Green product, in my opinion, is suitable for Islam. We call it halalan thayyiban, which means it does not destroy nature and the human body. - ARN company.

b. *Organisational environmental values*

The owners-managers of batik SMEs endeavour to internalise the ethical norms and beliefs in their organisational strategy. They consciously understand that businesses should protect the planet's ecosystem and not get caught up in short-term interests, such as excessive extraction of natural resources or use of environmentally damaging substances. This corporate environmental ethics policy can also be a competitive advantage as an intangible resource and capability to seize new opportunities amidst the widespread discourse about the urgency of handling climate change (Singh et al. 2019). For the green strategy and policies to be successful, environmental training should then be conducted for the employees to develop the employee's green competency (El-Kassar and Singh 2019). ARN, GLB, FLO, and MTE owners' statements substantiate this phenomenon.

> Welfare is not only about money, but also comfortability. The money would be useless if we do not have clean water, clean air, and a healthy environment. - ARN company

> We can use plants as raw material for the product without harming the environment. Using chemical or synthetic material will harm the environment because it cannot be decomposed. It will have a bad impact in the future. - GLB company

> We built this company to promote the beauty of Indonesian culture. We cannot do that by offering products that harm nature. Hence, we choose to be an environmentally friendly company. - FLO company

> We are committed to producing environmentally friendly products only. - MTE company

External factors

a. *National and international regulations*

Some batik SMEs admit that national and international regulations force them to change the way they operate their business to be greener. The government encourages these changes in a persuasive way, for example, by providing training programmes and financial support related to green innovation of colouring techniques for textile products. A hard approach is also implemented by improving standards in the Indonesian textile industry since 2015 due to suspicions of environmental damage because of chemical dyes. International regulations such as the 1995 Geneva convention regarding eco-labelling for clothing and textiles also encourage SMEs' batik not to use chemical dyes, especially if they want to export batik to Europe.

According to Peng et al. (2021), there are two types of international regulations in the context of adopting green innovation: command or control-based and market-driven. If the former results in production ban sanctions and company closures when they break the rules that make it more efficient in the short term for change, the latter uses some instruments, such as environmental taxes, grants, training and subsidies, to encourage companies to innovate more flexibly. Both of these ways can act as a catalyst to gain new external knowledge from regulators so that a company can reduce organisational inertia and reconfigure its organisational systems and routines to create green product innovation and green process innovation (Zhang et al. 2019). The owner or manager of SMEs becomes a key person in monitoring the constant changes in environmental regulations and leverages the learning capabilities of its employees to face either command or market-driven types. The FLO, DRB and BIX companies outline the changes in their business operations following soft and hard pressure from regulators as follows.

> As a batik SME, we were asked to reduce pollution by the government and meet the criteria as an environmentally safe company. The government supported us in fulfilling the regulations. In 2003, our company was invited to join a training program about environmentally safe production processes, and the government introduced us to natural colouring techniques for textile products. Since then, I have started using natural colouring techniques for textile products. - FLO company

> The government persuaded us to accommodate the idea of green products due to the regulation about pollution issues caused by the textile production process. In 2012, the SMEs Association I joined was supported by a grant from the Ministry of Cooperatives and Trade [Indonesia] to learn about natural dye techniques. The training interested me in natural dye techniques, and I decided to diversify the product and accommodate the idea of green products. Therefore, my company can comply with the regulation about environmental protection. - DRB company

> In 2009, the Indonesian government issued regulations that forbid us to use dangerous chemical dyes [The regulation about protection and management of the environment: Law No. 32, 2009]. Furthermore, the Geneva convention in 1995 about eco-labelling and other environmental requirements for the developing country for clothing and textile had restrictions to sell products that contain dangerous chemical dyes to Europe. - BIX Company

b. *Public community's norms*

In addition to the national and international regulations, we also identified that there are certain public norms that may affect the adoption behaviour. The consciousness not to harm neighbours or the wider public community is entrenched in some batik SMEs driving them to avoid chemical dyes. The public community demands also arise, for instance, by proposing batik SMEs adopt certain special technology, such as waste treatment plants. If they fail to meet societal expectations, social sanctions will apply, eventually leading to a bad reputation. The presence of such SMEs batik is a curse for the community where it operates.

Guoyou et al. (2013) stated that the power of community stakeholders can vary depending on income. Those who live in poor areas may not have the power to influence company policies to uphold environmental ethics, while those who are rich may do so. Therefore, the environmentalist civil society formed various non-governmental organisations (NGOs) to consolidate themselves so that their aspiration for a green policy would be more heard by company owners (Schweitzer and Meng 2022). The following quotes express how the public pressures batik SMEs to go green.

> In the beginning, our company used chemical dyes for producing batik. We did not have a waste treatment plant. Our chemical waste polluted the neighbour's water reservoir. Local government and our neighbours directed us to manage our waste. - BPL company

> Previously, we considered the green products unprofitable. We do not have a waste treatment plant because it is costly. However, now we have the waste treatment plant and improve our ability to make green products because the people around us asked for it. - BRJ company

c. *Trend and market demand*

Based on the in-depth interview with SME owners, we found that the uniqueness and aesthetics of batik produced by SMEs with natural dyes attract the attention of domestic and foreign markets. The production costs of batik with safe materials for nature are relatively higher compared to that with chemical-based materials. However, the prospective demand for batik is expected to be able to cover the costs as expressed confidently by the batik SME owners below.

Huang et al. (2016) stated that market pressure can be caused by increased customer awareness of environmental issues, customer preference for environmentally friendly products, customer ongoing concern for a company's environmental behaviour and customers seeking green suppliers. Various countries that are committed to sustainability in the world campaign to build customer awareness about sustainability which can eventually accumulate into market pressure to encourage companies to be more responsible by carrying out green innovations (Cuerva et al. 2014).

> We decided to use safe material for nature, even though the cost would be higher than chemical-based material. We have an advantage in the uniqueness of the products. Customers interested in natural products are willing to buy our batik even if the price is higher than the conventional one. - MTE company

> We have loyal customers who specifically asked for natural dye products. Our customers said the batik, which is coloured by leaves and flowers, is very beautiful, and they feel good about not contributing to environmental damage. The aesthetic of the products and their pro-environmental view are their reasons to buy. - GLB company

> Based on my experience, the demand for green products can be categorised as prospective, the same level as a synthetic-material product. - ECJ company

> Most of our customers are foreigners. They asked us to produce batik, which is safe for humans and the environment. They are willing to pay more for its uniqueness. The uniqueness is not only about the products but also the process. They would like to learn about how these green products were created. - SKB company

In sum, the green innovation adoption process is influenced by the religious, social and environmental values of SME owners. They seek to take others' safety and the sustainability of the environment as a priority in their strategic decision-making considerations to produce batik. Apart from internal drivers, the decision of some batik SME owners to accommodate green innovation as their strategy, either switch to green products or do diversification, seems to be their effort to attain legitimacy and fitness to society's requirements. The SMEs modified their production process due to demands and pressures.

Furthermore, knowledge is identified as an essential factor that supports green innovation adoption (Rogers 2003) as indicated by this study. The SME owners are exposed to the knowledge of green innovation products and processes obtained from various key stakeholders such as the government, civil society, communities, suppliers and customers. Albort-Morant et al. (2016) refer to this process as relational learning activities that allow SMEs to develop dynamic capabilities, for example, reconfiguring their production systems and organisations to become greener. Moreover, to gain legitimacy from the Indonesian government, batik SMEs in Indonesia can learn and apply knowledge of the Green Industry Standard (GIS) set by the Ministry of Industry No. 39 of 2019. GIS regulates various environmentally friendly batik production standards, such as the use of batik wax, the limit value for natural and synthetic dyes, types and quantities of maximum energy and water consumption, packaging and waste management (Indrayani et al. 2020).

To further optimise the utilisation of knowledge about green production systems, the batik SMEs need to implement knowledge management, starting from creation, acquisition, sharing and application (Abbas and Sağsan 2019). Knowledge creation and acquisition go into the R&D activities of SMEs which can be intensified to discover new technologies and efficient processes in the batik industry to achieve economic, social and environmental sustainability. The newly acquired knowledge can then be shared with all employees and external parties for application. The application process requires a smooth transition of organisational changes and should consider how strong the absorptive capacity of a batik SME that wants to adopt new knowledge or technology regarding cleaner production (Zhang et al. 2020).

Factors that are affecting the decision of SMEs to go green are depicted in Fig. 3.1.

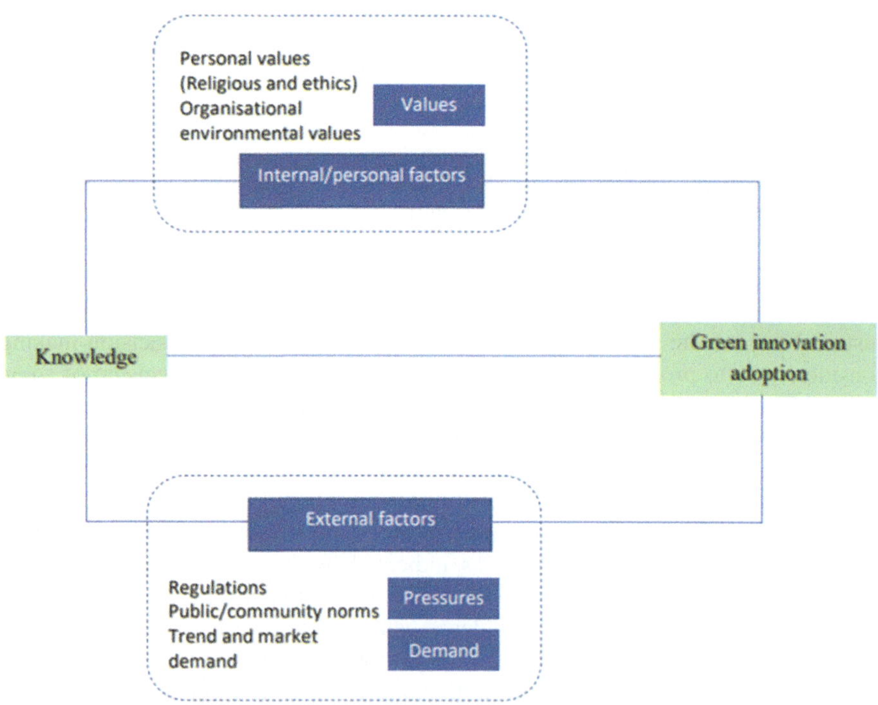

Fig. 3.1 Green innovation adoption factors (*Source* author's proposed framework)

3.5 Conclusion, Limitations, and Future Research Avenues

This study focused on the adoption of green innovation among SMEs in Indonesia from the batik sector. This business sector heavily used non-natural materials, which contributed to environmental issues. Exploring the type of the adoption of green innovation and the factors that affect their adoption becomes crucial to provide a deeper understanding of this field. Additionally, the use of institutional theory provides a better picture of what drives SMEs to adopt green innovation. The sense of legitimacy and recognition and attempts to mimic or become isomorphism are considered as the sources of green innovation adoption among SMEs in the batik sector in Indonesia.

The findings of the study based on in-depth interviews with 11 SME owners in the batik sectors in Indonesia conclude that there are still variations of adoption of green innovation among the SMEs, from purely green, the combination of brown and green and the shifting from brown to green. The variations are factors that determine SMEs' decision to adopt green innovation, which is mainly derived from internal and external factors. The internal factor is categorised as personal religious values, personal environmental ethics and organisational environmental values. On the other hand, external factors are defined as national and international regulations, public community norms, trends and market demands. Knowledge about green products and production obtained relationally from stakeholders, either voluntarily learned or forced, becomes the provision for SMEs to continue to maintain green innovation or change production systems from brown to green.

Participants of this research are limited to SMEs in the textile industry (batik). In fact, green practices do not only occur in the context of the textile industry. The need for doing green business may happen in other sectors such as cosmetics, food and beverages, furniture, etc. In the future, it is relevant to explore green innovation adoption in other sectors, such as food and beverages, as the most considerable number of Indonesian SMEs. The dynamic process of knowledge in the green innovation adoption process can also be explored further, for instance, by using the socialisation, externalisation, combination and internalisation (SECI) framework (Nonaka and Takeuchi 1995).

References

Abbas J, Sağsan M (2019) Impact of knowledge management practices on green innovation and corporate sustainable development: a structural analysis. J Clean Prod 229:611–620

Abdelzaher DM, Abdelzaher A (2015) Beyond environmental regulations: exploring the potential of "'Eco-Islam'" in boosting environmental ethics within SMEs in Arab markets. J Bus Ethics 145:357–371

Aboelmaged M, Hashem G (2019) Absorptive capacity and green innovation adoption in SMEs: the mediating effects of sustainable organisational capabilities. J Clean Prod 220:853–863

Aguilera-Caracuel J, Ortiz-de-Mandojana N (2013) Green innovation and financial performance: an institutional approach. Organ Environ 26(4):365–385

Albort-Morant G, Leal-Millán A, Cepeda-Carrión G (2016) The antecedents of green innovation performance: a model of learning and capabilities. J Bus Res 69(11):4912–4917
Arfi WB, Hikkerova L, Sahut JM (2018) External knowledge sources, green innovation and performance. Technol Forecast Soc Chang 129(January):210–220
Boohene R, Gyimah RA, Osei MB (2020) Social capital and SME performance: the moderating role of emotional intelligence. J Entrep Emerg Econ 12(1):79–99
Bruton GD, Ahlstrom D, Li HL (2010) Institutional theory and entrepreneurship: where are we now and where do we need to move in the future? Entrep: Theory Pract 34(3):421–440
Capasso M, Hansen T, Heiberg J, Klitkou A, Steen M (2019) Green growth–a synthesis of scientific findings. Technol Forecast Soc Chang 146(June):390–402
Chekima B, Azizi S, Syed W, Wafa K, Igau OA, Chekima S, Sondoh SL (2015) Examining green consumerism motivational drivers: does premium price and demographics mattern to green purchasing? J Clean Prod 112(4):3436–3450
Chen YS (2008) The driver of green innovation and green image-green core competence. J Bus Ethics 81(3):531–543
Cheng CCJ (2020) Sustainability orientation, green supplier involvement, and green innovation performance: evidence from diversifying green entrants. J Bus Ethics 161(2):393–414
CNN Indonesia (2022) https://www.cnnindonesia.com/nasional/20220613155729-293-808397/freeport-indonesia-raih-penghargaan-sustainable-business-awards. Accessed 4 Dec 2022
Creswell JW, Poth CN (2018) Qualitative inquiry & research design: choosing among five approaches, vol 1. Sage Publications
Cuerva MC, Triguero-Cano Á, Córcoles D (2014) Drivers of green and non-green innovation: empirical evidence in Low-Tech SMEs. J Clean Prod 68:104–113. https://doi.org/10.1016/j.jclepro.2013.10.049
Dimaggio PJ, Powell WW (1983) The iron cage revisited: institutional isomorphism and collective rationality in organizational fields. Am Sociol Rev 48(2):147–160
Du X, Jian W, Zeng Q (2014) Corporate environmental responsibility in polluting industries: does religion matter? J Bus Ethics 124:485–507
Ebrahimi P, Mirbargkar SM (2017) Green entrepreneurship and green innovation for SME development in market turbulence. Eurasian Bus Rev 7(2):203–228
El-Kassar A, Singh SK (2019) Green innovation and organizational performance: the influence of big data and the moderating role of management commitment and HR practices. Technol Forecast Soc Chang 144(July):483–498
Elkington J (1998) Cannibals with forks. Capstone Publishing
Firmansyah A (2020) The study of soga brown colour visual digitization from classical batik of yogyakarta. In: IOP conference series: materials science and engineering, vol 924(1), pp 1–6
Gleim MR, Smith JS, Andrews D, Cronin JJ (2013) Against the green: a multi-method examination of the barriers to green consumption. J Retail 89(1):44–61
Guoyou Q, Saixing Z, Chiming T, Haitao Y, Hailiang Z (2013) Stakeholders' influences on corporate green innovation strategy: a case study of manufacturing firms in China. Corp Soc Responsib Environ Manag 20(December):1–14
Horbach J, Rammer C, Rennings K (2012) Determinants of eco-innovations by type of environmental impact-the role of regulatory push/pull, technology push and market pull. Ecol Econ 78:112–122
Huang XX, Hu ZP, Liu CS, Yu DJ, Yu LF (2016) The relationships between regulatory and customer pressure, green organizational responses, and green innovation performance. J Clean Prod 112:3423–3433
Indrayani L, Triwiswara M, Evtriyandani (2020) Prinsip efisiensi energi untuk mewujudkan industri batik yang berkelanjutan (sustainable industry). In: Prosiding seminar nasional industri kerajinan dan batik, pp 1–10
Journeault M (2016) The Integrated Scorecard in support of corporate sustainability strategies. J Environ Manage 182:214–229
Jun W, Ali W, Bhutto MY, Hussain H, Khan NA (2021) Examining the determinants of green innovation adoption in SMEs: a PLS-SEM approach. Eur J Innov Manag 24(1):67–87

Kostova T (1997) Country institutional profiles: concept and measurement. Acad Manag Proc 1:180–184

Kusumawati N, Rahmadyanti E, Sianita MM (2021) Batik became two sides of blade for the sustainable development in Indonesia. In Green chemistry and water remediation: research and applications. Elsevier Inc.

Levi PA, Hunga AIR, Sidabalok H (2021) Reviewing batik with natural coloring based on the clean production theory with an ecofeminist perspective (Case Study: Putri Kawung Batik Community in Klaten). Salasika: Indones J Gend Women Child Soc Incl Stud 4(1):1–12

Li D, Zheng M, Cao C, Chen X, Ren S, Huang M (2017) The impact of legitimacy pressure and corporate profitability on green innovation: evidence from China top 100. J Clean Prod 141:41–49

Liu F, Lai K, Cai W (2021) Responsible production for sustainability: concept analysis and bibliometric review. Sustainability 13:1–27

Miles MB, Huberman AM, Saldana J (2014) Qualitative data analysis: a methods sourcebook. Sage Publications

Nonaka I, Takeuchi H (1995) The knowledge-creating company. Oxford University Press

Oliver C (1991) Strategic responses to institutional processes. Acad Manag Rev 16(1):145–179

Peng H, Shen N, Ying H, Wang Q (2021) Can environmental regulation directly promote green innovation behavior ?—based on situation of industrial agglomeration. J Clean Prod 314(June):128044

Qi G, Jia Y, Zou H (2021) Is institutional pressure the mother of green innovation? Examining the moderating effect of absorptive capacity. J Clean Prod 278:123957

Raya AB, Andiani R, Siregar AP, Prasada IY, Indana F, Simbolon TGY, Kinasih AT, Nugroho AD (2021) Challenges, open innovation, and engagement theory at craft SMEs: evidence from Indonesian batik. J Open Innov: Technol Mark Complex 7(121):1–24

Rogers EM (2003) Diffusion of innovations. Free Press

Ruef M, Scott WR (1998) A multidimensional model of organizational legitimacy: hospital survival in changing institutional environments. Adm Sci Q 43(4):877–904

Rui Z, Lu Y (2020) Stakeholder pressure, corporate environmental ethics and green innovation. Asian J Technol Innov 29(1):70–86

Schweitzer F, Meng Y (2022) How collaborating with NGOs makes green innovations more desirable. Bus. Soc., 1–38

Singh SK, Chen J, Giudice MD, El-Kassar A (2019) Environmental ethics, environmental performance, and competitive advantage: role of environmental training. Technol Forecast Soc Chang 146(May):203–211

Takalo SK, Tooranloo HS, Parizi ZS (2021) Green innovation: a systematic literature review. J Clean Prod 279:122474

Ullah R, Ahmad H, Rehman FU, Fawad A (2021) Green innovation and Sustainable Development Goals in SMEs: the moderating role of government incentives. J Econ Adm Sci (a-head-of-print)

Wang CY (2019) Building a network for preserving intangible cultural heritage through education: a study of Indonesian batik. Int J Art Des Educ 38(2):398–415

Widjojo H, Gunawan S (2020) Indigenous tradition: an overlooked encompassing-factor in social entrepreneurship. J Soc Entrep 11(1):88–110

Yin RK (2018) Case study research and applications: design and methods. Sage

Zhang J, Liang G, Feng T, Yuan C, Jiang W (2019) Green innovation to respond to environmental regulation: how external knowledge adoption and green absorptive capacity matter? Bus Strateg Environ 29(1):39–53

Zhang Y, Sun J, Yang Z, Wang Y (2020) Critical success factors of green innovation: technology, organization and environment readiness. J Clean Prod 264:121701

Towards Adopting Innovative Quick Response (QR)-Enabled Contactless Transaction Payment: The Malaysian MSMES' Entrepreneurial Perspective in COVID-19 Setting

4

Lennora Putit and Zahariah Sahudin

Abstract

Studies have shown that human nature is continuously evolving, and consumers are inevitably affected when it comes to adopting a new set of consumption behaviour. In the recent global pandemic notably termed as COVID-19, most consumers were fearful and forced to rely on new technological method of payments in the quest to acquire a safer method to buy goods and services. Entrepreneurs within the Micro, Small and Medium Enterprise (MSME) sector are no exception to such a phenomenon. The COVID-19 pandemic has given a huge socio-economic impact on many nations worldwide including Malaysia. According to the Malaysian Institute of Economic Research report, Malaysia's GDP development would decline by 2.61% in 2020 (Bernama 2020). Such an indicator is alarming enough for most businesses, resulting in the need to enhance buyers' spending power and stimulate socio-economic growth. One of the potential mechanisms is to analyse customers' needs by leveraging on their technological and socio-psychological well-being issues considering COVID-19 fearful circumstances. This could then potentially impact their acceptance of alternative cashless transaction payment mechanisms from MSME business owners and respective stakeholders (Chin et al. 2022; Amer et al. 2020). The use of mobile contactless payments is ascending throughout the recent COVID-19 occurrence and Movement Control Order (MCO) in Malaysia, with certain business competitors perceiving an increasing number of support for such financial transaction payment adoption (Mokhtar 2019; The Star 2020). A contactless payment is an innovative programming application that utilises electronic gadgets such as personal computers, or portable mobile smartphones for online

L. Putit (✉) · Z. Sahudin
Faculty of Business and Management, Universiti Teknologi MARA, Puncak Alam, Selangor, Malaysia
e-mail: lennora633@uitm.edu.my

A. Rafiki et al. (eds.), *Open Innovation in Small Business*, Contributions to Environmental Sciences & Innovative Business Technology,
https://doi.org/10.1007/978-981-99-5142-0_4

exchanges (Uduji 2019). It is an optional business strategy for merchandise or services' payments, which include the use of Quick Response (QR)-based mobile payment platforms. Adjustments could potentially be made towards many consumers' 'new normal' spending transaction payment procedure including propensity to use contactless payments and consider online platforms as an alternative shopping channel for necessities (Chin et al. 2022). Although gaining increased global acceptance, Malaysian MSMEs however, are still sceptical towards embracing contactless payment methods as a means of undertaking their business transactions (Aji et al. 2020). As such, it is thus pertinent to provide valid and explanatory reasoning for the potential adoption of QR-enabled contactless payments. Considering the pre-, during and post-COVID-19 pandemic fear, addressing these challenges could potentially be a novel approach towards investigating the extrinsic stimuli of MSME entrepreneurs and its implications on the nation's economic system. Drawing from the seminal Technology Acceptance Model and Fear Motivating Theory, a proposed model that explains the use of QR-enabled contactless payments based on psychosocial and technological perspectives is developed that could potentially result in significant national economic and societal implications. A mixed method-based research design would be employed to gather the necessary data in addressing the identified issues amongst the targeted population of interest. Several potential contributions could potentially be observed such as technological diffusion, fear and need for personal hygiene preservation in proximity transactions, as well as ecological benefits such as promoting reduction of carbon footprint associated with printing and disposing of paper and plastic-based banknotes, respectively.

Keywords

Innovation • Adoption • Contactless transaction payment • Quick response • MSME entrepreneurs • Technology acceptance model • Fear motivating theory • COVID-19

4.1 Introduction

Prior research has illustrated that human nature is continuously evolving. Consumers in general are affected when it comes to embracing a contemporary brand of consumption behaviour. After the recent occurrence of a worldwide pandemic most famously known as the coronavirus, consumers driven by fear are compelled to rely on up-to-date technological means of payments in search of more secure ways to procure goods and services. Entrepreneurs and stakeholders in the SME sector who conduct both business-to-business (B2B) and business-to-consumer (B2C) transactions within the SME sector are no exception to such a phenomenon.

The coronavirus, termed COVID-19 pandemic, has given a huge socio-economic impact on many nations worldwide including Malaysia. According to the Malaysian Institute of Economic Research report, Malaysia's GDP development will decline by 2.61% in 2020 (Bernama 2020). Such an indicator is alarming enough for most businesses resulting in the need to enhance buyers' spending power and stimulate socio-economic growth. One of the potential mechanisms is to leverage on addressing the technological and socio-psychological well-being from the perspectives of both business owners and stakeholders (Chin et al. 2022; Amer et al. 2020).

The use of contactless payments and reception of e-wallets are on the ascension throughout the COVID-19 occurrence and Movement Control Order (MCO) in Malaysia, with certain competitors perceiving the increasing number of supporters for such payment adoption (The Star 2020). Numerous adjustments could be made in customers' spending transactions that are the 'new normal', and that included the propensity to use cashless/contactless/e-wallet payments amongst most consumers at large, and also consider an alternative form of online shopping for necessities (Chin et al. 2022). A contactless payment is an innovative programming application that utilises electronic gadgets such as personal computers or portable cell phones for online exchanges (Uduji et al. 2019). It is an elective strategy for merchandise or services' payment. Contactless payments vary from 'pay wave' based credit and debit cards, as well as QR and Near-Field Communication (NFC)-based mobile payment platforms (Karjaluoto et al. 2019).

4.2 Problem Statement

There have been a number of contactless payment systems available in the Malaysian market such as AliPay, Touch' n Go, Boost and Grab Pay. This system has also been applied to pin based debit and credit cards from many commercial banks such as Maybank, CIMB, Public Bank and RHB Bank upon the announcement made by the Central Bank of Malaysia to replace all ATM cards to become PIN-based cards in 2016 (Bank Negara Malaysia 2016). The PIN-based cards of CIMB Bank, Public Bank, Maybank, RHB Bank and other banks were accepted for contactless payment functions in many major retail stores in Malaysia that included AEON BIG, TGV Cinemas, McDonald's, Golden Screen Cinemas and Starbucks (Joifin 2018). Nonetheless, the usage is still equivocal amongst MSMEs businesses, and if not seriously considered, it may impede the advancement of the MSME business ecosystem (Amer et al. 2020). Although gaining increased global acceptance, Malaysian MSMEs however, are still sceptical towards embracing contactless payment methods as a means of undertaking their business transactions (Aji et al. 2020).

Despite the rapid growth observed in mobile technologies' usage nowadays amidst the global pandemic state, the application of NFC-enabled mobile devices such as the aforementioned contactless or e-wallet payment to undertake financial transactions of goods/services is still new to many consumers in Malaysia (Amer et al. 2020; Balachandran and Tan 2015) but at the same time, have seen a potential acceptance towards such adoption (The Star 2020). Studies (e.g. Amer et al. 2020; Karjaluoto et al. 2019) have shown that contactless forms of payments are slowly gaining acceptance. At present, only 18% of Malaysians are using contactless payment methods as a means of doing transactions, mainly due to the majority's preferential need to hold cash (Shields 2018). Hence, the development of a new model is pertinent to provide valid and explanatory reasoning for the potential adoption of contactless payments, to which both technological and psychosocial factors as well as fear of COVID-19 infection could potentially be a novel approach towards investigating the extrinsic stimuli of SMEs entrepreneurs and its implications to a nation's economic system.

Following the above concerns, this proposed study generally seeks to investigate the influencing factors affecting contactless payment amongst MSME entrepreneurs amidst the recent and post-COVID-19 global pandemic. Specifically, it aims to explore the degree to which perceived ease of use, perceived usefulness, trust, social influence, fear and convenience impact SMEs' attitude towards contactless payment implementation. In turn, the study intends to assess whether personal attitude influences SMEs' intention towards such usage in the current COVID-19 epidemic.

Several research issues are identified as stated below:

- What are the SME entrepreneurs' current level perception of the contactless payment scheme within the recent and post COVID-19 global pandemic?
- To what extent do perceived usefulness, perceived ease of use and social influence affect contactless payment adoption amongst the SME entrepreneurs?
- Why do trust and fear influence contactless payment adoption amongst SME entrepreneurs?
- How does convenience influence contactless payment adoption amongst SME entrepreneurs?

4.3 Literature Review

4.3.1 An Overview of MSMEs in Malaysia

Malaysia's economy has gradually transitioned from a conventional economy to one that is reliant on the digital form. A financial system built on digital computer technology is referred to as a 'digital economy', in which transactions are made in digital trading platforms using the Internet (Jaafar 2018). As a response, the global realm, not excluding the members of the Association of South East Asian Nations (ASEAN), is undergoing a noteworthy economic and social transition that is being fuelled by technology via a digital economy.

The digital economy—evidently one major force behind business, job formation and advancement—is increasingly linked to economic prosperity and competitiveness in this rapidly changing landscape. In the context of developed and emerging economies, small and medium-sized businesses (SMEs) represent a significant source of employment, technological innovation and competitive advantage (Zahariah et al. 2020). SMEs are essential for fostering the financial growth of a nation. Furthermore, the very notion of SMEs has also cemented itself as any country's economic foundation. Malaysia, a nation that is steadily progressing, is no outlier. MSMEs in Malaysia, for example, are currently expanding proactively to improve and restructure the businesses to better align with Industry 4.0 and the principles brought about by the revolution (Lee and Wing Hooi 2017).

Even so, a majority of these MSMEs face financial difficulties as they attempt to innovate and restructure their establishments for the advent of Industry 4.0. For starters, developing nations' SMEs must pay a high price to embrace e-commerce infrastructure (Ahmad 2015). The difference between MSMEs, SMEs and larger businesses is typically higher. Therefore, this might have an impact on SMEs' long-term survival and competitiveness. Lee and Wing Hooi (2017) mentioned in their prior study that SME conferences have been conducted on e-commerce adoption possibilities such as cashless payment implementation in the Industry 4.0 era, and how the prospective revenue generated from this could reach the billion-dollar mark. That being said, SMEs' faith and stance towards such technological acceptance are still limited.

One of the major challenges is the minimal leveraging of companies' website, and corporate image (Ang et al. 2017). Furthermore, Lee and Wing Hooi (2017) further established that, although numerous+ MSMEs have commercial websites, they lack knowledge on the actual application of doing online business transactions, in particular provision of mobile and other forms of NFC-enabled mode of business payment. The websites are typically solely employed for communication and product and service promotion.

In addressing the aforementioned concerns, the review of literature is discussed, argued and synthesised as follows.

4.3.2 Contactless Payment

In today's business, the ability for organisations to combine diverse types of payment systems with smart mobile phone devices can provide consumers in the marketplace with a multitude of payment solutions. Examples include physical retail merchant services, fixed automated teller machines, electronic commerce (e-commerce) and mobile commerce (m-commerce) services. Albeit many past research on the development and growth of mobile business services, contactless or mobile payments (m-payments) have to a certain extent achieved varying degrees of success (Slade et al. 2013). The market introduction of contactless technology on a massive scale is pertinent in any type of payments sector. With regards to the greater magnitudes of control and convenience the customers could achieve, as well as a higher overall investment return for merchants, contactless payments have shown to benefit both B2B and B2C businesses and retail consumers (Brett 2017).

What is contactless payment? Contactless payment can be defined as the performance of financial transactions for goods and services, forgoing the prerequisite of exchanging physical cash, writing a cheque or entering a Personal Identification Number or PIN (Brett 2017). To reiterate, credit and debit cards, key fobs, smart cards, as well as other similar equipment that enable near-field communication (NFC) payments are the most prevalent forms of the system's presence. According to Nagashree et al. (2014a, b), radio-frequency identification (RFID) and other non-contact interconnectivity technologies were the precursors of NFC. In order to facilitate contactless transfers, NFC's relative simplicity and interoperability with current Bluetooth and Wi-Fi standards offer quite an appealing approach NFC-enabled contactless cards were created primarily for commercial application in the private sector (Trutsch 2014). Contactless card providers tout a number of advantages for both customers and retailers in order to encourage the utilisation of their cards.

Through the eyes of the consumer, this cutting-edge technology permits the swift, simple, and secure remittance of cash for products and services and has enhanced consumer convenience in carrying out daily transactions such as tolls incurred for transportation, parking services, grocery shopping and refilling of fuel. Perhaps the continuing global disease pandemic, particularly COVID-19 or the coronavirus, may also be blamed for the probable surge in contactless payment acceptance. After weeks of unremitting dread, there appears to be a decrement in the coronavirus infection rate, and plans are underway to start loosening lockdown restrictions in certain developed nations, which is offering a brief glimmer of hope to many. However, many developing nations disagree, warning that the issue may just be only getting started and that the toll from a significant COVID-19 epidemic is predicted to be considerably worse than in any advanced economy (Mohamed 2020). The COVID-19 disaster poses a consequential threat to many emerging regions in the world, and this has triggered many nations, including Malaysia, and global consumers at large, to be more fearful of its microbial infections that could possibly lead to severe medical fatality and death. In addition, a significant percentage of consumers avoid going shopping in person altogether out of fear of contracting an infectious disease and instead practise excessive social seclusion and other necessary hygiene precautions.

To better understand intention and actual behavioural adoption, the Protection Motivation Theory (Rogers 1975) as well as the Technological Acceptance Model (Davis 1989) as well as similar prior foundational studies along with pertinent literary frameworks are justifiably warranted. The Theory of Planned Behavior (TPB) (Ajzen 1991) posits an individual's intention is the factor that predicts behaviour the most closely, and the Theory of Reasoned Action (TRA) (Fishbein and Ajzen 1975) proceeds to explain that when intentions and behaviour are measured at the same level of specificity, they are thought to be strongly correlated. In response to these worries, a number of aspects are further investigated. In the wake of the COVID-19 epidemic, these variables may decide how subtly business consumers' propensities for adopting contactless payment transactions may change.

4.3.3 Behavioural Intention

Several researchers discovered that it is challenging to gauge actual adoption precisely if it is reviewed as a sole factor (Schuitema et al. 2013). Due to this, a majority of the research studies on the technology have focused on the adoption prediction's reliance on the variable of behavioural intention (Irani et al. 2009). Because behavioural intention is viewed to possess immediate and significant impact on actual behaviour, it is regarded as a predictor of behaviour (Fishbein and Ajzen 1975).

Ajzen (1991) asserts that intention is vital in regulating the catalyst components that would ultimately influence actual behaviour. The likelihood that a behaviour will actually be carried out will increase with the strength of the intention to deal with the behaviour. While Mutahar et al. (2017) employed intention as a factor of the mobile banking adoption amongst the customers in Yemen, Putit et al. (2020) and Putit and Muhammad (2015) concentrated on intention to explore the adoption of web-based travel transactions These prior research discovered that the actual adoption of emerging technology-related behaviours depends significantly on intentions. The intention to embrace e-payment possesses a significant beneficial impact on the real-life implementation of e-payment, according to Jusoh and Teng's (2019) research. In the light of these observations, intention is examined to gauge the targeted respondents' prospective acceptability of the studied contactless payment technology.

4.3.4 Attitude

It is an established fact that attitude towards usage has been identified as a variable which influences subsequent behavioural patterns or the reason for the intention that at last prompts specific conduct. In TAM, attitude towards usage is alluded to as the evaluative impact of a positive or negative sentiment of people in playing out a specific behaviour (Ajzen and Fishbein 2000). A prior study by Singh et al. (2020) discovered that the enhancement of behavioural intention towards e-wallet is reliant on the user's positive attitude. This study's comparison findings indicate that generation Y's inclination to use Sakuku in Indonesia is influenced by their attitude about behaviour (Jerry and Maulana 2017). Conversely, Shroffet al. (2011) discovered that there is no correlation between attitude and behavioural intention.

However, Teoh et al. (2020) observed that Malaysians have a positive outlook and are willing to give it a try given that this technique optimises payment speed while also being more user-friendly. Nevertheless, because the study primarily focused on a specific sampled population of interest that was not representative, it could not be generalised. According to Venkatesh and Davis's (2000) Technology Acceptance Model (TAM), behavioural intention was found to be influenced by both perceived usefulness and perceived ease of use, but they also highlighted the necessity for the development of an attitudinal component. Taking into account the aforementioned assertions, it could be hypothesised that attitude has the ability to affect the embracing of contactless payment technology by MSME entrepreneurs.

4.3.5 Perceived Usefulness and Perceived Ease of Use

The proportion to which a person views a certain paradigm as supporting their success in an activity is known as perceived usefulness (PU) (Mathwick et al. 2001). On the other hand, perceived ease of use (PEOU) is referred to how much development is straightforward (Zeithaml et al. 2002). Both factors have an impact on an information system's attitude (Venkatesh and Davis 2000). Alaeddin et al. (2018) supported the contention, to which there was a significant impact of perceived ease of use and perceived usefulness on attitude towards the implementation of technology. Moreover, Wijayanthi (2019) claimed that attitude towards utilisation could be influenced by perceived helpfulness. This case can likewise be reinforced by a comparative examination of these constructs undertaken by Jin et al. (2020).

Isrososiawan et al. (2019) discovered through their research the fact in which perceived ease of use creates a favourable and noteworthy influence on mobile payment implementation. As indicated by Davis (1989), who initially developed the Technology Acceptance Model (TAM), it is expected that PU and PEOU can impact attitude towards utilising technology. In any case, the assumption is unsound because Ajibade (2018) contended that factors outside the control of the potential users can influence the link between attitude towards using innovation and actual use. In this study, PU and PEOU use comprise a noteworthy impact on attitudes, which is hypothesised to be influencing variables in the intention to embrace contactless remittance of money amongst SMEs in Malaysia.

4.3.6 Trust

Vulnerability can be exposed to trust, which is a distinguishing climax of both social and financial cooperation significance (Pavlou 2003). Trust and social influence were mentioned separately in TAM2 and TAM3. Previous studies by Jin et al. (2020) have demonstrated a strong correlation between social and behavioural factors influencing acceptance of mobile wallets. According to Lai (2017), TAM2 and TAM3 were not significantly relevant when taking into account the novel idea of a single-stage e-payment system. Customers lack the confidence to implement contactless transactions (Aslam et al. 2017). However, Jerry and Maulana (2017) recommended that a variety of elements should be taken into account by trust, particularly those that have an impact on the implementation of cashless payments. In the light of this, it is hypothesised that trust may influence SMEs' attitudes towards the technological applications of cashless payments in their commercial dealings.

4.3.7 Convenience

Convenience refers to a person's inclination for advantageous products and services as far as user assessment of service encounters (Hsu and Chang 2013). Furthermore, prior research by Chang et al. (2012) on PDA-based English learning demonstrated essential constructive outcomes of perceived convenience on attitude towards the implementation of personal digital assistants. Another similar analysis revealed that attitudes towards using Moodle are unquestionably influenced by perceived convenience (Hsu and Chang 2013). Additionally, a study by Zarandona et al. (2019) discovered that mobile payment applications are easy to use and that both sexes agreed that using mobile devices to create purchases is convenient. In addition to convenience and usefulness, Lee and Jun (2007) argued that elements impacting selection goals should be taken into consideration by TAM. In any case, convenience is a major factor in choosing a single-stage e-payment system, according to Lai and Zainal (2015). The single-stage e-payment system's Technological Acceptance Model was extended in terms of convenience (Lai 2017). This study relies on the hypothesis that convenience influences SMEs' attitudes towards contactless payments.

4.3.8 Fear

For a long period of time, fear has been acknowledged as a potent stimulus that taught people the ways to respond in life-threatening situations. Given that the fear is brought on by environmental incentives, namely, threat, it is regarded as an extrinsic motivator. The protective motivation theory provides an explanation for this phenomenon in psychological investigations (Rogers 1975). The protection motivation theory (PMT) makes an effort to explain the driving force behind a specific behaviour that individuals exhibit in response to a perceived threat. The PMT states that humans' threat-avoiding behaviour is based on the following variables: (1) the perceived seriousness of a threat, (2) our confidence in our capacity to respond to the situation, (3) the likelihood that the threat will materialise and (4) the likelihood that a response (fight or flight) will be successful.

Based on this theoretical assumption, customers may discover that their fear of contracting microbes is a powerful motivator for them to be proactive about practising hygiene in their regular activities. Self-care, eating, socialising, travelling and—interestingly—making purchases at stores are amongst these activities. Even though the COVID-19 pandemic has negatively impacted the economy by way of business closures and social restrictions, vital services are still experiencing high amounts of daily visits from paying customers; this is evident in places such as the retail marketing industry. Even if a consumer wears a face mask, they still risk getting sick if they come into touch with contaminated objects like banknotes and card terminals as they are checking out. Thus, patrons are therefore encouraged to choose contactless forms of payment that prevent direct physical contact with these contact surfaces if they are concerned about the possibility of infection and are well aware of this issue.

Considering the aforementioned review of past literature, this study proposes the development of a number of hypotheses.

H1—Perceived usefulness impacts attitude towards transactions of contactless remittance amongst MSME entrepreneurs.
H2—Perceived ease of use impacts attitude towards transactions of contactless remittance amongst MSME entrepreneurs.
H3—Social Influences impact attitude towards transactions of contactless remittance amongst MSME entrepreneurs.
H4—Trust impacts attitude towards transactions of contactless remittance amongst MSME entrepreneurs.

4.4 Research Methodology

The proposed research is a combination of exploratory, descriptive and causal design as it is intended to model SME entrepreneurs' socio-psychological and technological attributes that can potentially result in the use of contactless payments technology. With this in perspective, the study would be applying both categories of research methods. Firstly, a qualitative approach is applied via focus group interviews amongst target respondents to explore the challenges and issues facing the potential usage of contactless payments in their SME business transactions. Besides, it is aimed to determine the relevancy of constructs identified for this study based on the extant literature reviews. Secondly, a quantitative approach is subsequently applied by conducting a survey to examine the contribution of each attribute derived from the initial qualitative phase.

4.4.1 Qualitative

In this data collection approach, 15 selected respondents across diverse ethnic and education backgrounds will be gathered via physical or virtual presence (using Zoom/WEBEX/Microsoft Teams/Google Meet platforms). Firstly, open-ended questions are developed to explore their business transactions experience and opinions are recorded to indicate their stance on the contactless payments in the current COVID-19 pandemic. Issues on resistance and receptiveness of such transactions are discussed, Secondly, the identification of these factors is further explored to ensure relevancy when considering shifting to a new form of business payment transactions. Thirdly, potential expectations on the current policies and regulations of the relevant authorities are to be probed on whether it could result in a hindrance towards using contactless forms of payments. At the end of the session, participants are also probed further in providing suggestions on how to overcome the challenges faced by SME entrepreneurs in performing B2B and B2C transactions that are high in microbial transmission risk.

Thematic content analysis is used in data analysis, where specific themes are developed to match the replies provided by the focus group respondents. Correct themes or attributes can be identified for use in the following phase of data collecting utilising this technique. The results from this focus group discussion will be validated by the panels of academics who are also experts in the area of study. The identified themes will be discussed further by these committees. The construction of items to evaluate each characteristic then follows after the final set of attributes has been decided. This activity involves the adoption of existing item measurements from published research work and relevant suggestions from the panels of experts. ATLAS T1 statistical software will be utilised in transcribing the data and thematic content analyses, as reiterated, will be applied in deriving potential themes from the qualitative data generated.

4.4.2 Quantitative

The quantitative data collection stage will be carried out continually until the final agreed-upon instrument has been determined. This stage involves validation of the newly created/adopted instrument to measure the consumers' psychological attributes that lead to the use of contactless payments technology, through exploratory questionnaire-based survey. Three hundred respondents will be purposively selected amongst SME players throughout various regions within the Peninsula Malaysia. The analysis will be conducted using Structural Equation Modelling (SEM) via SmartPLS4.0 statistical software. Through SEM, measurement and structural model of the proposed attributes can be determined.

Upon the initial data cleaning procedures, descriptive analysis will be carried out to determine the respondents' profiles. Thereafter, exploratory factor analysis using SPSS will be undertaken, followed by confirmatory factor analysis using SEM. Descriptive analysis will also be used to gauge the respondents' scores on each attribute. All the processes are meant to develop a validated newly developed instrument to assess consumers' psychological attributes that lead to the use of contactless payments technology so the SME entrepreneurial attributes can be assessed and empirically linked to business success. The entire data collection stage will address both research questions and hypotheses accordingly.

4.5 Expected Contributions

Several potential contributions could be generated from this study. In addition to technological diffusion, fear and need for personal hygiene preservation in close proximity transactions, it could also bring ecological benefits such as promoting reduction of carbon footprint associated with printing and disposing of paper and plastic-based banknotes. This research is also expected to contribute towards policy making framework that could assist the relevant government agencies in developing best practice policies with regards to contactless payment adoption, especially amongst commercial players such as financial service providers, and that in turn subsequently can assist major or small and medium-scale (SMEs) businesses to adopt such payment scheme for ease of transactional activities via offline or online technological platforms. Both alternative business and payment platforms are particularly sought after in the aftermath of COVID-19 pandemic setting.

4.6 Conclusion

This proposed study is developed in an attempt to understand how perceived ease of use, perceived usefulness, trust, social influence, fear, and convenience impact SME entrepreneurs' attitude and intention towards contactless payment implementation in the midst of the contemporaneous global epidemic. Through the seminal Fear Motivating Theory and Technology Acceptance Model, a theoretical model that explains the use of contactless payments based on psycho-social and technological perspective is a novel approach that could potentially result in significant and economic and societal implications. A mixed method-based research design would be employed to gather the necessary data in addressing the identified issues amongst the targeted population.

Acknowledgements The authors would like to thank the Faculty of Business and Management, Universiti Teknologi MARA, Puncak Alam Campus, Selangor, Malaysia, for its support in conducting this research.

References

Ahmad SZ (2015) An empirical study of factors affecting e-commerce adoption among small- and medium-sized enterprises in a developing country: evidence from Malaysia. https://doi.org/10.1080/02681102.2014.899961

Aji HM, Berakon I, Md Husin M (2020) COVID-19 and e-wallet usage intention: a multigroup analysis between Indonesia and Malaysia. Cogent Bus Manage 7(1):1804181

Ajibade P (2018) Levels of utilization and alignment of business information technology in the small and medium-sized enterprises (SMEs) in Nigeria and South Africa. Doctoral Thesis, University of Fort Hare

Ajzen I (1991) The theory of planned behavior. Organ Behav Hum Decis Process 50:179–211

Ajzen I, Fishbein M (2000) Attitudes and the attitude-behavior relation: reasoned and automatic processes. Eur Rev Soc Psychol 11(1):1–33

Alaeddin O, Altounjy R, Zainudin Z, Kamarudin F (2018) From physical to digital: investigating consumer behaviour of switching to mobile wallet. Polish J Manage Stud 17(2):18–30. https://doi.org/10.17512/pjms.2018.17.2.02

Amer A, Ibrahim I, Othman F, Jani SHBM (2020) Implication of cashless payment operation system among SMEs in the services sector. J Talent Develop Excellence 12(2s):2024–2033

Ang C-L, Tahar RM, Murat R (2017) An empirical study on electronic commerce diffusion in the Malaysian shipping industry. Electron J Inform Syst Develop Countries. https://doi.org/10.1002/j.1681-4835.2003.tb00092.x

Aslam W, Ham M, Arif I (2017) Consumer behavioral intentions towards mobile payment services: an empirical analysis in Pakistan. Trziste=Market 29(2):161–176

Balachandran D, Tan GW (2015) Regression modelling of predicting NFC mobile payment adoption in Malaysia. Int J Model Oper Manage 5(2):100–116. https://doi.org/10.1504/IJMOM.2015.072671

Bank Negara Report (2016)

Bernama (2020). Impact of pandemic on economy and recovery policy. Retrieved from

Brett J (2017) Dissertation Cardiff metropolitan university, U.K. http://hdl.handle.net/10369/8745

Chang CC, Yan CF, Tseng JS (2012) Perceived convenience in an extended technology acceptance model: mobile technology and English learning for college students. Australas J Educ Technol 28:809–826

Chin AG, Harris MA, Brookshire R (2022) An empirical investigation of intent to adopt mobile payment systems using a trust-based extended valence framework. Inf Syst Front 24:329–347. https://doi.org/10.1007/s10796-020-10080-x

Davis FD (1989) Perceived usefulness, perceived ease of use, and user acceptance of information technology. MIS Quart 319–340

Fishbein M, Ajzen I (1975) Belief, attitude, intention, and behaviour: an introduction to theory and research. AddisonWesley, Reading, MA

Hsu H, Chang Y (2013) Extended TAM model: impacts of convenience on acceptance and use of Moodle. https://www.bernama.com/en/features/news.php?id=1829686

Irani Z, Dwivedi YK, Williams MD (2009) Understanding consumer adoption of broadband: an extension of the technology acceptance model. J Oper Res Soc 60(10):1322–1334

Isrososiawan S, Hurriyati R, Dirgantari PD (2019) User mobile payment behavior using technology acceptance model (TAM): study of "dana" e-wallet users. Jurnal Minds: Manajemen Ide Dan Inspirasi 6(2):181. https://doi.org/10.24252/minds.v6i2.11274

Jaafar SS (2018) SME contribution to Malaysia's economy rose to 37% in 2017. The Edge. Retrieved from https://www.theedgemarkets.com/article/sme-contributionmalaysias-economy-rose-37-2017

Jerry M, Maulana V (2017) Factors affecting intention to use "SAKUKU" e-wallet of generation Y in Indonesia

Jin CC, Seong LC, Khin AA (2020) Consumers' behavioural intention to accept of the mobile wallet in Malaysia. J Southwest Jiaotong Univ 55(1). https://doi.org/10.35741/issn.0258-2724.55.1.3

Joifin F (2018) What is contactless payment. Retrieved from https://www.comparehero.my/credit-card/article/what-is-contactless-payment

Jusoh Z, Teng YZ (2019) Perceived security, subjective norm, self-efficacy, intention, and actual usage towards E-payment among UPM students. J Educ Soc Sci 12(2). ISSN No: 2289-9855

Karjaluoto H, Shaikh AA, Leppäniemi M, Luomala R (2019) Examining consumers' usage intention of contactless payment systems. Int J Bank Market 38(2):332–351. https://doi.org/10.1108/IJBM-04-2019-0155

Lai PC, Zainal AA (2015) Consumers intention to use a single platform e-payment system: a study among Malaysian internet and mobile banking users. J Internet Bank Commerce 20(1):1–13

Lai P (2017) The literature review of technology adoption models and theories for the novelty technology. J Inform Syst Technol Manage 14(1). https://doi.org/10.4301/s1807-17752017000100002

Lee TM, Jun JK (2007) The role of contextual marketing offer in Mobile commerce acceptance: comparison between Mobile Commerce users and nonusers. Int J Mobile Commun 5(3):339–356

Lee J, Wing Hooi L (2017) Crucial for SMEs to catch up on technology. Star Online. Retrieved from https://www.thestar.com.my/business/smebiz/2017/10/23/crucial-for-smes-to-catchupon-technology/

Mathwick C, Malhotra N, Rigdon E (2001) Experiential value: conceptualization, measurement and application in the catalogue and internet shopping environment. J Retail 77(1):39–56. https://doi.org/10.1016/S0022-4359(00)00045-2

Mohamed AE (2020) Chairman, president Barack Obama's global development council COVID-19: what are the biggest concerns for the developing world? Retrieved: 22 April 2020. https://www.weforum.org/agenda/2020/04/developing-worldcovid19. World Economic Forum
Mutahar AM, Norzaidi MD, Ramayah T, Putit L (2017) Examining the effect of subjective norms and compatibility as external variables on Tam: mobile banking acceptance in Yemen. Sci Int (Lahore) 29(4):769–776
Nagashree RN, Rao V, Aswini N (2014a) Near field communication. Int J Wireless Microwave Technol (IJWMT) 4(2):20–30
Nagashree RN, Rao V, Aswini N (2014b) Near field communication. Int J Wireless Microave Technol 2:20–30. Published Online March 2014b in MECS, http://www.mecs-press.net. https://doi.org/10.5815/ijwmt.2014.02.03
Pavlou PA (2003) Consumer acceptance of electronic commerce: integrating trust and risk with the technology acceptance model. Int J Electron Commer 7(3):101–134
Putit L, Muhammad N (2015) Predictors of web-based transactions: examining key variations within an intra-national culture. Adv Sci Lett 21(6):2068–2072
Putit L, SuKi AA, Nurhana AI, Afiqah MZ, Nurul Afif AM, Nurul Amirah MN (2020) Website attributes' influence on customer loyalty and travel service purchase: a case of online virtual community. ICEBA 2020 Proceeding
Rogers RW (1975) A protection motivation theory of fear appeals and attitude change. J Psychol 91(1):93–114
Schuitema G, Anable J, Skippon S, Kinnear N (2013) The role of instrumental, hedonic, and symbolic attributes in the intention to adopt electric vehicles. Transp Res Part a: Policy Pract 48:39–49
Shroff RH, Deneen CC, Ng EM (2011) Analysis of the technology acceptance model in examining students' behavioural intention to use an e-portfolio system. Austr J Educ Technol 27(4). https://doi.org/10.14742/ajet.940
Singh N, Sinha N, Liébana-Cabanillas FJ (2020) Determining factors in the adoption and recommendation of mobile wallet services in India: analysis of the effect of innovativeness, stress to use and social influence. Int J Inf Manage 50:191–205. https://doi.org/10.1016/j.ijinfomgt.2019.05.022
Slade EL, Williams MD, Dwivedi YK (2013) Mobile payment adoption: classification and review of the extant literature. Mark Rev 13(2):167–190
Teoh TM, Hoo C, Lee T (2020) E-wallet adoption: a case in Malaysia
The Star Online (2020) Covid-19 outbreak steepens adoption curve of e-wallets in Malaysia. Retrieved from https://www.thestar.com.my/news/regional/2020/04/28/covid-19-outbreak-steepens-adoption-curve-of-e-wallets-in-malaysia
Trutsch T (2014) The impact of contactless payment on spending. Int J Econ Sci 3(4)
Uduji JI, Okolo-Obasi EN, Asongu SA (2019) The impact of e-wallet on informal farm entrepreneurship development in rural Nigeria. Electron J Inform Syst Develop Countries 85(3):e12066
Venkatesh V, Davis FD (2000) A theoretical extension of the technology acceptance model: four longitudinal field studies. Manage Sci 46(2):186–204
Wijayanthi I (2019) Behavioral intention of young consumers towards e-wallet adoption: an empirical study among Indonesian users. Russian J Agric Socio Econ Sci 85(1):79–93. https://doi.org/10.18551/rjoas.2019-01.09
Zahariah MZ, Jusoh AZ, Ros Intan SM, Putit L (2020) Drivers of E-commerce adoption amongst small & medium sized enterprises (SMEs) in the services sector. J Int Bus Econ Entrepreneurship 5(1): 1–15
Zarandona J, Cariñanos-Ayala S, Cristóbal-Domínguez E, Martín-Bezos J, Yoldi Mitxelena A, Cillero IH (2019). With a smartphone in one's pocket: a descriptive cross-sectional study on smartphone use, distraction, and restriction policies in nursing students. Nurse Educ Today 82:67–73
Zeithaml VA, Parasuraman A, Malhotra A (2002) An empirical examination of the service quality-value-loyalty chain in an electronic channel. Working Paper. University of North Carolina

Digital Strategies and Policy Approach for Small Medium Micro Business Development in Indonesia

5

Ahmad Budi Setiawan, Karman, Ari Cahyo Nugroho, Amri Dunan, and Bambang Mudjiyanto

Abstract

The utilization and implementation of Information and Communication Technology (ICT) for Micro-, Small and Medium-sized Enterprises (M-SMEs) is a widely discussed topic. MSMEs in Indonesia face a multitude of challenges that are not limited solely to the adoption and sophistication of technology. Furthermore, ICT needs to be developed in a way that increases its competitiveness and robustness, to effectively drive sustainable development and bridge the digital divide that exists in countries like Indonesia. The Indonesian government has conducted a public briefing through Law No. 19 of 2016 and Presidential Regulation No. 74 of 2017 on information and electronic transactions and the National Electronic-Based Trading System Roadmap, respectively. These two frameworks were created to emphasize the importance of leveraging ICT to achieve economic growth. Stakeholders are promoting ICT firm growth through formal measures. A review of government initiatives is needed for success.

A. B. Setiawan (✉) · A. C. Nugroho · A. Dunan · B. Mudjiyanto
Institute of Governance, Economy and Community Welfare, The National Research and Innovation Agency of Indonesia, Jakarta, Indonesia
e-mail: ahma109@brin.go.id

A. C. Nugroho
e-mail: aric001@brin.go.id

A. Dunan
e-mail: amri003@brin.go.id

B. Mudjiyanto
e-mail: bamb065@brin.go.id

Karman
Institute of Social Sciences and Humanism, The National Research and Innovation Agency of Indonesia, Jakarta, Indonesia
e-mail: karm001@brin.go.id

A. Rafiki et al. (eds.), *Open Innovation in Small Business*, Contributions to Environmental Sciences & Innovative Business Technology,
https://doi.org/10.1007/978-981-99-5142-0_5

This research aims to fill gaps to reach desired outcomes, to understand the adoption and spread of innovation among MSMEs, and to help them thrive in a rapidly evolving global market. The study involves experts who provide guidance and insights. The findings show that MSMEs need support in implementing and deploying ICT, due to low technology literacy and inadequate technical and operational management. To make the best use of resources, ICT policies for MSMEs should embrace both proactive and reactive methods and foster strategic alliances for human capital innovation. Policymakers should prioritize a flexible environment, breaking down sectoral barriers, and enhancing collaboration between universities, research institutions, industries, and MSMEs.

Keywords

Digital strategies • Policy • Micro-small and medium enterprises • Development • Indonesia

5.1 Introduction

The adoption and spread of ICT innovations among MSMEs have become increasingly complex and critical. Initially, the focus was on streamlining company procedures for efficiency, but now it is centered on competitiveness and resilience through global networks. This aligns with the World Summit on Information Society's strategic agenda, which views ICT as a driver for sustainable development by strengthening global connections, promoting economic productivity, and reducing the digital divide through knowledge-based societies. To promote the adoption of information and communication technology (ICT) and bridge the digital divide, the Indonesian government has enacted Law No. 19 of 2016 regarding information and electronic transactions and Presidential Regulation No. 74 of 2017 on the National Electronic-Based Trading System Roadmap aim to promote the economic benefits of ICT through public awareness and support. The focus has shifted from simply increasing efficiency to fostering competitiveness and resilience in a rapidly changing and interconnected global market. The use of ICT as a catalyst for sustainable development aligns with the goals of the World Summit on Information Society and strengthens interconnectivity, drives productive economic activities, and promotes knowledge-based societies. This effort requires a comprehensive approach to enhance technology literacy and foster good corporate governance practices.

The literature review at the micro-organizational level underscores the significance of ICT-based innovation in shaping organizational strategies. The classic themes are consistently emphasized (Galliers and Baets 1998; Broadbent et al. 1999; Broadbent and Siong Neo 1999; Kappelman et al. 2016). The use and utilization of ICT as an enabler at the organizational level, particularly for micro, small and medium-sized businesses, transform the way it creates added value and the number of strategic benefits it generates. Adopting ICT among MSMEs also presents complex organizational challenges, particularly in managing operational activities and maintaining efficiency, effectiveness, and business competitiveness (Shailer et al. 2000).

Previous studies indicate that adopting and spreading ICT innovations in MSMEs can change the face of industrial competition and emphasize the need for MSMEs to craft a winning innovation strategy in the highly competitive global market. The effective use of ICT at the organizational level is key to maximizing value and strategic advantages for MSMEs. But it also poses organizational challenges, particularly in optimizing operational processes to achieve efficiency, effectiveness, and competitiveness (Utomo 2000). Effective technology-based innovation strategies are key to enhancing global competitiveness for micro, small and medium-sized enterprises (MSMEs) through the advancement of product, process, and institutional innovation. The expected values in these strategies play a central role in this pursuit. Supervision is necessary to ensure the proper implementation of technology-based innovation strategies, especially when these strategies are related to public policies. The policies aim to ease the space for freedom to create and innovate in the digital economy era, explore the benefits of information and communication technology in society in a sustainable and equitable manner, and understand the impact of people's behavior while preserving the sovereignty of the nation and the state.

The use and utilization of ICT by MSMEs is a pressing concern that requires empirical studies for evaluating communication and information policies. Policymakers should adjust ICT policies if needed to reflect changing circumstances and developments. The public's interest in ICT covers not just e-commerce, but also other valuable pursuits. Hence, policymakers must monitor the impact of ICT on business productivity. This study investigates human capital development variables as key factors that both aid and hinder the diffusion and adoption of innovation by business actors and the mastery of technology and connectivity in global business networks. The study explores patterns and creates strategies for MSME development, with ICT as a driving force. These strategies are a part of the plan.

5.2 Literature Review

This section outlines the relevant theory for the discussed topics. The literature review presents an argument that (1) explains MSME innovation strategy, (2) highlights MSME innovation initiatives, (3) emphasizes the need for technology-based innovation framework for MSME competitiveness, and (4) examines Information Technology adoption patterns in MSMEs.

5.2.1 Micro, Small and Medium Enterprises Innovation Strategy

Indonesian Micro, Small and Medium Enterprises (MSMEs) must improve and strengthen their innovation capacity to sustain growth in a disruptive era. To strengthen innovation capacity, MSMEs must cultivate productive resources and integrative capabilities to respond effectively to environmental changes (Liao et al. 2009). It indicates the need for MSMEs to foster human capital with relevant skills and competencies to respond to changes in business models triggered by the adoption of technology-based innovations (see van Dam 2017). The application of technology-based innovations transforms organizational processes by disrupting various types of work and the scope of conventional productive activities.

Christensen et al. (2016) said that disruptive innovation is an organizational challenge in mastering technology and adapting to market changes. MSMEs must address preconditions proportionately to avoid confusion and manage technology and organizational practices effectively. Technology complexity, seen as an enabler, helps MSMEs overcome barriers of distance, time, and place to compete. Using technology-based innovation is now critical for SME's competitiveness, even though it has become a necessary part of organizational life in an uncertain environment.

Given MSMEs' limited resources, they should formulate competitiveness and resilience strategies in stages. The first stage is to increase business actors' sensitivity to changes, then have the courage to take proactive, risk-taking decisions. MSMEs with a proactive attitude, backed by sufficient data and analytical skills, gain bargaining power among stakeholders. Thus, MSMEs' decision-making speed becomes increasingly important with public encouragement to realize their role as a "bumper" in maintaining national economic stability and driving productive growth.

5.2.2 Innovation Initiatives Among MSMEs

The awareness of MSME actors toward information technology sophistication has a positive impact on enhancing the innovation ecosystem. Over 4.7 million MSME actors have utilized digital technology for business purposes in various formats such as e-commerce and marketplace platforms, according to data from the Coordinating Ministry for Economic Affairs (2019). The growth of business actors will continue with incentives such as expediting public services and boosting business activities. However, MSME actors play a limited role in promoting export activities, with MSME exports accounting for only 15.7% of total non-oil and gas exports in Indonesia.

Classic obstacles to competitiveness and resilience, such as technology and innovation mastery, human capital capacity, legal compliance, access to funding sources and markets and product development, and exceptional service creation, continue to require integrated handling from the organization's technical management and public policy level (Yoshino and Hesary 2016). Despite the limitations inherent in MSMEs, the current situation and macro conditions offer gaps and growth potential for accelerating economic growth through information technology-based innovations.

SMEs can increase their absorptive capacity for innovation. MSMEs' flexibility in decision-making can serve as a strength to explore innovation opportunities that involve external knowledge sources. Lichtenthaler (2009) stated that absorptive capacity, or the absorption of technology and market knowledge, is necessary for initiating innovation and improving organizational learning. MSMEs generally possess the absorbent step and adaptive behavior for organizational creativity in response to environmental changes.

Absorptive capacity involves the ability to master external knowledge through a sequential process, starting with exploring, transforming, and exploiting learning resources to achieve organizational excellence. When creating MSME growth strategies, it is essential to choose the appropriate innovation areas and use information technology as an enabling factor to promote innovation in goods, processes, and institutional factors. The sectoral uniqueness of MSMEs means that they need to choose a variety of innovation arenas instead of following a single, linear path, leading to variations in the innovation process and its stages.

The diversity of the innovation process highlights the unique behavior of MSMEs, particularly in absorbing external knowledge for innovation and organizational development. Cultural compatibility, for instance, is now a critical factor in the creative industries, particularly in fashion and digital application devices. Other factors that play a role in the diffusion of innovation, such as relative superiority, complexity, visualization, and ease of trying, become supporting elements in the innovation diffusion and adoption process (see Rogers 1995). It prompts MSMEs to fully understand the quality of interactions and business network development to avoid pseudo-egocentric forces. Therefore, MSMEs must master knowledge sources in technology and market behavior to start innovation. Sari and Kusumawati (2022) also argue that strengthening the capacity to absorb knowledge also motivates MSME players to explore innovation opportunities, such as aligning activities with innovation policy support, developing business activity networks, and enhancing interactions with educational institutions and other incubation agents. Reinforcing the innovative capacity of MSME actors plays a coherent role in developing their ability to absorb external knowledge for MSME development.

5.2.3 The Urgency of MSMEs to Adopt the Basic Framework for Technology-Based Innovation

Adopting technology-based innovation in micro, small and medium-sized enterprises (MSMEs) requires a framework that starts with ideas, continues with concept preparation, prototype development, market acceptance testing, and ends with final utilization. This results in increased economic value for users. MSMEs must orchestrate innovation sources and relate them to public needs to benefit from these innovations. Synchronizing MSME activities with government policies to accelerate the Information and Communications Technology (ICT) adoption and diffusion process will boost the growth and development of productive businesses. MSMEs must take advantage of opportunities and relate to innovation centers and agents of change to develop and accelerate their business activities.

Literature studies have shifted the discussion of ICT adoption in MSMEs from technical–operational issues to managerial and strategic ones. The adoption of ICT aims to accelerate the diffusion of innovations that promote the creation of new values in society's socio-economic life through best practices in various fields. The big challenge is how to turn the potential of information technology-based innovation from the macro policy level into concrete actions at the micro-organizational level. Aligning strategy, structure, system, and human capital is crucial for small and medium business actors to develop their innovation capacity.

ICT-based innovation can transform organizational strategy and generate added value and strategic benefits. However, it also raises organizational challenges for MSMEs, especially in stimulating operational activities and achieving efficiency, effectiveness, and competitiveness. Research has revealed the potential for ICT adoption and diffusion to change competitiveness, highlighting the importance of formulating innovation strategies for MSMEs in Indonesia in the global era. The innovation strategy anticipates the emergence of new values through product innovation, process innovation, and institutional innovation.

All of this certainly requires a separate guarding process, especially if it is related to the implementation of public policies that lead to:

(1) Promoting creativity in the digital economy era by removing barriers to creative freedom.
(2) Examining the fair and sustainable utilization of information and communication technology in society.
(3) Assessing the behavior of individuals while preserving national and state sovereignty in history.

The ICT policy implementation may require correction if adjustments are necessary to keep up with changing circumstances to support business growth and acceleration (Utomo and Dodgson 2001). Both central and regional governments are encouraging businesses to leverage the benefits of ICT to improve productivity through initiatives such as the "a million.id domains" by the Ministry of Communication and Informatics, the "1000 digital start-ups" program, and the "MSMEs Go Online." The Ministry of Cooperatives and MSMEs, and the Ministry of Villages, Development of Disadvantaged Regions, and Transmigration have also launched "*Gebyar UKM*" to promote ICT innovation. The MSME Design Centre, which relies on digital technology, has been established by the Ministry of Trade. However, there is a need to thoroughly review and reassess current models and approaches to promote ICT adoption among MSMEs. The wide range of socio-demographic and psychographic backgrounds and types of MSME activities highlights the need for alternative policies that account for unique contextual and organizational factors. The unequal ICT adoption among MSMEs poses new challenges for formulating public policies that provide equal opportunities for businesses to succeed in the digital economy era. It's important to understand that not all MSMEs are equally prepared to take advantage of technological innovations. The policy for using ICT should align with the readiness level of businesses, considering the patterns of adoption and diffusion. The goal is not only to create marketplaces but also to build the capabilities of MSMEs to compete in an increasingly open, cross-border, and dynamic global market environment.

5.2.4 Innovation Adoption Patterns

The adoption pattern describes the logical flow that has been adhered to by business actors in orchestrating their productive resources and innovation opportunities. Understanding adoption patterns is relevant in determining strategic priorities and allocating resources for organizational development. The primary key lies in the ability to observe adoption patterns and then apply an appropriate mentoring model in response to the complexity of the MSME development model. This empirical study explores various perspectives and the need to accelerate MSMEs through cross-disciplinary arguments in economics, business, anthropology, design, and engineering. This exploratory research, with the participation of specialists who investigate crucial topics from diverse domains, is designed to provide more thorough and complete insights that are valuable for relevant public officials.

In general, the development and acceleration of business activities carried out with the assistance of information and communication technology (ICT) follow the management process flow. ICT can provide creative potential and open new doors to improve the efficiency and effectiveness of business operations. It's just that the form and model of ICT application cannot be equated between one business activity and another. As a result, public policies relating to ICT development for MSMEs can only be one policy for some. Utomo (2000) divides ICT functions and roles into several stages:

(1) ***efficiency***. ICT become a central aspect of business operations. It is a management tool to help achieve business objectives more efficiently. For instance, ICT is used for customer relationship management to connect with consumers or for quality control to ensure product excellence;
(2) ***effectiveness***. ICT takes on a more central role, a tool for management to help achieve business objectives more effectively. The use of ICT to bind consumers is an example of implementation at this stage (customer relationship management) or ICT to ensure production quality (quality control);
(3) ***competitiveness***. Information and communication technology plays a critical and strategic role and becomes more interactive. Its strategy must align with the business strategy, to map consumer behavior, create business models, compile sensitivity analyses, build more interactive product designs, market control, and supply chains, among other things.

Concerning optimizing the process of ICT-based innovation diffusion, Utomo and Dodgson (2001) explained that every series of activities at each of these three

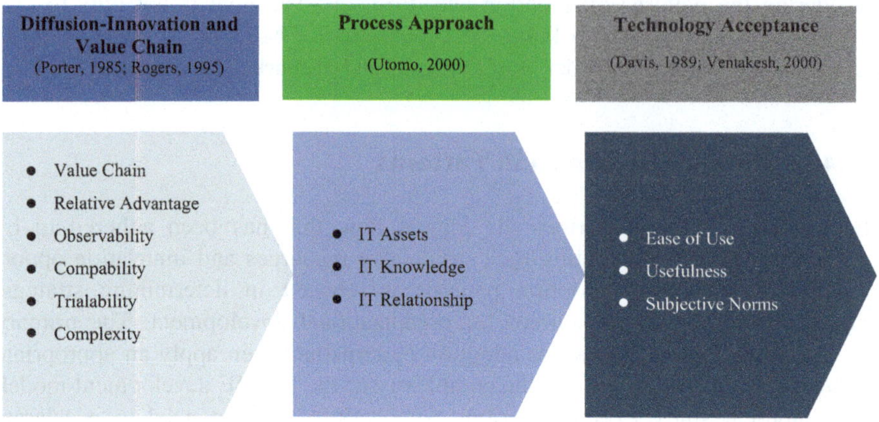

Fig. 5.1 Interrelationships between interests in Technology Adoption. *Source* Utomo, H & Dogson (2001)

stages required three main prerequisites: ICT assets (infrastructure), ICT knowledge (human resources), and ICT relationships (connectivity). The utilization of adequate ICT infrastructure is to optimize its use. The existence of human capital equipped with good ICT can increase the ability to absorb environmental changes in a better direction in line with changes that occur in the business environment and global markets. Strengthening the ICT ecosystem that can maintain the harmony of entity interests that supports good interactions and networks that mutually reinforce interests between entities is a new requirement for organizations operating in the era of disruption (see Fig. 5.1).

In particular, the patterns and prerequisites for using ICT, as described above, are the theoretical frameworks as an essential reference in developing policies, programs, and activities for developing MSME productivity. Thus, the literature framework in this exploratory study will provide a foundation for empirical exploration of how ICT acts as an enabler to encourage the growth of export-oriented and highly competitive MSMEs. From a public policy perspective, we hope that the findings of this empirical study provide useful information about innovation strategies and optimizing the use of information and communication technology resources to boost MSME productivity and economic growth in Indonesia.

We carry out this kind of program in various countries through different programs. Australia, for example, issued a policy on the Building on IT Strength (BITS) Incubator program, the National Technology Online Program (ITOL), Business Entry Point, and Online Business Resources Facility (OECD 2002). A similar program is run through the Wales Information Society (WIS) and the UK Online for Business. In Austria, a similar program is like "Let's eBiz." Finland and the Netherlands also have similar programs called "eASKEL" and "Netherland Go Digital" (OECD 2002). Meanwhile, Sweden, Luxembourg, and Ireland have similar initiatives known as "SVEA," "APSI/CRP-HT Guide," and "PRISM initiative," respectively (OECD 2002)..

5.3 Methodology

This study accommodates the possibility of the dynamics of the MSME context occurring during the study process. Therefore, we chose Action Research in this study to provide space and flexibility in examining problems at a practical level to build patterns and action on technology-based strategic innovations for business development. To answer the questions in this research, we use multiple case studies to strengthen arguments and reveal them in depth by explaining the uniqueness of sectoral contexts and alternative development strategies. By remembering and considering the limitations inherent in the study approach and method, the selection of observed objects, sources, duration of observation, and the degree of attachment to the object during the research process has its urgency in obtaining the results we expect. Action Research procedures include:

1. Pre-survey and, at the same time conducting mapping regarding ICT facilitation in the study target areas by involving cross-sectoral experts.
2. Workshops and Focus Group Discussions to formulate change and development agendas.
3. Introduction of innovations.

The readiness of business actors to adopt technology is a change that requires separate monitoring at each stage. Mastery of the detailed aspects of the use of ICT with the escort of technical aspects. While optimizing the use of technology in business activity units requires supervision of managerial aspects. The basic model that will be used in this research to control change is schematically presented in Fig. 5.2.

5.4 Result and Discussion

5.4.1 Micro-SME Situation and Conditions

The situation of MSMEs in Indonesia is unique, not only because of the business sector, but also because of other factors such as business ownership, geographic work area, and business maturity. The existence of MSMEs plays a critical role in economic development by reducing potential volatility in the elements that comprise gross domestic product through employment and increasing the volume of activities that encourage growth in consumption and investment in the real sector. The awareness of MSME actors in utilizing ICT sophistication is also a positive indicator.

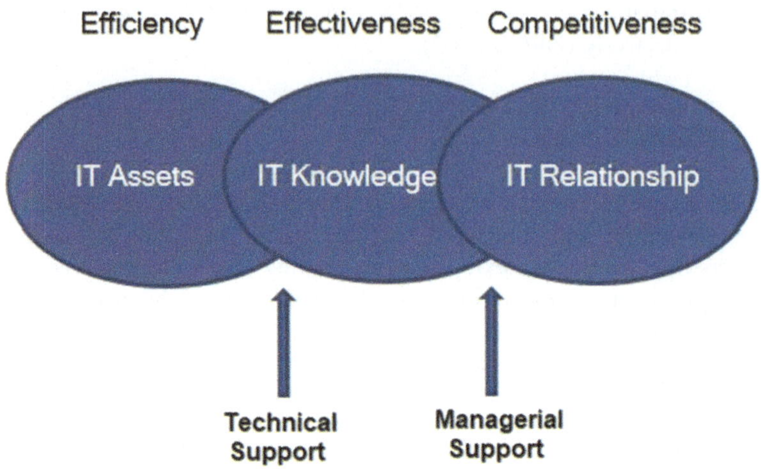

Fig. 5.2 Process approach. *Source* Results of Analysis and Processing of Researchers

According to the results data of the 2016 Central Bureau of Statistics Economic Census, 26.71 million business units in Indonesia fall into the MSME category. By absorbing 96.7% of the total workforce, this category of business actors contributes 61.41% of GDP or IDR 6,228 trillion. Meanwhile, MSME actors contributed 56.2% of total national investment in terms of investment contribution (Central Bureau of Statistics 2016). Furthermore, according to data from the Central Bureau of Statistics (2017), more than 4.7 million players used digital technology for business purposes in various formats (e-commerce and marketplace platforms). The number of business actors continues to grow because of various incentives such as faster public services to facilitate expansion of business activities.

However, the role of MSME actors in promoting export activities remains limited. MSME exports account for only about 15.7% of total non-oil and gas exports in this country (Coordinating Ministry for the Economy 2019). The development of business management behavior is relevant to efforts to increase MSMEs' competitiveness and resilience in the face of changes in the global business environment. Concerning the potential strength of MSMEs and the amount of their contribution to the national economy, scrutinizing patterns and developing innovation strategies has its urgency to implement. MSMEs become more competitive globally as information and communication technology is adopted and spread.

The central and regional governments have been trying to optimally encourage MSME actors by utilizing ICT sophistication to increase the productivity of business activities. The Ministry of Communication and Informatics of Indonesia launched a series of effective programs and activities through the one million domains.id program; 1000 digital startups; and UMKM Go Online, all of which increased the quality of national independence through innovation and technology.

Meanwhile, the Ministry of Cooperatives and SMEs and the Ministry of Villages, Development of Disadvantaged Regions, and Transmigration initiated various "*Gebyar UKM activities,*" including the significance of ICT-enabled innovation. In addition, the Ministry of Cooperatives and SMEs also launched an Integrated Business Service Centre to encourage business activities' growth through incubation and mentoring. Another effort that is also working in the MSME area is the Ministry of Trade launching the MSME Design Center, which also relies on digital technology; At the same time, the Creative Economy Agency rolled out a program in the form of an ICT facility package for the community.

The proper steps illustrate the seriousness of policymakers in encouraging the growth of productive business activities through technology-based innovation. The government's efforts to accelerate the use and utilization of ICT among MSMEs still require a comprehensive review and re-examination so that the models, approaches, and impacts can be entirely achievable.

5.4.2 Strategy for Innovation and Development of MSME Resilience

The development and application of information and communication technology have resulted in various business models for MSME actors who add value to the organization. The process of sophistication will continue in line with the increasingly open business environment in this country and increasing public awareness of the function of information technology. It indicates that there is still a lot of space available for developing and implementing technology-based innovations to increase competitiveness and resilience in various fields and sectors of the productive economy. The three main pillars that trigger competitiveness, and resilience in business activities are:

(1) Essential triggering factors include stable macroeconomic conditions, public health, and physical infrastructure.
(2) Efficiency trigger factors include the availability of quality human resources and sufficient support from money market and capital market players, as well as technology readiness for adoption; and
(3) Innovation and sophistication factors create a variety of innovative products and processes, which in turn lead to increased productivity.

The first and second trigger factors are undeniably necessary conditions for a solid competitive position. However, they are somewhat resistant to being able to face influences, challenges, and even pressure both from within and outside the country. With the existing supporting resources, there is no other choice but to encourage activities that trigger competitiveness from research and development activities so that they can spread out immediately to the public for the domestic market and the export market. Diffusion or spread of technology is also largely determined

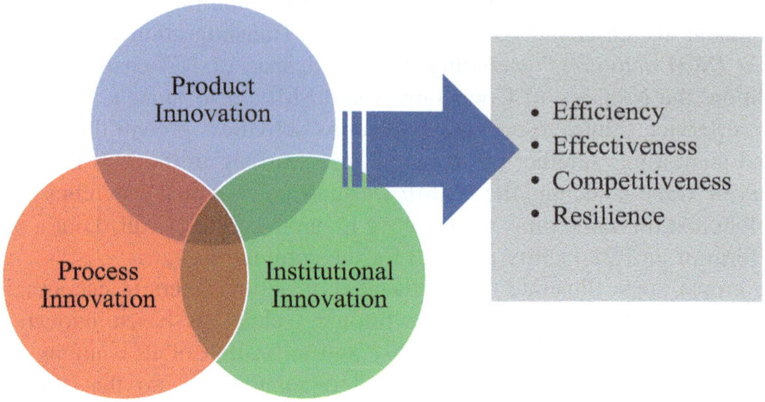

Fig. 5.3 Strategy and impact of ICT innovation for MSMEs. *Source* Rogers (1995)

by the strength of recognizing the characteristics of adopting innovation. Rogers (1995) reveals innovation adoption as perceived by users/beneficiaries as follows: relative advantage; compatibility; complexity; the ability of tested or trialability; and the ability of observed (see Fig. 5.3).

In-depth discussions regarding the formulation of business development strategies in general so far have only focused on the creative efforts taken to achieve revenue streams only. Serious efforts to improve the quality of interaction between MSME actors and interested parties are the preconditions needed for the expected added value creation to be realized. On the one hand, the variety of innovations with various business models is indeed the main attraction for potential development in business development. However, on the other hand, it becomes a separate obstacle if the provision of physical and supporting infrastructure needed for business development, such as access to innovation and business assistance, is not or has not been fully available. The challenge faced then is optimizing business networks (formal and non-formal) that can support the need for innovation development and at the same time become part of strengthening the business chain which is the focus of MSMEs.

5.4.3 Collaborative Strategy in MSME Development

To support the independence of MSMEs, we need to formulate a collaborative strategy to address a key challenge in MSME development. The flexibility and compatibility in resource management make MSMEs a buffer and shock absorber for the community's economic activities, particularly when business actors make choices due to limited resources. Rather than seeking short-term profits through unlimited exploitation, MSME actors prioritize long-term viability over opportunistic behavior.

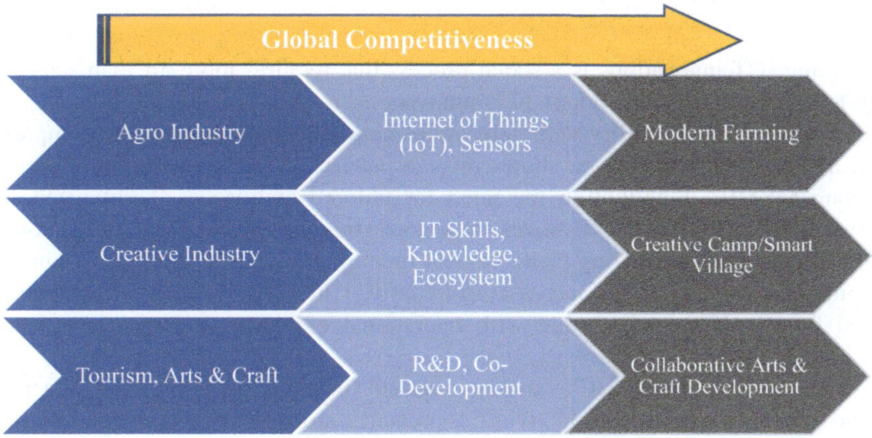

Fig. 5.4 Innovation area for MSME development. *Source* Utomo (2000)

The traditional reason, insufficient financial resources, can impede SMEs that have just begun their business. For small and medium enterprises, the criterion for the level of resource adequacy is difficult to understand. The time availability has almost focused to test the concept and develop business processes. That is why these business actors will take bold ways by relying on alternative funding sources to get around limited funds and to bridge the desire of entrepreneurs to realize innovation through their ideas with limited resources. Hence, it is urgent to examine potential areas to develop an innovation strategy before moving on to a larger scale (see Fig. 5.4).

The support for collaborative strategies for MSMEs should ideally be able to move in a direction that leads to increased prosperity and sustainability. Management reorientation regarding the use of natural resources has strategic value for social welfare, as mandated in the Indonesian constitution. To innovation, it is possible to provide and facilitate the synergy between academia, industry, and government to help and realize a national innovation system.

The intervention of external factors into the environment of MSME activities is not easy to avoid, especially with the opening-up of investment opportunities. Collaborative efforts among MSMEs are strategic social capital to build an industrial foundation. Efforts to encourage industrialization in Indonesia are already in a fast-moving position. The need to innovate and achieve competitive advantage certainly requires adequate readiness, especially the provision of human resources that are reliable and equipped with superior competence to maintain industry sustainability. The institutional synergy between industry and public education partners is a source of strength for this nation to survive and have a positive attitude in facing competition.

5.5 Conclusion

Adopting ICT innovations is a social process that requires more art than science. MSME actors will only adopt ICT innovations if they see relative advantages, observability, compatibility with current business practices, manageable complexity, and testability. Evidence shows diverse patterns of ICT innovation adoption among MSMEs, highlighting the need for tailored approaches. Unlike general trends, MSMEs' use of ICT moves from downstream to upstream, with increased public connectivity boosting technological literacy. However, a shortage of human capital makes it difficult to increase upstream ICT literacy. MSMEs follow a standard flow in ICT use, starting with operational efficiency and moving toward competitiveness.

Therefore, MSMEs' ICT strategy should balance proactive and reactive approaches to optimize productivity. A proactive strategy aims to expand served market segments. MSMEs need to enhance their human capital in managing ICT's technical and operational aspects to improve ICT capabilities. To compete, MSMEs can adopt a reactive approach to maintain their market position and enhance their competitiveness. Labor-intensive industries like crafts, clothing, and furniture focus on customer connectivity to survive competition. Digital creative industries, on the other hand, benefit from a proactive strategy for business sustainability. MSMEs can optimize limited resources by forming strategic alliances in ICT use.

Community-based MSMEs can particularly benefit from alliances based on human capital strength to increase innovation capacity. Public policies can accelerate downstream MSME innovation results by encouraging stakeholder participation in research, funding, operations, and marketing. The tourism, garments, crafts, and hospitality industries have the potential to drive socio-economic development. Emphasizing local creativity, a solid production base, and skilled labor is key to developing MSMEs and supporting national economic growth and equity. Public policies for MSME development should also consider innovation through co-creation among stakeholders to increase productivity and sustainability.

Acknowledgements We are deeply grateful to all those who made this research possible. Our sincere thanks address to the Research Institute of Informatics Application and Public Communication, Ministry of Communication and Informatics of Indonesia, our previous place of work before we move to the National Research and Innovation Agency of Indonesia, for their funding and support throughout the research process. We also appreciate the collaboration with the Directorate of Business Development and Incubation, University of Gadjah Mada under leadership of Mr. DR. Hargo Utomo, MBA, M.Com and extend our gratitude to the scholars there who kindly shared their valuable time and expertise with us. This research would not have been successful without the collective effort and support of everyone involved.

References

Broadbent M, Weill P, St Clair D (1999) The implication of information technology infrastructure for business process redesign. MIS Q 23(2):159–182

Broadbent M, dan Siong Neo B (1999) Strategic context and patterns of IT infrastructure capability. J Strat Inf Syst 8(2):157–187

Central Bureau of Statistics (2016) Business/company registration result data–economic census 2016. Central Bureau of Statistics of Indonesia, Jakarta

Central Bureau of Statistics (2017) Indonesia economic report 2017. Central Bureau of Statistics of Indonesia, Jakarta

Chiliya N, Chikandiwa C, Afolabi B (2011) Factors affecting small micro medium enterprises (SMMEs) adoption of e-commerce in the Eastern Cape Province of South Africa. Int J Bus Manag 6(10):28–36

Christensen MC, Altman JE, McDonald R, Palmer J (2016) Disruptive innovation: intellectual history and future paths. Harvard business review, working paper 17-057

Coordinating Ministry for Economic Affairs (2017) Indonesia economic outlook 2017: increasing competitiveness to support exports. Coordinating Ministry for Economic Affairs of Indonesia, Jakarta

Coordinating Ministry for Economic Affairs (2019) Indonesia economic outlook 2019: increasing competitiveness to support exports. Coordinating Ministry for Economic Affairs of Indonesia, Jakarta

Van Dam NMH (2017) The 4th industrial revolution & the future of jobs, 1st ed, e-Book. ISBN 978-87-403-1883-8

Davis FD (1989) Perceived usefulness, perceived ease of use and user acceptance of information technology. MIS Q 13:319–339

Davis FD, Bagozzi RP, Warshaw PR (1989) User acceptance of computer technology: a comparison of two theoretical models. Manage Sci 35:982–1003

Dodgson M, Gann D, Salter A (2008) The management of technological innovation. Oxford University Press, Oxford

Felin T, Hesterly SW (2007) The knowledge-based view, nested heterogeneity, and new value creation: philosophical considerations on the locus of knowledge. Acad Manag Rev 32(1):195–218

Galliers RD, dan Baets WRJ (1998) Information technology and organizational transformation: innovation for the 21st century organization. Wiley, Chichester

Hassan R, Nhemachena C (2008) Determinants of African farmers' strategies for adapting to climate change: multinomial choice analysis. Afr J Agric Resour Econ 2(1):83–104

Kacira M, Sase S, Okushima L, Ling PP (2005) Plant response-based sensing for control strategies in sustainable greenhouse production. J Agric Meteorol 61(1):15–22

Kappelman L, Johnson V, McLean E, Torres L (Mar 2016) The 2015 SIM IT issues and trends study. MIS Q Exec 15(1)

Korner O, Van Straten G (2008) Decision support for dynamic greenhouse climate control strategies. Comput Electron Agric 60(1):18–30

Liao J, Kickul RJ, Ma H (2009) Organizational dynamic capability and innovation: an empirical examination of Internet firms. J Small Bus Manage 47(3):263–286

Lichtenthaler U (2009) Absorptive capacity, environmental turbulence, and the complementarity of organizational learning processes. Acad Manag J 52(4):822–846

OECD (2002) OECD small and medium size enterprise outlook. www.oecd.org

Perry BD, Grace D, Sones K (2013) Current drivers and future directions of global livestock disease dynamics. Proc Natl Acad Sci 110(52):20871–20877

Rogers EM (1995) Diffusion of innovations. Free Press, New York

Sari NTP, Kusumawati A (2022) Literature review: the efforts to strengthening of micro, small and medium-sized enterprises (MSME) in Indonesia. Asian J Manag Entrep Soc Sci 02(01):98–115. ISSN: 2808 7399. https://ajmesc.com/index.php/ajmesc/article/view/52

Shah Alam S, Ali MY, Jani MF (2011) An empirical study of factors affecting electronic commerce adoption among SMEs in Malaysia. J Bus Econ Manag 12(2):375–399

Shailer G, Stevens K, Utomo H (May 2000) EFTPOS impacts on branch banking: an extra organisational analysis. Aust J Inf Syst 7(2)

Utomo H, Dodgson M (2001) Contributing Factors to the diffusion of IT within small and medium-sized firms in Indonesia. J Glob Inf Technol Manag 4(2):22–37

Utomo H (2000) The diffusion of information technology innovations within small and medium-sized firms in Indonesia. PhD Dissertation, Unpublsihed, Australian National University, Canberra

Venkatesh V (2000) Determinants of perceived ease of use: integrating control, intrinsic motivation, and emotion into the technology acceptance model. Inf Syst Res 22:342–365

Yoshino N, Hesary F-T (2016) Major challenges facing small and medium-sized enterprises in Asia and solutions for mitigating them. Asian Development Bank (ADBI) working paper series, no 564, April 2016

Zhang N, Wang M, Wang N (2002) Precision agriculture—a worldwide overview. Comput Electron Agric 36(2–3):113–132

Social Media as an Open Innovation: Deciphering Its Relationship with Firm Performance, Compatibility, and Security Concern

6

Rosman Mohamed Hussin, Lennora Putit, and Geetha Subramaniam

Abstract

Intense competition, unpredictability, and volatility are just some of the hallmarks of the current market scenario. The advent of COVID-19 is a prime example of the "perfect storm", representing the sudden and unexpected environmental changes a firm can face. It has propelled digitalization into the spotlight for organizational change, warranting the implementation and integration of specific external resources of open innovations such as social media. This adaptive course of action is taken so that firms can continue to operate functionally but safely; amid movement restrictions, stringent Standard Operating Procedures, and Work From Home working arrangements. Especially for Small and Medium Enterprises (SMEs), social media is a cost-effective and easy-to-use open innovation resource with a multitude of potential benefits if and when utilized properly and effectively. While knowledge on the subject matter is still evolving, information regarding social media's direct and indirect relationship with firm performance remains an important and pertinent opportunity for onward investigation. The assertion of Malaysia's own Ministry of Finance (MOF) emphasizes this point, as it considers digitalization a means to make SMEs more resilient and competitive, a mitigating measure to address the issue related to SME mortality. This chapter describes the association of social media and firm performance operating under challenging circumstances; and certain elements that can facilitate or impede the decision to adopt social

R. M. Hussin · L. Putit (✉)
Faculty of Business and Management, Universiti Teknologi MARA (UiTM), 40450 Shah Alam, Selangor, Malaysia
e-mail: lennora633@uitm.edu.my

G. Subramaniam
Faculty of Education, Language, Psychology and Music, SEGi University, 47810 Petaling Jaya, Selangor, Malaysia

A. Rafiki et al. (eds.), *Open Innovation in Small Business*, Contributions to Environmental Sciences & Innovative Business Technology,
https://doi.org/10.1007/978-981-99-5142-0_6

media. The conceptual framework positions social media adoption as an intermediary construct to examine deployment factors and synthesize their effects on firm performance. Based on the literature review, two factors have been identified and are singled out as antecedents: compatibility and security concern. The Resource-Based View is the theory premise for this research, where social media is slated as an essential resource to elevate competitive advantage. The discussions and conclusion presented in this chapter are intended to offer an updated perspective from an academic and real-world practice standpoint.

Keywords

Social media • Innovation • Small and medium enterprises (SMEs) • Entrepreneurship • Security concern • Compatibility • Firm performance

6.1 Introduction

At the end of a specific business period, the success or failure of a firm is determined by comparing the actual results against the set goals. The utilization of traditional accounting to assess firm performance by looking into financial indicators has been sufficient and widely practiced in the past (Yip et al. 2009). Nonetheless, recent trends reveal the examination of both the financial and non-financial indicators for a more holistic approach (Falahat et al. 2020; Kamalrulzaman et al. 2021).

To attain peak performance, firms are under continuous pressure to adapt to do things better, faster, and at lower costs. The formulation of a conceptual framework presented in this chapter is based on this reality, which includes the key constructs of Compatibility, Security Concern, Social Media Adoption, and Firm Performance. The firms scrutinized will be those that fall within the Malaysian Small and Medium-sized Enterprises (SMEs) classification. SMEs are drivers of a nation's economy, a common consensus shared among numerous researchers (Lutfi et al. 2022; Tran et al. 2021). This sentiment is reinforced by the World Bank (2022) projections that by 2030, 600 million jobs will be required to meet global demand. Since about 7 out of 10 employment opportunities are created by SMEs, safeguarding their survival and ensuring SME growth is vital. However, due to the nature in which they operate, SMEs are typically vulnerable and prone to suffer severe drawbacks, exposing them to difficult challenges to mortality. High failure rates heighten concerns as they can result in a substantial loss of resources from the consequence of the failure (Abdullah et al. 2019).

In realization to the importance of SMEs for Malaysia, a national policy coined as Dasar Keusahawanan Nasional (DKN 2030) was launched in 2019, primarily to facilitate SMEs with five core objectives to be achieved through six strategic thrusts (SME Corp. 2019). The policy called for specific SME achievements targeted by 2030, such as 50% Gross Domestic Product (GDP) contribution and 80% of the jobs created nationwide (MEDAC 2019). Given the historically high failure rates associated with SMEs, these are ambitious goals. Nevertheless, the Malaysian

Ministry of Finance acknowledged that digitalization, such as social media adoption, can amplify resilience and competitiveness (MOF 2020) therefore, can move SMEs positively toward the goals outlined in DKN 2030. During an Asia–Pacific Economic Cooperation (APEC) ministerial meeting, Malaysia reiterated the need for member countries to urge SMEs to make the digitalization transition, which is deemed essential for their survival (APEC 2020). Firms that utilize social media are more likely to abandon archaic business strategies in favor of innovative and modern practices (Parveen et al. 2016).

Firm performance is included in the research framework because measuring performance is essential for determining performance levels. By mediating the constructs of compatibility and security concern, social media adoption, on the other hand, represents the deployment of a tool and how it can affect firm performance. A literature search reveals a lack of recent research in Malaysia that jointly looks into both social media and firm performance, with those available only from 2019 or prior. Furthermore, considering that we are currently in a recovery period post-COVID-19 aftermath, this opens up a window of opportunity to provide a fresh and updated outlook on research in this area.

6.2 Theoretical Foundation

A well-established strategic management theory known as Resource-Based View (RBV), which Wernerfelt (1984) originally introduced and Barney (1991) later made popular, stipulates that important firm resources, be they tangible or intangible, can impact firm performance (Mikalef and Gupta 2021). This intended study attempts to extend and update previous findings by applying RBV as its theoretical foundation.

Employing open resources innovatively and leveraging their specific integrating competencies can enable a business with average resources to achieve outstanding results (Luo and Bu 2018). Utilizing these public resources efficiently and rapidly can give firms a distinct competitive advantage (Gao et al. 2018). However, scholars point out that even though social media is practically free to use, allocating sufficient and dedicated resources to manage it effectively requires costs (Lepkowska-White and Parsons 2019).

From this context, RBV is used to evaluate social media adoption as a potent resource, devising a framework for examining social media adoption and firm performance linkages. In parallel, social media adoption will be investigated as a facilitator through its placement within the framework as a mediating construct. RBV has been utilized in the past to analyze the correlation between social media and performance outcomes (Fang et al. 2021; Oyewobi et al. 2020), but scholars also insist that more studies are needed to fully comprehend the benefits and drawbacks social media can offer (Olanrewaju et al. 2020). The anticipated outcome from this study may bring some clarity on whether social media is, in fact, a significant strategical resource compared to other, more traditional resources.

6.3 Development of a Proposed Conceptual Framework

Through a literature review exercise, compatibility, security concern, social media adoption, and firm performance are identified and singled out as the main components of the proposed conceptual framework. The following chapter sections present a detailed description and explanation for each relationship.

6.3.1 The Compatibility and Firm Performance Relationship

Scholars in the past have regarded compatibility as a firm's characteristic in a multi-dimensional context; with cultural, operational, and technical aspects making up the three main components of organizational compatibility (Rajaguru and Matanda 2019; Srimarut and Mekhum 2020). According to Sarkar et al. (2001), compatibility can promote effective cooperation between business partners, while incompatibility can result in a difficult and hostile working environment. For example, suppliers, clients, or partners may be turned off from joining a network that is not the same as theirs. So when a firm subscribes to an incompatible network, it may be forced to subscribe to another more common network, thus incurring additional costs. These costs are not limited to only switching costs alone but also other costs related to maintaining the connections already inside the present network (Blouch et al. 2021). Therefore, because of incompatibility, both users and firm performance are affected. Given these considerations, the proposed hypothesis is:

H1a: Compatibility will positively affect Firm Performance.

6.3.2 The Security Concern and Firm Performance Relationship

A firm's security breach can trigger a significant decline in its market value, placing them in a precarious situation (Goel and Shawky 2009). A data breach can damage a firm's public reputation and customer confidence as it reflects the inadequacy of security measures and internal controls, revealing a tell-tale sign that the firm is vulnerable to the risk of cyber-attacks (Martin et al. 2017). Nonetheless, it is also argued that openness regarding security deficiencies can positively mitigate concerns and increase trust because it shows a firm's honesty and continued commitment to protecting personal information (Gordon et al. 2010). In an era of a fast-moving data environment, identifying and addressing security concerns is critical to determine how they can affect performance so that firms can fortify their cyber resilience against unwelcomed and illegal intrusion. The following hypothesis is deduced in line with this reasoning.

H1b: Security Concern will negatively affect Firm Performance.

6.3.3 The Compatibility and Social Media Adoption Relationship

Compatibility is one of the five innovation characteristics presented in the book Diffusion of Innovations by Rogers (2003). Compatibility is expressed as the conflict between modern and traditional methods (Jia et al. 2022). From the context of this intended study, compatibility refers to how it will affect firm performance and social media, if its adoption is compatible with the existing business operation and practices. Regarding social media, Tajudeen et al. (2018) state that the greater the compatibility, the higher the chance it will lead to adoption. They further argued that based on its open accessibility and simplicity of use, social media would be highly compatible with most firms. Nevertheless, the decision to adopt social media is dependent on whether a firm sees the benefits behind it; for example, SMEs would exploit social media if they perceived its deployment as cost-effective. Typically, compatibility is regarded as a crucial criterion for innovation acceptance (Ainin et al. 2015; Syaifullah et al. 2021), although there is also the discovery of conflicting results (Bouargan et al. 2020; Ramdani et al. 2009). Due to these inconsistencies, further investigation is warranted. As a result, the hypothesis below is proposed:

H2a: Compatibility will positively affect Social Media Adoption.

6.3.4 The Security Concern and Social Media Adoption Relationship

Even though social media can be a dynamic instrument for building and establishing a solid presence in the market, experts cautioned that its adoption comes with various risks, such as reputational damage, unwanted negative publicity, and legal and financial implications (Ekandjo et al. 2018). Unscrupulous hackers can infiltrate firm vulnerabilities illegally in a variety of methods, including cross-site scripting, phishing, malware, clickjacking, and profile cloning (Jain et al. 2021), raising a security concern and hampering adoption ambitions for social media. However, while these threats are real and serious, they can be easily mitigated by imposing a stringent cyber security compliance regimen within the firm, such as strict access control and password security, installation of malware scanners, and implementation of firewalls (Rajasekharaiah et al. 2020). In addition, the benefits of effective social media adoption such as extensive market reach, two-way interaction, and customized content capability (Wibawa et al. 2022) may outweigh the risks. The hypothesis derived from the explanation above is as follows:

H2b: Security Concern will negatively affect Social Media Adoption.

6.3.5 The Social Media Adoption and Firm Performance Relationship

Several studies have shown that adopting social media improves performance, signifying that social media has various potential advantages (Ainin et al. 2015; Oyewobi 2021; Qalati et al. 2022). Businesses demonstrating dedication towards building trust on social media are better positioned to capitalize on this resource (Mahmoud et al. 2020). But then, merely exploiting technology such as social media to create and sustain relationships with consumers and business partners is inadequate because it needs to be deployed effectively to enhance firm performance. The influence social media can wield on firm performance remains a pertinent topic for onward examination and discourse. Based on the justifications mentioned above, the following hypothesis proposes:

H3: Social Media Adoption will positively affect Firm Performance.

6.3.6 The Mediation Function of Social Media Adoption

A myriad of researchers have examined the motivations for social media adoption on numerous occasions (Aronica et al. 2021; Olaleye et al. 2021; Sugandini et al. 2022; Trawnih et al. 2021; Yasa et al. 2021). Findings were diverse, including (but not limited to) the apparent correlation between social media adoption and firm performance or the fact that some firms benefited tremendously from the decision to adopt social media. Generally, a firm's resolve to adopt social media can be attributed to an adaptive reaction to the shift in the business scenario, to stay at par or outperform its competitors. Taking advantage of social media within an organizational context could enable firms to apply strategies that improve multiple performance indicators, such as profitability and growth (Tajvidi and Karami 2021). Seamless social media integration as a resource could elevate firm performance since factors such as compatibility alone might not be enough to improve it. Moreover, firms particularly SMEs, require effective tools to steer their businesses toward their set goals and increase their chances of success. The hypotheses below are proposed based on the merits presented:

H4a: Social Media Adoption mediates the relationship between Compatibility and Firm Performance.

H4b: Social Media Adoption mediates the relationship between Security Concern and Firm Performance.

6.3.7 Research Conceptual Framework

Figure 6.1 depicts an illustrated rundown of all the connections between the constructs based on the descriptions and explanations presented in Sects. 6.3.1–6.3.6.

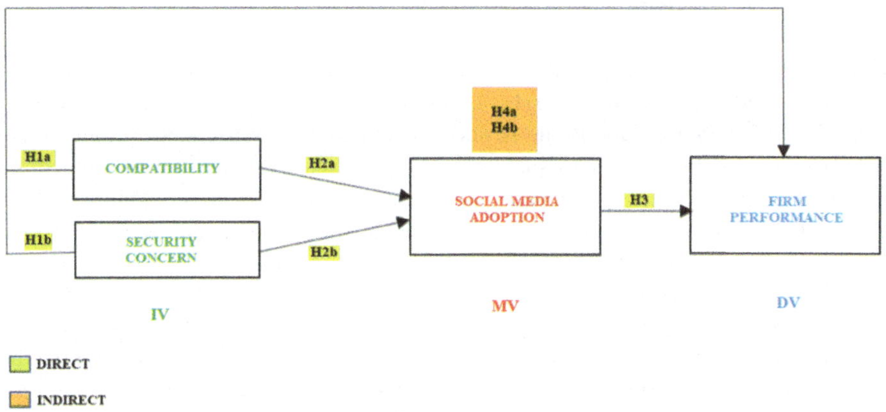

Fig. 6.1 The author's proposed conceptual framework

6.4 Research Methodology

As the first step, a search was undertaken for scholarly literature on Scopus (https://www.scopus.com/) to determine the relationships of SMEs, social media, and firm performance, including information regarding factors that can encourage or inhibit social media adoption. "Firm performance", "social media", and "small medium enterprises" formed the primary initial keywords used, either individually or as a combined variation search string. Once the relevant articles were identified, the full text was retrieved from the corresponding sources, such as Emerald Insight (https://www.emerald.com/insight/) and ScienceDirect (https://www.sciencedirect.com). Additionally, various online working papers and reports released by government and non-government organizations are also obtained to support a more thorough investigative research, with only those closely related to the subject matter set aside for further evaluation before being referenced and cited.

From the preceding literature review exercise, compatibility, and security concern were found to be relevant constructs to be examined and therefore, included in the conceptual research framework as antecedents. Search emphasis then turned to "compatibility" and "security concern", specifically in respect to their association with the adoption of social media and firm performance; and their indirect links to firm performance via the mediating role of social media adoption.

In line with previous suggestions and research on SMEs (Ghobakhloo and Ching 2019; Kull et al. 2018), this research will target a single respondent holding a top management post of manager and above in a Malaysian SME. Each construct will be scrutinized using a 7-point Likert scale, with the instrument based on adapted and adopted items; such as compatibility from Tajudeen et al. (2018) or Ramayah et al. (2011)'s firm performance items. Data acquired will undergo a screening and cleaning process in preparation for it to be analyzed using SmartPLS, the outcomes of which will be presented and explained.

6.5 Discussion

In the current volatile marketplace, firms must accept the fact that they will have to face unpredictably shifting market conditions as well as a hostile business environment and fierce competition. A sudden and unexpected change might make it harrowing for businesses to sustain a competitive advantage unless immediate remedial action is taken to adapt to counteract the adverse effects (Martín-Rojas et al. 2017). The recent unprecedented and devasting emergence of the COVID-19 pandemic, where firms needed to comply with strict Standard Operating Procedures (SOPs) and adapt to new ways of conducting business, such as "Work From Home" (WFH), is a perfect example of a crisis that compelled organizations to prioritize digital transformation. Adopting social media as a resource within the confines of a firm's strategic and operational framework is an avenue to embrace this digital transformation.

Palalic et al. (2020) observed that of the 70% of consumers conducting product research, 60% decided to share the information on their social media. On the homefront, findings in a poll of internet users (MCMC 2020) reinforce this point, whereby the number of users sharing content on social media platforms showed an increase from 73.8% in 2018 to 86.5% in 2020. Furthermore, the same survey also found that there has been a notable upward trend in online social media activity in the country, increasing from 85.6% in 2018 to 93.3% in 2020. With 30.25 million users of social media, Malaysia ranked 2nd in the world in early 2022 for "active social network penetration" (Kemp 2022; Statista 2022). All these findings demonstrate social media's potential and capability to influence decisions. It is worthwhile, therefore, considering the post-COVID-19 pandemic setting and using the conceptual framework described in this chapter to examine the direct and indirect consequences of social media adoption, as evidenced by the rising statistical trends.

6.6 Conclusion

This chapter puts forward a conceptual framework and presents an overview of the relationships between the constructs in the framework proposed, comprising compatibility, security concern, social media adoption, and firm performance. It laid out arguments and justifications to scrutinize Malaysian SMEs, using the carefully selected constructs that can affect social media adoption and firm performance; while outlining how social media adoption can fit into a mediating role. More precisely, the research intent encompasses three primary goals; firstly, to determine the compatibility and security concern effect on firm performance; secondly, to determine the consequence of compatibility and security concern on social media adoption; finally, to investigate if there is any indirect link between compatibility and security concern with firm performance, through social media adoption. In conclusion, the research aims to enhance current literature by providing fresh beneficial viewpoints that can be applied to managerial practices to achieve notable performance gains, as well as relevant information to policymakers so it can be used to formulate more effective policies and strategies to assist Malaysian SMEs overcome the obstacles they frequently face.

References

Abdullah NAH, Ahmad AH, Zainudin N, Rus RM (2019) Predicting financially distressed small- and medium-sized enterprises in Malaysia. Glob Bus Rev 20(3):627–639. https://doi.org/10.1177/0972150919837053

Ainin S, Parveen F, Moghavvemi S, Jaafar NI, Shuib NLM (2015) Factors influencing the use of social media by SMEs and its performance outcomes. Ind Manag Data Syst 115(3):570–588. https://doi.org/10.1108/IMDS-07-2014-0205

APEC (2020) Going digital is necessary for small businesses to survive: Minister. https://www.apec.org/Press/News-Releases/2020/1023_SMEMM

Aronica M, Bonfanti RC, Piacentino D (2021) Social media adoption in Italian firms. Opportunities and challenges for lagging regions. Papers Reg Sci 100(4):959–978. https://doi.org/10.1111/pirs.12606

Barney J (1991) Firm resources and sustained competitive advantage. J Manag 17(1):99–120. https://doi.org/10.1177/014920639101700108

Blouch R, Bakhsh K, Shakeel W, Khan MM, Yasmeen A (2021) Linking firms' strategies with firm performance: mediating role of network value in telecom industry. Int J Product Perform Manag. https://doi.org/10.1108/IJPPM-02-2020-0047

Bouargan M, Halim RAHA, Husaini NH, Jusniah NA, Ahmad NHMH, Almunawar MN (2020) The utilization of social media by small and medium food vendors in Brunei Darussalam. Int J Asian Bus Inf Manag 11(1):142–163. https://doi.org/10.4018/IJABIM.2020010109

Ekandjo TAT, Jazri H, Peters A (2018) Online social networks risks to organisations. ACM Int Conf Proc Ser. https://doi.org/10.1145/3283458.3283465

Falahat M, Ramayah T, Soto-Acosta P, Lee YY (2020) SMEs internationalization: the role of product innovation, market intelligence, pricing and marketing communication capabilities as drivers of SMEs' international performance. Technol Forecast Soc Chang 152:119908. https://doi.org/10.1016/j.techfore.2020.119908

Fang GG, Qalati SA, Ostic D, Shah SMM, Mirani MA (2021) Effects of entrepreneurial orientation, social media, and innovation capabilities on SME performance in emerging countries: a mediated-moderated model. Technol Anal Strateg Manag 34(11):1326–1338. https://doi.org/10.1080/09537325.2021.1957816

Gao H, Tate M, Zhang H, Chen S, Liang B (2018) Social media ties strategy in international branding: an application of resource-based theory. J Int Mark 26(3):45–69. https://doi.org/10.1509/jim.17.0014

Ghobakhloo M, Ching NT (2019) Adoption of digital technologies of smart manufacturing in SMEs. J Ind Inf Integr 16:100107. https://doi.org/10.1016/j.jii.2019.100107

Goel S, Shawky HA (2009) Estimating the market impact of security breach announcements on firm values. Inf Manag 46(7):404–410. https://doi.org/10.1016/j.im.2009.06.005

Gordon LA, Loeb MP, Sohail T (2010) Market value of voluntary disclosures concerning information security. MIS Q Manag Inf Syst 34(3):567–594

Jain AK, Sahoo SR, Kaubiyal J (2021) Online social networks security and privacy: comprehensive review and analysis. Complex Intell Syst 7(5):2157–2177. https://doi.org/10.1007/s40747-021-00409-7

Jia Q, Lei Y, Guo Y, Li X (2022) leveraging enterprise social network technology: understanding the roles of compatibility and intrinsic motivation. J Enterp Inf Manag 35(6):1764–1788. https://doi.org/10.1108/JEIM-05-2021-0225

Kamalrulzaman NI, Ahmad A, Ariff AM, Muda MS (2021) Innovation capabilities and performance of Malaysian agricultural SMEs: the moderating role of strategic alliance. Int J Bus Soc 22(2):675–695. https://doi.org/10.33736/ijbs.3751.2021

Kemp S (2022) Digital 2022: Malaysia. https://datareportal.com/reports/digital-2022-malaysia/

Kull TJ, Kotlar J, Spring M (2018) Small and medium enterprise research in supply chain management: the case for single-respondent research designs. J Supply Chain Manag 54(1):23–34. https://doi.org/10.1111/jscm.12157

Lepkowska-White E, Parsons A (2019) Strategies for monitoring social media for small restaurants. J Foodserv Bus Res 22(4):351–374. https://doi.org/10.1080/15378020.2019.1626207

Luo Y, Bu J (2018) Contextualizing international strategy by emerging market firms: a composition-based approach. J World Bus 53(3):337–355. https://doi.org/10.1016/j.jwb.2017.01.007

Lutfi A, Alsyouf A, Almaiah MA, Alrawad M, Abdo AAK, Al-Khasawneh AL, Ibrahim N, Saad M (2022) Factors influencing the adoption of big data analytics in the digital transformation era: case study of Jordanian SMEs. Sustainability (switzerland) 14(3):1802. https://doi.org/10.3390/su14031802

Mahmoud MA, Adams M, Abubakari A, Commey NO, Kastner ANA (2020) Social media resources and export performance: the role of trust and commitment. Int Mark Rev 37(2):273–297. https://doi.org/10.1108/IMR-02-2019-0084

Martín-Rojas R, Fernández-Pérez V, García-Sánchez E (2017) Encouraging organizational performance through the influence of technological distinctive competencies on components of corporate entrepreneurship. Int Entrep Manag J 13(2):397–426. https://doi.org/10.1007/s11365-016-0406-7

Martin KD, Borah A, Palmatier RW (2017) Data privacy: effects on customer and firm performance. J Mark 81(1):36–58. https://doi.org/10.1509/jm.15.0497

MCMC (2020) Internet users survey 2020. In Malaysian communications and multimedia commission (MCMC). https://www.mcmc.gov.my/skmmgovmy/media/General/pdf/Internet-Users-Survey-2018-(Infographic).pdf

MEDAC (2019) Dasar Keusahawanan Nasional 2030. https://www.medac.gov.my/admin/files/med/image/portal/Dasar-Keusahawanan-Nasional-(DKN)-2030.pdf

Mikalef P, Gupta M (2021) Artificial intelligence capability: conceptualization, measurement calibration, and empirical study on its impact on organizational creativity and firm performance. Inf Manag 58(3). https://doi.org/10.1016/j.im.2021.103434

MOF (2020) Economic outlook 2021. In: Ministry of Finance Malaysia. https://belanjawan2021.treasury.gov.my/pdf/economy/2021/economic-outlook-2021.pdf

Olaleye BR, Hamdan S, Awwad RI (2021) The mediation-moderation effect on the nexus between social media adoption, customer satisfaction, social media competence and SMEs performance during COVID-19 pandemic. Int J Data Netw Sci 5(4):803–814. https://doi.org/10.5267/j.ijdns.2021.x.001

Olanrewaju AST, Hossain MA, Whiteside N, Mercieca P (2020) Social media and entrepreneurship research: a literature review. Int J Inf Manag 50:90–110. https://doi.org/10.1016/j.ijinfomgt.2019.05.011

Oyewobi LO, Adedayo OF, Olorunyomi SO, Jimoh R (2021) Social media adoption and business performance: the mediating role of organizational learning capability (OLC). J Facil Manag 19(4):413–436. https://doi.org/10.1108/JFM-12-2020-0099

Oyewobi LO, Olorunyomi OS, Jimoh RA, Rotimi JOB (2020) Impact of social media usage on performance of construction businesses (CBs) in Abuja, Nigeria. J Financ Manag Prop Constr 26(2):257–278. https://doi.org/10.1108/JFMPC-05-2020-0042

Palalic R, Ramadani V, Mariam Gilani S, Gërguri-Rashiti S, Dana L (2020) Social media and consumer buying behavior decision: what entrepreneurs should know? Manag Decis 59(6):1249–1270. https://doi.org/10.1108/MD-10-2019-1461

Parveen F, Jaafar NI, Ainin S (2016) Social media's impact on organizational performance and entrepreneurial orientation in organizations. Manag Decis 54(9):2208–2234. https://doi.org/10.1108/MD-08-2015-0336

Qalati SA, Ostic D, Sulaiman MABA, Gopang AA, Khan A (2022) Social media and SMEs' performance in developing countries: effects of technological-organizational-environmental factors on the adoption of social media. SAGE Open 12(2):1–13. https://doi.org/10.1177/21582440221094594

Rajaguru R, Matanda MJ (2019) Role of compatibility and supply chain process integration in facilitating supply chain capabilities and organizational performance. Supply Chain Manag 24(2):301–316. https://doi.org/10.1108/SCM-05-2017-0187

Rajasekharaiah KM, Dule CS, Sudarshan E (2020) Cyber security challenges and its emerging trends on latest technologies. IOP Conf Ser Mater Sci Eng 981(2):022062. https://doi.org/10.1088/1757-899X/981/2/022062

Ramayah T, Samat N, Lo MC (2011) Market orientation, service quality and organizational performance in service organizations in Malaysia. Asia-Pacific J Bus Adm 3(1):8–27. https://doi.org/10.1108/17574321111116379

Ramdani B, Kawalek P, Lorenzo O (2009) Predicting SMEs' adoption of enterprise systems. J Enterp Inf Manag 22(1):10–24. https://doi.org/10.1108/17410390910922796

Rogers EM (2003) Diffusion of innovations, 5th ed. Free Press

Sarkar M, Echambadi R, Cavusgil ST, Aulakh PS (2001) The influence of complementarity, compatibility, and relationship capital on alliance performance. J Acad Mark Sci 29(4):358–373. https://doi.org/10.1177/03079450094234

SME Corp (2019) DKN 2030. https://www.smecorp.gov.my/index.php/en/policies/2015-12-21-09-09-50/about-dkn2030

Srimarut T, Mekhum W (2020) The impact of compatibility on the process integration of the supply chain in improving firm performance: the mediating effect of information technology capability. Int J Supply Chain Manag 9(1):155–167

Statista (2022) Active social network penetration in selected countries and territories as of January 2022. https://www.statista.com/statistics/282846/regular-social-networking-usage-penetration-worldwide-by-country/

Sugandini D, Effendi MI, Istanto Y, Arundati R (2022) Social media adoption on SMEs in Indonesia: TOE model. Lecture notes in networks and systems, vol 486, pp. 793–804. https://doi.org/10.1007/978-3-031-08087-6_55

Syaifullah J, Syaifudin M, Sukendar MU, Junaedi J (2021) Social media marketing and business performance of MSMEs during the COVID-19 pandemic. J Asian Financ Econ Bus 8(2):0523–0531. https://doi.org/10.13106/jafeb.2021.vol8.no2.0523

Tajudeen FP, Jaafar NI, Ainin S (2018) Understanding the impact of social media usage among organizations. Inf Manag 55(3):308–321. https://doi.org/10.1016/j.im.2017.08.004

Tajvidi R, Karami A (2021) The effect of social media on firm performance. Comput Human Behav 115:105174. https://doi.org/10.1016/j.chb.2017.09.026

Tran Q, Doan AT, Tran T (2021) What are the drivers of SMEs' financial performance? The interaction of intellectual capital and ownership. Aust Econ Pap 61(4):751–777. https://doi.org/10.1111/1467-8454.12239

Trawnih A, Yaseen H, Al Adwan AS, Alsoud AR, Jaber OA (2021) Factors influencing social media adoption among SMEs during Covid-19 crisis. J Manag Inf Decis Sci 24(6):1–18

Wernerfelt B (1984) A resource-based view of the firm. Strateg Manag J 5(2):171–180. https://doi.org/10.1002/smj.4250050207

Wibawa BM, Baihaqi I, Nareswari N, Mardhotillah RR, Pramesti F (2022) Utilization of social media and its impact on marketing performance: a case study of SMEs in Indonesia Int J Bus Soc 23(1):19–34. https://doi.org/10.33736/ijbs.v23i1. https://doi.org/10.33736/ijbs.4596.2022

World Bank (2022) Small and medium enterprises (SMEs) finance. https://www.worldbank.org/en/topic/smefinance

Yasa NNK, Giantari IGAK, Sukaatmadja IPG, Sukawati TGR, Ekawati NW, Nurcaya IN, Rahanatha GB, Astari AAE (2021) The role of relational and informational capabilities in mediating the effect of social media adoption on business performance in fashion industry. Int J Data Netw Sci 5(4):569–578. https://doi.org/10.5267/j.ijdns.2021.8.011

Yip GS, Devinney TM, Johnson G (2009) Measuring long term superior performance. The UK's long-term superior performers 1984–2003. Long Range Plan 42(3):390–413. https://doi.org/10.1016/j.lrp.2009.05.001

7 Designing Value Proposition for Increasing Business Competitiveness: A Case Study of Startup Business in Indonesia

Astri Ghina, Grisna Anggadwita, Ramo Palalic, and R. Nadya Dewi

Abstract

The recent pandemic has caused many negative effects on businesses across the world. However, at the same time, it has opened new business opportunity windows leading to accelerated and robust competition. The negative consequences of COVID-19 were observed primarily in service sectors and digital agencies. As agencies use media and technology, they are pushed to fierce competition. In order to sustain their businesses, digital agencies sought better ways for innovation that will lead to sustainable competitive advantage. In this context, the study aims to identify the value proposition design of digital agencies identified as startups in Indonesia by developing a value map and making it fit customer profiles. The research employs qualitative methods using a case study approach. It focuses on one of the cases of a digital agency that failed due to the unclear business model and the difficulty of creating a value map and customer profile. The results of the study reveal that there is a discrepancy between each value created and consumer complaints and expectations. This research contributes to the theory of better understanding the base of the value proposition. At the same time, the study proposes a new value for a startup business to increase business competitiveness.

Keywords

Business competitiveness • Customer profile • Startup • Value map • Value proposition design

A. Ghina · G. Anggadwita (✉) · R. N. Dewi
School of Economics and Business, Telkom University, Bandung, Indonesia
e-mail: grisnaanggadwita@telkomuniversity.ac.id

R. Palalic
Department of Management, Sultan Qaboos University, Muscat, Oman

A. Rafiki et al. (eds.), *Open Innovation in Small Business*, Contributions to Environmental Sciences & Innovative Business Technology,
https://doi.org/10.1007/978-981-99-5142-0_7

7.1 Introduction

Startup businesses play a significant role in the global and national economy, including Indonesia, which is considered one of the emerging economies. Also, startups contribute to creating new jobs in society by providing broad employment opportunities for productive workers in Indonesia . However, due to COVID-19 in early 2020, startup businesses have experienced a significant decline. The social restriction policy required every society to adapt to new online business activities. COVID-19 triggered and accelerated the development of information and communication technology by establishing online activities in various fields. Thus, business actors are required to be able to utilize this technology media as a means of doing online business through promotions or digital marketing. The high interest of business people in digital marketing has driven the rise of digital agency service offerings in Indonesia, causing intense competition among these service providers. It encourages business actors to develop good business models to ensure the sustainability of their businesses (Stubbs and Cocklin 2008).

The business model determines how an organization produces value for customers, convinces them to pay for that value, and shares this value with the customers (Blyler and Coff 2003; Amit and Zott 2001; Zott and Amit 2007). Chesbrough and Rosenbloom (2002) define a business model as a heuristic logic that integrates technical potential into creating economic value. Al-Debei and Avison (2010) define a business model as an abstract representation of an organization. Wibawa and Baihaqi (2014) suggest that a business model innovation is essential for firms to survive in a dynamic market environment. According to Teece (2010), a "good" business model generates an attractive value proposition for customers, achieves reasonable cost and risk structures, and enables enterprises to produce and deliver products and services to capture significant value. Thus, the value proposition design is the core of the firm business model (Osterwalder et al. 2014).

Kaplan and Norton (2001) state that the value proposition is "the essence of strategy" and begins by providing differentiated value (Hammonds 2001). Lehmann and Winer (2008) highlight how a value proposition represents a company's core strategy statement. Payne and Frow (2005) propose three aspects of the value creation process: recognizing what value the firm delivers to its consumers, determining what value the company receives from its customers, and engaging in this co-creation successfully (e.g., Prahalad and Ramaswamy 2004; Vargo and Lusch 2004; Payne et al. 2008). Consequently, customer value relates to the customers and the firm in a way that makes the concept particularly appealing in comprehending the success of the company's best practices. However, the term value proposition is usually used in a very informal and trivial way rather than being developed strictly from a strategic point of view (Lanning 2003). Most businesses need an established value proposition (Payne and Frow 2014).

The COVID-19's consequences were not all negative. Instead, it has created various business opportunities, including digital marketing services. However, such a wave of opportunities brought intense competition, which caused several of them to fail. This paper examines the value proposition design for digital marketing services in Indonesia by identifying the value map and adjusting it to the customer profile. This research uses a case study approach to examine a digital marketing service in Indonesia that wishes to redevelop its business to avoid past failures.

This paper is structured as follows. First, it presents the background and problem statement of the study. Second, it briefly reviews the relevant literature on the value proposition concept. Third, it explains the research of this study. Third, it explores a case study by illustrating an alternative approach for establishing a value proposition in the digital marketing services industry. Finally, the paper discusses and concludes the significance of value proposition creation in comprehending the desires and needs of the target market. Additionally, it discusses research limitations and future studies.

7.2 Literature Review

The COVID-19 pandemic has encouraged open innovation in various organizations, including startups and micro, small, and medium-sized businesses (MSMEs). Startups and small businesses benefit from entrepreneurial activity and have a comparative advantage in penetrating market niches, allowing them to form symbiotic relationships with large corporations (Dana et al. 2013; Dana 2014). Open innovation is defined as "a distributed innovation process that relies on a deliberate flow of managed knowledge across organizational boundaries, using monetary and non-monetary mechanisms aligned with the organization's business model to guide and motivate knowledge sharing" (Chesbrough and Bogers 2014). A business model is a set of specific activities—activity systems—designed to satisfy a perceived market need by defining how a company is connected to its ecosystem (e.g., agents, customers, suppliers, and institutions) (Sahut et al. 2020). Innovation can generate a new business model, revise an existing one, or configure and integrate multiple business models (Dana et al. 2022).

Business models are used to define the value that firms deliver to their consumers and partners ("value proposition"), how this value is generated by integrating different resources and activities ("value creation"), and how the value provided can be monetized by companies ("capturing value") (Gassmann et al. 2014; Johnson et al. 2008). Whittington et al. (2020) analyzed business models in three aspects, namely, value proposition and customer segments (value creation), resources and activities (value configuration), and revenues and costs (value capture). The essence of the business model is how companies deliver value to customers, persuade customers to be willing to pay for that value, and generate significant profits for the company (Teece 2010). This business model is the key to the success of the sustainability of a business; innovation will only be practical with a suitable business model (Chesbrough 2010).

Business model development is essential for sustainable business success; thus, creating a value proposition that explains how a company's products and services assist customers in completing their tasks more effectively than other alternatives is crucial (Christensen et al. 2005). This development can be done by utilizing open innovation, where the business model is developed with a user-center. When the value proposition created meets user needs, it is necessary to adjust the resources and activities in the business (Hienerth et al. 2011).

Lanning and Michaels (1988) define a value proposition as a promise of value to customers that combine benefits and pricing. Competitive advantage is attained by giving a superior value proposition to a selected group of customers. Because of the importance of the value proposition concept, businesses must adequately define it in order to obtain a competitive advantage in their target markets. To do this, they must have a deep understanding of the benefits that customers value the most. The value proposition is acknowledged as an integral component of a company's business model (Morris et al. 2005; Lindgardt et al. 2009; Voelpel et al. 2004). In this context, Vargo and Lusch (2004) developed many fundamental premises that highlight the significance of the value proposition as an offer of value to customers and the customer's role as a co-creator of value. The value co-creation process between companies and customers can increase trust and direct feedback, stimulating creativity in the design of value propositions (Leckel et al. 2020).

A value proposition is a transparent and credible expression of the experience that consumers will receive from a business. The value proposition component consists of ability, impact, and cost. Ability is what a business can do to meet consumer needs. The impact is how much our capabilities can help consumers. Cost is how much consumers have to pay for the advantages of our capabilities and impact.

Similarly, a value proposition is a tool that can tell us whether we are likely to run a business successfully (Barnes et al. 2009). In other words, the value proposition is a blueprint for business units that enables businesses to convey genuine value to their target consumers. Value helps business units to become customer-centric, not just branding or marketing stories (Barnes et al. 2017).

The value creation process is divided into two things, namely, activities carried out within the organization and activities carried out outside the organization (with external stakeholders). Value creation with external partners can fill the resource gap owned by startups in capturing new opportunities (Jabeen et al. 2022). The concept of open innovation is in line with the value creation process with external stakeholders, where the business model will be user-centered (Bogers et al. 2020). However, the literature linking innovation to business models is still early (Jabeen et al. 2022).

Value proposition design can begin the development of user-centered business model innovations. The value proposition design is the first step in linking open innovation with business model development. Value proposition design is finding a value proposition that customers want and making it available. It can be done through the iterative search for what customers want, which is crucial to create a high-value proposition for customers (Osterwalder et al. 2014). In recent years, value proposition design has become increasingly significant as a method that assists businesses in systematically building their value propositions (Osterwalder et al. 2014). The value customers gain from an organization is the concept of advantages that increase the customer's offering, which can be incorporated into a value proposition (e.g., Lanning and Michaels 1988).

7.2.1 Customer Profile (Problem Mapping)

The customer profile describes specific consumer segments in a more structured business model. First, there are customer jobs describing the things consumers try to get done at work or in their daily lives. This settlement must satisfy consumers and be taken from their perspective to be seen from functional, social, and emotional jobs. The second is customer pain. It describes anything that annoys the consumer, and this pain represents a risk, potentially leading to a bad outcome.

Moreover, the third, customer gains, describes the results and benefits consumers want. Some benefits are required, expected, or desired by the consumer. The benefits include functional utility, social gain, positive emotion, and cost. The way to do problem mapping is to identify the customer jobs (prioritizing the customer jobs according to the rank of essential and insignificant), customer pains (prioritizing the customer pain according to the rank of extreme and moderate), and customer gains (prioritizing customer gains according to the rank of essential and nice to have) (Osterwalder et al. 2014).

7.2.2 Value Map (Solution Mapping)

The value map describes the features of a particular value proposition in a structured and detailed business model. The first three parts are products and services helping consumers complete functional, social, emotional, and basic needs. Creating such products and services must be interconnected with consumer segments and support consumers in carrying out the role of buyers. Second, pain relievers describe how these products and services can reduce or even eliminate consumer pain. The third is gain creators, which describes how products and services create consumer profits. By outlining what results and benefits consumers expect, including functional utility, social benefit, positive emotions, and cost-effectiveness. There is no need to address every advantage, but focusing on consumer-relevant things, products, and services can make all the difference to consumers. The way to do solution mapping is to map the products and services, pain relievers, and gain creators. Then, ranking those mapping by order of importance (Osterwalder et al. 2014).

7.2.3 Problem–Solution Fit

Problem mapping occurs when a business has evidence that consumers care about specific jobs, pains, and gains. Solution mapping means the business has designed a value proposition to address those jobs, pains, and gains. Achieving problem–solution fit occurs when the business has evidence that customers get excited about value propositions that address important customer jobs, alleviate extreme pains, and create essential customer gains. Consumers expect a lot from products and services, even though they realize that getting everything in a product and service is impossible. As a business owner, focus on the essential gains for consumers and create a difference. Consumers also have many pains. Organizations can only fulfill some of them. Business owners focus on the most extreme and unresolved pains. Consumers are judges and executioners of value propositions and will abandon products and services if they feel their needs are unmet. To check the fit, see and review the products and services, pain relievers, and gain creators one by one, and check whether those fit with customer jobs, pains, and gains. If the pain relievers and gain creators do not fit anything, the business still needs to create a value proposition for the customers (Osterwalder et al. 2014).

7.3 Research Methodology

7.3.1 Research Design

This study aims to validate the value proposition design in startup businesses, which is a digital marketing service providers, by identifying the compatibility between value maps and customer profiles. This study employs an interpretive methodology that enables researchers to comprehend complex phenomena by incorporating humans in fast-changing environments (Yin 1984; Wainwright 1997; Creswell and Poth 2016). Qualitative research allows researchers to understand complex phenomena that cannot be understood through quantitative methods (Malhotra 2006). Therefore, case study research is appropriate for an in-depth investigation of programs, events, activities, processes, and individuals (Creswell and Creswell 2017). The case study approach is expected to investigate and explore field-related descriptions of situations and facts. Lead by this research philosophy, this study applies qualitative methods.

7.3.2 Case Study Overview

This research was conducted at Bandung, an Indonesia-based startup specializing in digital marketing service providers. Because they do not wish to be identified, we use fictitious names for businesses and respondents. However, this still needs to diminish the insights gained from this study. We use the fictitious name "X-Creative" for the business name in this study. X-Creative was founded during the COVID-19 pandemic in 2020 due to the high demand for digital marketing services. Due to policies limiting mobility and social interaction, business actors must utilize technology to sell their products/services, making e-commerce media the preferred method. X-Creative assists these business actors in digitally marketing their firms, as they typically need help branding their businesses.

During the post-COVID era, X-Creative failed and was forced to stop operating due to its inability to compete effectively with other industry businesses and its weak value proposition. Some of the causes of this failure are service innovations established without a research or validation procedure and determining inappropriate value propositions that prioritized only affordable expenses. The founder of X-Creative desires to remodel his firm.

In order for X-Creative to provide the best possible service to its customers, it needs to do research or validation to make sure that the innovation it comes up with is tailored to the needs and desires of those consumers. It is crucial to avoid past failures and stay competitive with other digital agencies.

7.3.3 Setting and Participants

Samples were selected using a non-probability method and purposive sampling technique based on predetermined criteria that could represent the phenomenon under investigation. The criteria for informants in this study are the founders of X-Creative and business people as consumers of digital marketing services.

The primary data source in this qualitative research is the results of interviews and observations with the participating actors. The study employs semi-structured, in-depth, face-to-face interviews to collect the opinions and perceptions of informants regarding the significance of digital marketing services. Before each participant's interview, they were informed of the interview questions and the nature and goal of the study. Table 7.1 displays the number of participants in this study.

The interviews were conducted in Indonesian; thus, the interview findings were transcribed and translated, and double English translations were performed to guarantee that the meaning of the interview results was not altered. Interviews were conducted between January and February 2023 and lasted 10–20 min per interview. All interviews were recorded with the participant's consent. To ensure the

Table 7.1 The number of informants

Code	Gender	Position
N1	Female	Founder X-Creative
N2	Female	Consumer–Business owner
N3	Male	Consumer–Business owner
N4	Female	Consumer–Business owner
N5	Female	Consumer–Business owner
N6	Male	Consumer–Business owner
N7	Female	Consumer–Business owner

anonymity of informants, we employ a code or initials. The results of the interviews were sent to the informants for review to ensure the validity and reliability of the study results.

7.3.4 Data Analysis

This research uses descriptive and exploratory nature as its analysis technique. Data analysis was performed after data collection was completed within a specific timeframe (Sugiyono 2013). Qualitative data analysis is performed interconnected and continuously until it is completed and saturated. Data analysis consists of data reduction, display, and conclusion drawing/verification. Data reduction aims to summarize the data by focusing on important information related to this study. Data display can be done in brief descriptions, charts, relationships between categories, and flowcharts. Display data in this study using a ranking to arrange and prioritize the statement points of customer jobs, pains, and gains from most significant to least significant. Lastly, the conclusions are drawn from the validation findings of the value proposition design in terms of the value map and consumer profile's fit.

7.4 Result and Discussion

The research results are explained based on the theory of Value Proposition Design (Osterwalder et al. 2014), which includes customer profiles, value maps, and fit. This article will explain the results of each aspect in the following sub-sections.

7.4.1 Customer Profile (Problem Mapping)

The customer profile consists of customer jobs, customer pains, and customer gains. This article will summarize the findings in Tables 7.2, 7.3 and 7.4.

7.4.2 Customer Jobs

Customer jobs from six respondents are the priority issues that the respondents are trying to solve or priority needs that respondents are trying to meet when running a business in various fields. This article will map the results into functional, social, emotional, and supporting jobs. The summary of findings is in Table 7.2.

Based on Table 7.2, the findings were re-confirmed to the respondents to assess which ones were important and which were insignificant. The results of the rating of customer jobs by six respondents stated that all findings were significant. Those are sorted by importance: functional, emotional, supporting, and social jobs.

Table 7.2 The findings of customer jobs

Customer jobs	The findings
Functional jobs	Requires a third party to manage product marketing using social media
Social jobs	Desire to increase brand awareness of its business products; hence it looks more professional to consumers
Emotional jobs	Feel happy when there is a third party who can help manage their social media business because time is more efficient; hence respondents can focus on other aspects such as production and distribution
Supporting jobs	Desire to be involved in determining the marketing theme so that it still highlights the business identity (co-creator of value)

Table 7.3 The findings of customer pains

Customer pains	The findings	Code	Rating
Undesired outcomes/ problems	Feeling unable to self-manage their social media because they do not feel creative	Personal obstacle	Extreme
	Digital agencies only focus on content without considering the internal aspects of the business so that the marketing concept does not represent the business brand	Incompetent agency	Extreme
	Content created by digital agencies is monotonous because they use templates from internet sources, making them less attractive to consumers	Incompetent agency	Extreme
	The digital agency cannot fulfill the work according to the agreed time	Unachieved output	Extreme
	There is no significant increase in engagement with consumers on social media	Unachieved output	Extreme
Obstacles	Lacking time to get the job done professionally	Personal obstacle	Extreme
	Communication with digital agencies could be more effective, so there are many miscommunications	Incompetent agency	Extreme
Risks	Losing potential profit if they have to self-manage their social media business	Financial risk	Extreme
	The costs incurred do not match the benefits (too expensive)	High cost	Moderate

Customer Pains

Customer pains from six respondents included undesired outcomes, problems, or obstacles when they had to meet priority needs while running their business. These pains are explored based on each other's experiences when meeting their needs,

Table 7.4 The findings of customer gains

Customer gains	The findings	Code	Rating
Functional utility	Desire to be assisted by a third party who can provide product photo/video services	Complete service	Nice to have
	Desire to be assisted by a third party that provides endorsement services	Complete service	Nice to have
	Desire help from a third party who understands and can run an effective campaign strategy	Complete service	Nice to have
Social gains	There is an increase in brand awareness of the business	Achieving significant output	Essentials
	Increased positive feedback from consumers	Achieving significant output	Essentials
Positive emotions	Desire to be assisted by a responsive third party in communicating the contents	Professional services admin	Essentials
	Happy when assisted by creative third parties who can understand the marketing needs of their products	Professional services admin	Essentials
	Happy when marketing content is made on time as needed	Professional services admin	Essentials
Cost saving	Significant profit increase	Achieving significant output	Essentials

even from experience interacting with digital agencies that have helped them. The summary of findings is in Table 7.3.

Based on Table 7.3, this article codes the findings into several themes: personal obstacle, financial risk, incompetent agency, unachieved output, and high costs. These six themes were then re-confirmed to the respondents to assess which were extreme and which were moderate. The results of the rating by six respondents are sorted by level of pain as follows: personal obstacles (extreme), incompetent agency (extreme), output not achieved (extreme), financial risk (extreme), and high costs (moderate).

Customer Gains

Customer gains from six respondents include functional utility, social gains, positive emotions, and cost savings. Findings are summarized in Table 7.4.

Based on Table 7.4, this article groups the findings into several themes: complete service, significant output achievement, and professional service admin. These three themes were then re-confirmed to the six respondents to assess which ones were essential and which were excellent. The results of the rating by six

Table 7.5 Customer jobs, customer pains, and customer gains

Customer jobs	Customer pains	Customer gains
1. Requires a third party to manage product marketing using social media 2. Want to increase brand awareness of its business products; hence it looks more professional to consumers 3. Feel happy when there is a third party who can help manage their social media business because time is more efficient; hence respondents can focus on other aspects such as production and distribution 4. Desire to be involved in determining the marketing theme so that it still highlights the business identity (co-creator of value)	1. Personal obstacles (extreme) 2. Incompetent agency (extreme) 3. Unachieved output (extreme) 4. Financial risk (extreme) 5. The high-cost (moderate)	1. Achieving significant output (essentials) 2. Professional services admin (essentials) 3. Complete service (friendly to have)

respondents are sorted by level of gain as follows: achieving significant output (essentials), professional services admin (essentials), and complete services (friendly to have).

The results are summarized as priority customer jobs and the rating results of customer pains and gains in Table 7.5.

7.4.3 Value Map (Solution Mapping)

The value map consists of products and services, pain relievers, and gain creators from exploration with business owners. The products and services offered by X-Creative are social media management and advertising. Pain relievers describe the things from products and services that can relieve customer pains. Pain relievers from X-Creative are written agreements regarding the expected output, affordable prices, and custom services according to consumer needs. Gain creators explain explicitly how products and services produce benefits according to consumer expectations, including functional utility, social gains, positive emotions, and cost savings. Gain creators from X-Creative are guaranteed to be able to reach a specific target market audience so that it is more significant in increasing profits.

7.4.4 Problem–Solution Fit

Problem–Solution Fit is the stage where the value proposition can answer priority needs and benefits consumers want to experience based on customer jobs. The fit will be achieved when products and services match customer jobs, the pain relievers match customer pains, and the gain creators match customer gains. Value maps will help to make all consumer jobs easier (customer jobs), reduce or even eliminate problems that bother consumers (customer pain), and it can provide the benefits expected by consumers (customer gains). The summary of the Problem–Solution Fit from X-Creative is in Fig. 7.1.

The X-Creative's products and services, pain relievers, and gain creators can answer almost all of the points in customer jobs, customer pain, and customer gains. This study illustrates the Problem–Solution Fit for value proposition design for X-Creative in Fig. 7.2. If it fits, the check symbol will identify the fits, but the cross symbol will identify when it is not fit.

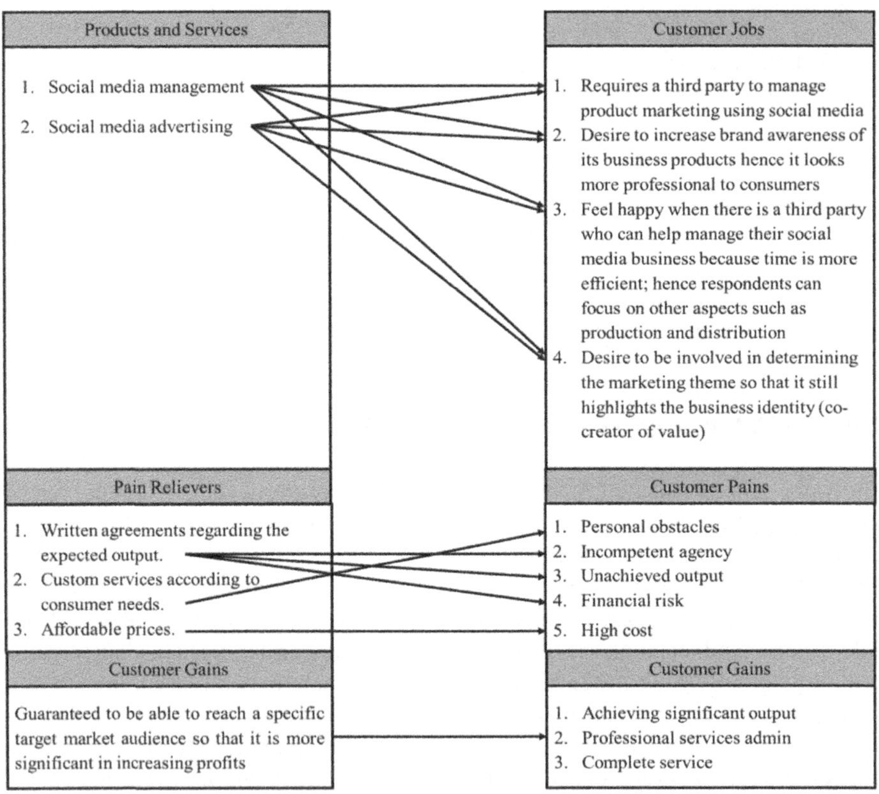

Fig. 7.1 Summary of the problem–solution fit from X-Creative

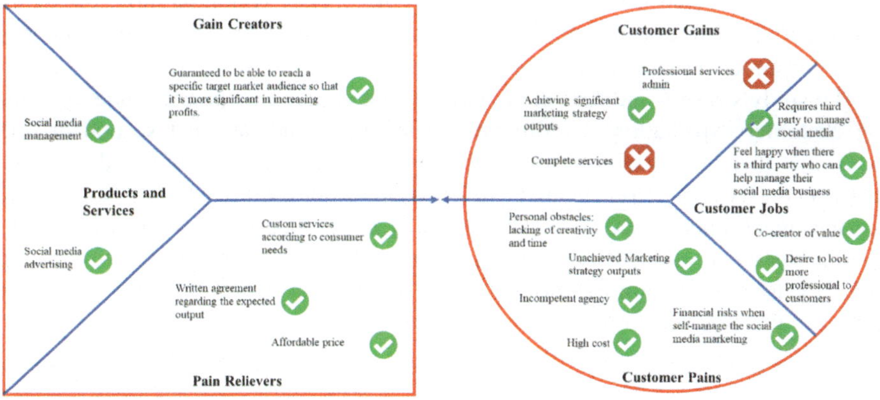

Fig. 7.2 Value proposition design X-Creative

Based on Fig. 7.2 X-Creative provides services that include social media management and advertising. The products and services fulfill customer jobs from respondents, which include the need for a third party to manage social media, feeling happy when others handle social media management tasks, co-creator of value, and desire to look professional by customers. The products and services meet all the customer jobs, illustrated with a green checklist icon (Fig. 7.2). According to Osterwalder et al. (2014), products and services which consist of offering lists have to solve jobs or meet the customer's basic needs. Empathizing with customer needs is crucial to startup success (Hougaard 2005). The available market is one of the primary factors in business success (Hartnett and Keisler 2008).

Pain relievers from X-Creative are custom services according to consumer needs, written agreements regarding the expected output, and reasonable prices. The pain relievers fulfill customer pains from respondents, including a lack of creativity and time (personal obstacles), unsuccessful marketing strategy outputs, and financial risks when forced to self-manage social media marketing. The pain relievers are fulfilling all the customer's pains. According to Osterwalder et al. (2014), pain relievers must be able to reduce or even eliminate customer pains in completing work. According to Barnes et al. (2009), businesses can generate profits if the institution recognizes the consumers as the primary stakeholders. In this case, pain relievers must accommodate the basic needs of their target consumers.

Gain creators from X-Creative are guaranteed to reach the target market specifically to increase profits significantly. The gain creators fulfill one of the customer gains, namely significant marketing strategy outputs. In this case, there is a gap where two aspects of customer gains exist that X-Creative cannot fulfill. This condition can be used as basic information to improve products and services. According to Osterwalder et al. (2014), the gain creators can address only some customer gains. The startup can focus on relevant gain creators, which make a difference and create meaningful customer experiences. Value is about the experience delivered to consumers, not what we do. Businesses only need to address some consumer desires, as doing so can lead to bankruptcy (Barnes et al. 2009).

The study's results indicate the importance of creating corporate value in meeting customer expectations. Additionally, the results indicate that several drivers determine the success of turning an idea into a business. Thus, the study implies that success is a function of talent focused on opportunities where products and services meet market needs. As a driver, the business focuses on creating and monetizing customer value (Kuenne and Danner 2017). An organization can develop ideas by looking at familiar products and services in the market and exploring opportunities relevant to its business (Drummond 2009).

7.5 Research Implications

This study has both academic and practical implications. Additionally, it provides broader insights regarding implementing value proposition design in startups by validating the value map and customer profile. The design of value propositions plays a significant role in engaging and building long-term customer relationships, thereby emphasizing the importance of the value proposition as the center point of the business model to enhance organizational competitiveness. This study can facilitate student learning in developing value propositions for startup businesses and broader development for other business scales such as micro, small, and medium-sized enterprises.

From a practical perspective, developing products and services can fulfill other aspects of customer gains. For example, to fulfill various services, X-Creative can collaborate with other parties in providing services so that it is more cost-efficient considering X-Creative's limited resources. To fulfill the expected benefits in the form of a professional admin, X-Creative can develop services using project management applications so that all processes will be more controlled to produce expected outputs. The project management dashboard can help strengthen gain creators, which provides guarantees regarding the ability to target specific target markets so that the interaction between X-Creative and its consumers will be more effective and efficient. It is necessary to conduct market research to understand pains and gains better; hence the pain relievers and gain creators owned by products and services can satisfy consumers (Pinson and Jinnett 2006). Startups have to create good ideas, which come from listening to their colleagues and environment; hence, the ideas generated will be different but a bit better (Southon and West 2002). Ideas can also be developed by looking at trends; for example, currently, there is a trend of using digital in running a business; hence we have to put technology elements into developing ideas (Pinson and Jinnett 2006). For the creative mind, developing new ideas is sometimes easy. The biggest challenge is turning ideas into products that meet market needs and whether they are technologically feasible and profitable (Strauss 2002).

7.6 Conclusion

This research contributes to the literature by better understanding the value proposition in digital marketing. This in-depth understanding is presented by a case study that highlights how startup companies engaged in digital marketing services can survive again to meet at least customer expectations. In doing so, new insights about customer needs for business models can be gained and help develop a successful sustainable business model.

This case study demonstrates that X-Creative has a value map since it offers social media management and advertising services. Customers believe digital agency services will assist with social media management and brand visibility among consumers. Based on the customer profile, the study has identified the following: customer pains, including consumer complaints about the lack of digital agency competence, the inability to achieve the desired output, as well as inappropriate costs and timelines; customer gains, including the presence of service quality that meets customer expectations, output achievement, service completion, and timeliness. Similarly, the identified value map identifies pain relievers, such as the existence of custom services, service quality that follows consumer desires, agreement regarding the output produced, and affordable costs; and gain creators, such as the availability of various services, the achievement of a specific target audience, and affordable costs. The analysis results demonstrate the fit between the customer profile and the value map, such as agreement on the output to be created, affordable pricing, and social media management services and ads (advertising).

This research has several limitations that can provide opportunities for further research. This study is limited to a single case study, limiting the generalizability of its findings to other case studies. Therefore, additional case study research is required to investigate different methods for building value propositions in various business domains. In addition, longitudinal studies are suggested to investigate companies' acceptance and utilization of value propositions by analyzing the company's development and implementation experiences.

Future research can also conduct empirical testing by analyzing the relationship between value propositions and other aspects, such as competitive advantage and organizational performance, to better understand the value proposition. The implementation and application of the value proposition concept at the segment and individual customer levels require additional investigation.

References

Al-Debei MM, Avison D (2010) Developing a unified framework of the business model concept. Eur J Inf Syst 19(3):359–376. https://doi.org/10.1057/ejis.2010.21

Amit R, Zott C (2001) Value creation in e-business. Strateg Manag J 22(6–7):493–520. https://doi.org/10.1002/smj.187

Barnes C, Blake H, Pinder D (2009 Creating & delivering your value proposition: managing customer experience for profit. Kogan Page Limited, United States

Barnes C, Blake H, Howard T (2017) Selling your value proposition: how to transform your business into a selling organization. Kogan Page Limited, United States
Blyler M, Coff RW (2003) Dynamic capabilities, social capital, and rent appropriation: ties that split pies. Strateg Manag J 24(7):677–686. https://doi.org/10.1002/smj.327
Bogers M, Chesbrough H, Strand R (2020) Sustainable open innovation to address a grand challenge: lessons from Carlsberg and the green fiber bottle. Br Food J 122(5):1505–1517. https://doi.org/10.1108/BFJ-07-2019-0534
Chesbrough H (2010) Business model innovation: opportunities and barriers. Long Range Plan 43(2/3):354–363. https://doi.org/10.1016/j.lrp.2009.07.010
Chesbrough H, Bogers M (2014) Explicating open innovation: clarifying an emerging paradigm for understanding innovation. New frontiers in open innovation. Oxford University Press, Oxford, pp. 3–28 (Forthcoming)
Chesbrough H, Rosenbloom RS (2002) The role of the business model in capturing value from innovation: evidence from Xerox Corporation's technology spin-off companies. Ind Corp Chang 11(3):529–555. https://doi.org/10.1093/icc/11.3.529
Christensen CM, Cook S, Hall T (2005) Marketing malpractice–the cause and the cure. Harv Bus Rev 83(12):74–86
Creswell JW, Poth CN (2016) Qualitative inquiry and research design: choosing among five approaches. Sage Publications
Creswell JW, Creswell JD (2017) Research design: qualitative, quantitative, and mixed methods approaches. Sage Publications
Dana LP (2014) Asian models of entrepreneurship—from the Indian Union and Nepal to the Japanese Archipelago: context, policy and practice, 2nd ed. World Scientific Publishing
Dana LP, Etemad H, Wright RW (2013) Toward a paradigm of symbiotic entrepreneurship. Int J Entrep Small Bus 5(2):109–126. https://doi.org/10.1504/IJESB.2008.016587
Dana LP, Salamzadeh A, Hadizadeh M, Heydari G, Shamsoddin S (2022) Urban entrepreneurship and sustainable businesses in smart cities: exploring the role of digital technologies. Sustain Technol Entrep 1(2):100016. https://doi.org/10.1016/j.stae.2022.100016
Drummond H (2009) How to be a successful entrepreneur. Replika Press Pvt Ltd., India
Gassmann O, Frankenberger K, Csik M (2014) The business model navigator: 55 models that will revolutionise your business. Pearson UK
Hammonds KH (2001) Michael Porter's big ideas. Fast Co 44:150–156
Hartnett R, Keisler K (2008) Small business, big opportunity. Sensis, Victoria
Hienerth C, Keinz P, Lettl C (2011) Exploring the nature and implementation process of user-centric business models. Long Range Plan 44(5/6):344–374
Hougaard S (2005) The business idea. Springer Berlin, Germany
Jabeen F, Belas J, Santoro G, Alam GM (2022) The role of open innovation in fostering SMEs' business model innovation during the COVID-19 pandemic. J Knowl Manag. Emerald Publishing Limited. ISSN 1367–3270
Johnson MW, Christensen CM, Kagermann H (2008) Reinventing your business model. Harv Bus Rev 86(12):50–59
Kaplan RS, Norton DP (2001) Transforming the balanced scorecard from performance measurement to strategic management: Part 1. Account Horiz 15(1):87–104
Kuenne C, Danner J (2017) Built for growth. Harvard Business Publishing Corporation, Massachusetts
Lanning MJ, Michaels EG (Jul 1988) A business is a value delivery system. McKinsey Staff Pap 41
Lanning M (2003) An introduction to the market-focused philosophy, framework and methodology called delivering profitable value. www.exubrio.com/services/white_papers/DPVIntro-eXubrio.pdfm. Accessed Dec 2022
Leckel A, Veilleux S, Dana LP (2020) Local Open Innovation: a means for public policy to increase collaboration for innovation in SMEs. Technol Forecast Soc Chang 153:119891. https://doi.org/10.1016/j.techfore.2019.119891
Lehmann DR, Winer R (2008) Analysis for marketing planning. Irwin, Burr Ridge, IL

Lindgardt Z, Reeves M, Stalk G, Deimler M (2009) Business model innovation. Boston Consulting Group
Malhotra NK (2006) Questionnaire design. The handbook of marketing research: uses, misuses, and future advances. Sage Publication, pp 83–94
Morris M, Schindehutte M, Allen J (2005) The entrepreneur's business model: toward a unified perspective. J Bus Res 58(6):726–735
Osterwalder A, Pigneur Y, Bernarda G, Smith A, Papadakos T (2014) Value proposition design: how to create products and services customers want. Wiley, London. ISBN 978-1-118-96805-5
Payne A, Frow P (2014) Developing superior value propositions: a strategic marketing imperative. J Serv Manag 25(2):213–227. https://doi.org/10.1108/JOSM-01-2014-0036
Payne A, Frow P (2005) A strategic framework for customer relationship management. J Market 69(4):167–176. https://doi.org/10.1509/jmkg.2005.69.4.167
Payne AF, Storbacka K, Frow P (2008) Managing the co-creation of value. J Acad Mark Sci 36:83–96. https://doi.org/10.1007/s11747-007-0070-0
Pinson L, Jinnett J (2006) Steps to small business startup. Kaplan Publishing, USA
Prahalad CK, Ramaswamy V (2004) Co-creation experiences: the next practice in value creation. J Interact Mark 18(3):5–14. https://doi.org/10.1002/dir.20015
Sahut JM, Dana LP, Laroche M (2020) Digital innovations, impacts on marketing, value chain and business models: an introduction. Can J Adm Sci/revue Canadienne Des Sciences De L'administration 37(1):61–67. https://doi.org/10.1002/cjas.1558
Southon M, West C (2002) The beermat entrepreneur: turn your good idea into a great business. Pearson Education Limited, Great Britain
Strauss SD (2002) The big idea. Dearborn Trade Publishing, USA
Stubbs W, Cocklin C (2008) Conceptualizing a "sustainability business model." Organ Environ 21(2):103–127. https://doi.org/10.1177/1086026608318042
Sugiyono D (2013) Metode penelitian pendidikan pendekatan kuantitatif, kualitatif dan R&D. Alfabeta, Bandung, Indonesia
Teece DJ (2010) Business models, business strategy and innovation. Long Range Plan 43(2–3):172–194. https://doi.org/10.1016/j.lrp.2009.07.003
Vargo SL, Lusch RF (2004) The four service marketing myths: remnants of a goods-based, manufacturing model. J Serv Res 6(4):324–335. https://doi.org/10.1177/1094670503262946
Voelpel SC, Leibold M, Tekie EB (2004) The wheel of business model reinvention: how to reshape your business model to leapfrog competitors. J Chang Manag 4(3):259–276
Wainwright D (1997) Can sociological research be qualitative, critical and valid?. TQR 3(2):1–17
Whittington R, Regnér P, Angwin D, Johnson G, Scholes K (2020) Exploring strategy text and cases, Pearson, UK
Wibawa BM, Baihaqi I (2014) Desain inovasi model bisnis untuk pengembangan bisnis Vaksin HydroVac. In: Konferensi nasional riset Manajemen VIII
Yin RK (1984) Beyond method: strategies for social research. Admini Sci Quart 29(2):321–323. https://doi.org/10.2307/2393196
Zott C, Amit R (2007) Business model design and the performance of entrepreneurial firms. Organ Sci 18(2):181–199. https://doi.org/10.1287/orsc.1060.0232

Comparing Critical Factors for Big Data Analytics (BDA) Adoption Among Malaysian Manufacturing and Construction SMEs

8

Ku Halim Ku Ariffin, Norizah Ahmad, Saravanan Paramasivan, Cut Kesuma Pahlufi, and Yossie Rossanty

Abstract

The study was conducted with the aim of exploring the factors that impact the implementation of BDA as well as the usage of BDA in two industries of interest, i.e., the manufacturing and construction industries. In particular, the study narrowed down the organizations in the research to only involve SMEs, thereby providing a fair base of comparison while truncating financial factors. The study adopted a modified technology-organizational-environmental (TOE) model. The study was conducted using content-based approaches. We expected that the study of SMEs in the manufacturing and construction industries may have positively and significantly influenced the strongest parameters for BDA adoption and ultimately contributed to enriching theoretical and practical development.

Keywords

Perceived big data analytics • Manufacturing • Construction • SMEs • Malaysia

8.1 Introduction

Currently, we are sitting at the boundary of the Fourth Industrial Revolution (IR4) and the Fifth Industrial Revolution (IR5). It has been succinctly elaborated how knowledge application itself has evolved from merely producing technology or an

K. H. K. Ariffin · N. Ahmad · S. Paramasivan · C. K. Pahlufi
Faculty of Applied and Human Sciences, Universiti Malaysia Perlis, Arau, Malaysia

Y. Rossanty (✉)
Department of Management, Universitas Pembangunan Panca Budi, Medan, Indonesia
e-mail: yosunpab@gmail.com

A. Rafiki et al. (eds.), *Open Innovation in Small Business*, Contributions to Environmental Sciences & Innovative Business Technology,
https://doi.org/10.1007/978-981-99-5142-0_8

innovation, right up to acquiring knowledge from data collected from a particular technology or innovation. There is a discernible migration from technology centric initiatives to data-centric initiatives, where the data is valued for the knowledge and information it provides. This is very true for IR4 and IR5. IR4 stepped up automation by exploring autonomous systems through machine learning , Big Data Analytics (BDA) and Artificial intelligence (AI). The culmination and application of cloud computing, IoT, and BDA inadvertently opened a bigger virtual market worth USD 11.5 trillion (as of 2016) in the form of a digital economy (Nagy et al. 2018). Lifestyles, business practices, and government delivery systems were drastically overhauled with elements of IR4, whereby phones are now getting "smarter," and this is inadvertently accompanied by the migration of physical infrastructure and services such as banking, shopping, delivery systems, and records into a virtual electronic IoT space that can be accessed anytime and anywhere, while also keeping the cost of providing such services to an extreme minimum (Surya Saputra 2011).

The inevitable change that IR4 has brought about can be clearly seen during the onset of the COVID-19 pandemic and its aftermath. According to Pew Research Center's report (2020), it was found that up to 71% of employees in the United States embraced remote working during the pandemic, a substantial increase from 20% before the pandemic outbreak, whereby 60% of them preferred working remotely even after the pandemic. This is also corroborated by statistics published by the Manpower Research and Statistics Department of the Ministry of Manpower Singapore (2021), which indicated that up to 49.3% of the country's workforce was working from home (WFH) during the COVID-19 pandemic and that 75% of these remotely working people would prefer to continue working remotely post-pandemic. As a matter of interest, it is also reported that Belgium expects to formalize remote working post-pandemic as a solution towards pollution eradication and traffic control.

The ability for many businesses within such countries to immediately adopt a remote-working norm effectively shows that businesses within these countries have the adequate infrastructure that supports IoT, cloud computing, and BDA incorporation within their business operations, and the high percentage of workers who prefer to continue working remotely indirectly does show a high level of readiness among workers to adopt elements of IoT, cloud computing, and BDA. Although these numbers do operationally indicate a country's overall responsiveness and its people's readiness to adopt the systemic application of IR4 elements in their work practices, they do not address the micro-variability that exists between industries in their degree of BDA implementation.

Since BDA is the common denominator that pervades in the creation of autonomous systems, the operationalization of IoT, the foundation of machine learning systems, and the creation of "smart" systems such as smart farming and smart cities, it is therefore within the interest of this study to explore further the variability that exists within industries and between industries in their BDA implementation or adoption.

BDA refers to the application of mathematics, physical analysis, and statistical analysis to large unstructured and/or unstructured datasets in order to obtain functional information that can be used for informed decision-making and as a precursor for secondary processing. The extent to which BDA is adopted in industries depends invariably on the type of industry, the resources available from an organizational support system, as well as the driving regulations and trends prevalent within the industry and in the country. The differences that arise from the functional application of BDA show the influence of which types of industries can affect BDA adoption tendencies. For instance, in the case of the manufacturing industry, BDA is instrumental in the areas of process and instrumentation optimization that can be used towards decisions that benefit work-cell optimization, improving yield, supply chain management, predictive maintenance, and product lifecycle management (Li et al. 2022). In the financial industry, BDA is instrumental in the areas of processing a large base of customer datasets in ways that can enhance efforts in sales and marketing, customer cultivation, risk management, internal management, and informed product and service development (Nobanee 2021).

This difference in the functional use of BDA in the manufacturing and financial industries subsequently leads to differential needs in system architecture for BDA implementation, whereby the former involves more intensive software-hardware interfacing and machine learning, and the latter involves a more intensive statistical algorithm to manage, isolate, and process an unstructured customer dataset (Stojanovic et al. 2015; Nobanee 2021).

In the construction industry, on the other hand, BDA is largely new in its proliferation. Much of the advances in the construction industry have been on computation speed, computation techniques, algorithm developments and to a larger extent, technology innovation. Despite the slow growth and emphasis of BDA within the construction industry, BDA has proven to be functional in the areas of predictive design and modeling, managerial and operational management as part of integrated cost–benefit analysis, project management, inventory and waste management, and equipment tracking through sensor usage (Bieh-Zimmert and Felden 2015). Therefore, it can be seen that the possible usage of BDA within the construction industry is diverse and encompasses greater usage potential applicable to different segments of the workforce, but only if BDA can be streamlined throughout an organization belonging to the said industry. Thus, the study is aimed at isolating and understanding factors that affect the adoption of BDA among small and medium enterprises (SMEs) in two industries in Malaysia, i.e., manufacturing and construction. These factors were examined by parameterizing them into three broader categories, such as organizational factors, technological factors, and environmental factors, as stipulated under a standard Technology-Organization-Environment (TOE) framework model, with the aim of comparing the quantified effects of these factors on BDA adoption within and between SMEs belonging to the manufacturing and construction industries.

The Malaysian Reserve reported that the construction industry in Malaysia is performing poorly on the productivity front. The construction industry faced the lowest productivity level compared to other industries in 2016, as reported by the Malaysian Productivity Report 2016–2017. In the same article, the Minister of Works, Datuk Seri Fadillah Yusof, was quoted as saying that the application of disruptive technology such as IoT, BDA, AI, and cloud computing poses a wealth of opportunities to enhance productivity and create more innovative data-centric services. In particular, the corroborative use of BDA within the construction industry is indeed seen as a solution towards optimizing value along a poorly performing construction industry's supply chain. Hence, this is the fundamental reason why this study is important.

Although digital transformation is inevitable in manufacturing, overhauling an entire assembly line or production line is extremely expensive (Estrada and Ruiz 2016). Therefore, many manufacturing companies have not been able to fully adopt the benefits of digital transformation within their organizations due to financial constraints. Aside from capital expenditure constraints, cybersecurity could be a barrier to BDA adoption in the manufacturing industry. Data that is generated through various interfaces, such as sensors, devices, and networks, is vulnerable to malicious exploitation by irresponsible parties. Many manufacturing companies work under the pretext of industrial design secrets and industrial patents; thus, the adoption of BDA is somewhat deterred due to the risk posed by a weak cyber-security infrastructure. Apart from that, it is also stated that another pertinent challenge in the adoption of BDA in the manufacturing industry is the niched technical expertise, skills, and training required, especially for process- and instrumentation-centric BDAs. Manufacturing companies are highly customized and personalized in their approach toward product and process design, development, and fabrication; therefore, most often, technical workers rarely migrate across manufacturing companies with different product and service interests. Thus, adequate skills also pose a big adoption challenge in the manufacturing industry. The combination of all these factors poses a foundational basis for this study, which will investigate and quantify the factors that affect BDA adoption among SMEs in manufacturing and construction in Malaysia.

- RQ1: How do TOE factors affect BDA adoption among Malaysian SMEs within the manufacturing and construction industries?

Accordingly, the purpose of this study is to quantify the effect of TOE factors on the degree of BDA adoption among Malaysian SMEs in the manufacturing and construction industries and to compare the extent of BDA adoption between these two industries. The significance of this study is to comprehend the driving factors behind the varying degree of BDA adoption among SMEs in the manufacturing and construction industries. Understanding how organizational, technological, environmental, and individual factors dynamically impact BDA adoption among SMEs can shed light on the aspects in which resources and assistance can be directed by private and public organizations in helping to accelerate the effective adoption of BDA among SMEs in such industries.

8.2 Literature Review

8.2.1 Big Data

Data has taken a different form today than it did in the past. Data at present is manifested in polymorphous form due to the influx of data coming in from all variable sources, an influx catalyzed by improvements in sensor and instrumentation technology as well as increasing awareness of data-backed decision-making processes (Riahi 2018). Polymorphous data in this very sense refers to the many shapes and forms in which data today is being received, whereby at first glance, they just appear to be a collection of random, non-systemic, and ambiguous numbers coming in at an unprecedented rate and volume (Margara and Rabl 2018). This is where the concept of Big Data lies. The sheer volume, retrieval speed, and degree of heterogeneity in the format of the data are defining characteristics of Big Data. Since the functionality of Big Data varies from one industry to another, the way Big Data is utilized or harvested in terms of the techniques and categories of technologies deployed, delineates the unique domain and perspectives of what Big Data is in respective industries (Abbasi 2016).

The classical characterization of what is Big Data had always boiled down to its volume, the speed of information influx, i.e., the velocity, and the multitude of format possibilities it can take, i.e., the variety. As a result, the three Vs—velocity, volume, and variety—have been the guiding principles of Big Data, based solely on descriptive observation of its characteristics (Venkatram and Geetha 2017; Gupta and Gupta 2015). However, in later years of its evolution, industrialists became increasingly aware of the importance of qualifying and assessing the harvested big data because not all incepted digital numbers are usable and some have simply been "polluted," "diluted," or "attenuated" due to interferences or disparities in sensor sensitivity (Holmes 2017). This realization came to light due to the parallelism drawn between the yield of Big Data processing and classical signal processing, which showed that information (whether in Big Data or signal form) can differ in its quality as well as its capacity to undergo secondary and tertiary processing (Falk and Sejdić 2018). Therefore, this brings us to the most recent two other Vs that were added into components of Big Data i.e., its veracity, which denotes the quality of data, and its value, which denotes the secondary utilization of data (Pendyala 2018; Atal and Mike 2018).

Volume refers to the amount of data produced, contained, and utilized within a particular system. With the ever-increasing sophistication of data acquisition technology and the capacity to store vast amounts of information, the volume factor of Big Data is also capitalized by increasing intent and importance with regards to its functions (Hartung 2018; Venkatram and Geetha 2017). Velocity is related to the frequency with which data is acquired, processed, and migrated after processing within a particular system. Better sensors and instrumentation have allowed for the acquisition of data at an unprecedented rate, and all this data is streamed in real time. As a matter of fact, the velocity of data is one of the primary reasons why logical processing techniques and artificial intelligence solutions are currently being sought and will be sought in the future for software-hardware interfaces. This huge amount of data coming in every second needs to be processed without constant human involvement (Dautav and Distefano 2017; Walker 2015).

In this sense, "variety" refers to the variability and multiplicity that exist in the types of data received on the back end of data acquisition. Due to the complexity posed by the multiplicity factor of the Big Data received, the processing methodologies and techniques associated with delineating these data have constantly been a subject of research and improvisation. Most importantly, the differential links that exist between these data subtypes have led to the discovery of alternate uses for these data, especially when differential causalities are established (Kockum and Dacre 2021; Nyamful and Agrawal 2017).Veracity is related to the challenge posed by the high rate of receipt of voluminous variable data into the system, i.e., how do we ascertain the quality and truthfulness of the data received? The process of ascertaining the quality and truthfulness of this Big Data is called "data veracity." The first part of the data processing technique is very much involved in filtering out inconsistencies, any duplications, abnormalities, and biases, in a process most often called filtering. During this process itself, the volatile component of a data set is also identified. Volatility refers to the differential rate of change of a trend or lifetime of data. Volatile data is more unpredictable and requires a different type of statistical metric to comprehend. As much as we have seen how quality is exhibited based on oddities that exist in the data, truthfulness and quality can also be adjudged based on the data source, data type, and even the methods used in handling those data (Pendyala 2018; Berti-Equille and Ba 2016; Zhang and Xiang 2015).

Data value is a more abstract component of Big Data, hence its recency. The other four Vs are often measurable at the point of data receipt, but in the case of value, this quality is not always straightforward. Data can be harvested based on its known useability, either from an economic or technological perspective. However, in many cases, the usability of some received data is unknown until additional processing is performed, and some of these usabilities may simply be byproducts of an already existing methodical big data processing. Therefore, in this sense, the value of Big Data is limited to the existent technology and techniques available at that moment for extracting its value. Aside from that, the applicability of data is heavily dependent on the requirements of the presiding time. A high value of data may be warehoused and kept on a backup server due to the limited knowledge of its potential use during its time of processing, and it may take a few years into the future for its value to be realized and potentially utilized. These kinds of latent values associated with Big Data often pose a challenge for even economists to ascribe an economic value to subtypes of Big Data.

8.2.2 Big Data Analytics (BDA)

The description of Big Data in the preceding section would have elucidated the importance of applying additional analysis to Big Data, a procedure now famously known as "Big Data Analytics" (BDA). Succinctly put, BDA is essential because new procedures, techniques, tools, hardware, and software are needed to accommodate the increased need for sensitivity for data acquisition, increased complexity and quantity of data aggregates, increased urgency for data utilization, and the increased need for a quicker programmable data processing system (Holmes 2017; Shi 2022). These essentialities have basically created a new breed of analytics with its own growing BDA ecosystem. The system that is put in place needs to be able to assist the processes involved in making sense of this Big Data to ensure a timely exploitation of this data in real-time. BDA requires an automatically prudent system capable of isolating and interpreting a high rate of incoming large volume datasets while also taking into account a potential volatility construct (Hussain and Roy 2016; Bala 2022). Hence, BDA can be described as a collection of processes and technologies that are involved in the collection, organization, and analysis of Big Data sets with the aim of extracting discernible useful patterns and information that can be of value to the user (Kumaraguru and Chakravarthy 2017). The end result of BDA is expected to assist and improve old, existential, and future problems for an organization or institution. There are four broad categorizations of types of analytics within the context of BDA: descriptive analytics, diagnostic analytics, predictive analytics, and prescriptive analytics.

Descriptive Analytics
At the first instance of data processing, this type of analysis centers and focuses on describing what is observed from the pool of unorganized, structured, or unstructured big data. Mining and organizing data into easily discernible categories, as well as assisting future processing of identifying trends, are two of the core functions of descriptive analytics. Descriptive statistics also provide statistical insight into the numeric data received, equipping the user with the ability to identify any possible aberrations and oddities within the large dataset, which may require immediate attention from an action point of view (Kalaian et al. 2016; Hodeghatta and Nayak 2017).

Diagnostic Analytics
This type of analysis focuses on determining the causality of a specific event, behavior, or oddity in a large dataset. Diagnostic analysis, as opposed to descriptive analytics, necessitates more processing power and a more complex set of techniques in order to solve a "lead-lag" problem from a large dataset. The analysis necessitates extensive spatial processing to cross-reference different datasets and, to a lesser extent, extensive temporal processing to cross-reference time series trends for the investigated event, behavior, or oddity (Shi 2022; Loshin 2013a, b; Elitzur et al. 2020).

Predictive Analytics
This type of analysis centers on and focuses on outlining the most likely event or set of events that may occur following the processing of a large dataset. Predictive analysis is a more advanced type of analysis that occurs after descriptive

and diagnostic analysis. Extensive temporal data needs to be investigated to understand historical datasets in an attempt to correlate these events with past datasets. Upon correlation, a probabilistic model is created to understand and forecast future behavior or events from the received dataset. Due to its complexity, most of the innovation in artificial intelligence (AI) has stemmed from the need to automate actions based on a pre-meditated response obtained from predictive analytics (Finlay 2014; Nair et al. 2017).

Prescriptive Analytics

This type of analysis is the most advanced compared to the other three. In descriptive, diagnostic, and predictive analysis, the datasets that are utilized, processed, and analyzed against are continuous data that are real and historical. However, in prescriptive analysis, a large dataset is simulated and differentially extrapolated against present and historical data to understand the probable outcome due to an action or interference made. Prescriptive analysis does not just use forward-looking models to predict an event; it also allows insight into the best possible solution to an anticipated problem in the future. For instance, ascertaining and predicting the extent of structural burden on a hydroelectric dam due to an increasing volume of water would be the application of predictive analytics, whereas identifying construction solutions and testing these solutions in a model with the intent of enhancing the structural stability of the dam bodes well within the domain of prescriptive analysis. As such, the boundary of prescriptive analysis is very wide and is only limited by computational capacity (Gonçalves 2017; Burk and Miner 2020; Deshpande et al. 2019).

8.2.3 Technology Assimilation and Adoption in Industry

Big Data and BDA are the cornerstones of Industry 4.0 and are precepted as the foundations for the impending Industry 5.0. At each industrial revolution paradigm, the rate at which the presiding technology evolves and gets replaced within the workings of an organization, as well as in an industry, is getting more rapid and is happening at an unprecedented rate (Baird et al. 2017; Girasa 2020). As such, organizations need to quickly adapt to the ever-changing demands of the industry by being sensitive to newly introduced technologies or processes that are considered standards of the time due to regulated compliance or due to innovative solution towards better product/service delivery to the consumers (Aslan 2021). This is what the concept of technology assimilation entails. In a nutshell, technology assimilation can be described as the manner and extent to which a new technology or innovation is adopted by a company. According to Rivera Vargas (2018), a more industrial interpretation of technology assimilation should include all dimensions and types of technologies and processes (Rouse 2019), and thereby, in this context, technology assimilation is defined as all processes involved within an organization from the moment of conceptualization or awareness of a technology or innovation, right up to all the steps undertaken to adopt the stated technology or innovation in parts of the organization or the full diffusion of the technology or innovation for full deployment within an organization (Fowler 1996; Ali and Soar 2018; Fang 2016).

The best example of the importance of technology assimilation and adoption can be seen in the current industrial status quo. Technology such as Big Data has become so fluid and evolving with potentially varying functionalities across many industries, and as a matter of fact, the rate at which technologies associated with Big Data and BDA are evolving is outpacing Moore's Law (Bhuvaneswari 2021), i.e., the law that compares technology innovation yield against current resources and demand (Bohr 2008). Hence, technology and innovation in their present form have a very short replacement-cycle life compared to the defining technologies of the past (Hempell and Zwick 2008). One example of this scenario would be smartphones. Due to accelerated hardware and software innovations caused by fierce competition, the replacement-cycle life of an innovated smartphone is estimated to be between 8 and 12 months. For a smartphone company to maintain its edge in the market, constant new innovations need to be sought, and in accordance with that, systems and processes need to be assimilated just as quickly to operationalize production (Kordic et al. 2017). The same could be said in the case of Kodak. According to research conducted by Yuzawa (2018), Kodak failed and eventually declared bankruptcy due to the organization's failure in incorporating and assimilating the concept of digitalization into its products, services, and technology. Kodak monopolized the photography and videography industries throughout the twentieth century, and during its peak years, Kodak even held a Fortune 500 position for decades (Fortune 500 official website). However, things took a calamitous turn for Kodak when digitalization started making waves as an emerging technology across all industries. Kodak's approach to digitalization was rather dismissive, and in just a few years.

Kodak could not respond to the rapidly evolving industry due to lag induced inelasticity, hence leading Kodak towards a permanent downfall (Li 2015). Indeed, according to Hall and Khan (2003), any assimilation and adoption process has two defining characteristics: the first is the inevitable slowness in disruptive technology adoption, and the second is the differing rate of adoption or acceptance of the technology across and within industries. Question then arises, if a successful monopoly such as Kodak can fall prey due to failure of disruptive technology adoption, what then would be the fate for SMEs these days if they do not catch up in time to adopt the current disruptive technology i.e., BDA and how differently these SMEs are adopting BDA (Verma and Chaurasia 2022)? Therefore, to examine the factors that can impact technology assimilation, readiness, and eventual adoption in a company, the technology-organization-environment (TOE) framework model (Przechlewski and Strzała 2009) will be examined in the following sections.

8.2.4 Technology-Organization-Environment (TOE) Framework Model

The TOE Model can provide a different perspective with regards to the adoption of BDA. The TOE model can be systematically used to understand the effect of an innovation or technology in an organization. TOE helps in differentiating a technology's inherent creative index, an organization's resource limitations and directions, and the presiding environmental trend with regards to innovation (Przechlewski and Strzała 2009; Leo Handoko et al. 2021). Given these three main domains in which technology adoption could be influenced, a more detailed exploration of these domains will provide the variables that could be examined for this study.

Technology Factors

There are five parameters within the domain of technological factors, and they are perceived benefits of technology, ease of use, system compatibility, data security, and data quality (Malik et al. 2021; Al-Balushi 2016). In the context of perceived benefits, the organization's weighted cost–benefit analysis on the aggregate benefits conferred by the technology is evaluated against overall cost savings, increased operational efficiency, and improved marketing strategies. If the organization perceives that there is no net benefit from the adoption of the technology, then it will lack the will and intent to assimilate and adopt the technology. Therefore, perceived benefit does stand as a significantt determinant in the aspect of technology adoption (Mohamad Shafi and Misman 2019; Ali 2021). On the other hand, ease of use refers to the relative simplicity required to install, implement, and operate the technology, and in the case of BDA, this would refer to the full deployment of Big Data infrastructure with its analytic tool. It is found that with a higher degree of simplicity in the implementation and installation of technology, there is a higher likelihood of the technology's adoption (Leo Handoko et al. 2021; Stadler et al. 2016). System compatibility, on the other hand, refers to a technology's suitability for a company's current infrastructure. In the case of BDA, the suitability and compatibility of BDA tools and systems with the company's information systems, such as enterprise resource planning (ERP) and customer relationship management (CRM), can be compared due to the inflated costs that could be incurred if the systems are incompatible (Catherine and Abdurachman 2018; Gwadabe and Arumugam 2021). Hence, for organizations at the SME level, this would be a decisive factor. In the case of data security, not all technologies would face this issue because not all adopted technologies would deal with sensitive data.

However, in the case of BDA, issues pertaining to the privacy of personal and industrial data, and the possibility for unlawful usage and hijacking of data can be a big deterrent towards Big Data adoption, and therefore, subsequently, BDA (Haldorai and Ramu 2022). If compliance and prudent security protocols are not in place, the adoption of Big Data and BDA would be counterproductive (Loshin 2013a, b). Lastly, data quality in this sense also specifically refers to technologies that utilize, interpret, or yield data, and in the case of BDA, these descriptions fit perfectly. Data quality is an essential factor in analytics because the acquisition of Big Data needs to correspond to its application in a useable manner once BDA is applied (Becker et al. 2015; Ganapathi and Chen 2016). For instance, a 10% useable dataset from Big Data would prove the data quality to be too poor to be integrated into an application, and this could affect the adoption rate across the organization as well as pose questions on the type and source of the data being harvested (Benbernou and Ouziri 2017).

Organization Factors

There are three parameters within the domain of organizational factors, which are management support, technological capacity, and financial competence (Malik et al. 2021). From the context of management support, the support refers to the belief held by the management in the process of adopting new technology, which is translated to the commitment and support to ensure the technology is adopted well within the organization (Ganguly 2022). If the management perceives BDA as important and very relevant to the organization, then there will be an increase in the uptake of BDA adoption in the organization (Edu and Agozie 2022). Technological capacity refers to the infrastructure readiness of the organization to immediately utilize and operationalize a new technology.

In the context of BDA, this would refer to all aspects of an organization's infrastructure to acquire, store, process, analyze, and manage Big Data. Therefore, this would include the areas of human resources, employee training, knowledge, and skills, as well as education with regards to their capacity to immediately handle Big Data and engage in BDA (Sayginer and Ercan 2020; Olszak and Mach-Król 2018; Gwadabe and Arumugam 2021). Financial competence, on the other hand, refers to the financial readiness of the organization to immediately liquidate and adopt a new technology. In the context of BDA, this would refer to an organization's willingness and capacity to invest in big data systems and BDA. The capital expenditure to overhaul or set up a Big Data and BDA system is expensive, and the same goes for the operational expenditure due to the requirement of highly personalized skills, analytics, and consulting requirements (Verma and Chaurasia 2022; Alalawneh and Alkhatib 2020; Riahi 2018). Hence, organizations need to be financially competent to enhance the likelihood of BDA adoption, whereby the opposite is also true: the burden of incurring additional costs can promulgate hesitance for BDA adoption.

Environment Factors

There are three parameters within the domain of environmental factors, which are competitor's adoption, partner's adoption, and government support (Malik et al. 2021; Al-Balushi 2016). Both competitor's adoption and partner's adoption refer to the trends that are existent within an industry the organization belongs to, as well as the supporting industries up and below the value chain that support the said organization. Adoption of a new technology by a competitor would demonstrate the changing dynamics within the industry, as well as the industry's positive outlook toward the technology (Shrafat 2018; Baker 2011). For example, Fujifilm, the competitor of Kodak, prospered in the era of digitalization due to its timely adoption and reaction towards modernizing its production line to yield products that embrace digitalized functions. Although Kodak did not follow the trend, it was forced to make a management decision when its competitors demonstrated a willingness to embrace digitalization. Hence, the same situation can be anticipated even for BDA. Within a specific industry setting, if a competitor had already embraced and utilized the functions of BDA, then an organization would be forced to strategically consider the likelihood of also adopting a similar approach to keep up with the competitive advantage (Ho and Chen 2018). The assimilation and adoption of technology by business stakeholders within an organization, such as suppliers and service provider partners, is referred to as partner adoption of new technology. Although a partner's adoption rate is reported to be less decisive compared to a competitor's adoption rate, in certain industries such as the manufacturing industry and the financial industry, this can prove to be equally instrumental (Prause 2019; Sima et al. 2020). This is because manufacturing industries and financial industries are formalized and are tasked with enhanced compliance regulations that can affect the areas of quality control, safety protocol, and data security standards, where the requirements to maintain a minimum standard can affect the processing infrastructure and methodology in place by partners that are along the value chain (Yadegaridehkordi et al. 2018; Ram et al. 2019). For example, in the context of BDA adoption, a bearing manufacturer may apply BDA to enhance quality control procedures or even to apply predictive analytics before a bearing is approved for a production line. The necessity for the latter may stem from the requirement of a hard-disk manufacturer to assess the feasibility of bearing design from BDA predictive models before orders are placed (Belhadi et al. 2019; Moktadir et al. 2019). Government support refers to encouragement or policies that are in place for an industry or within a country whereby a technology is either compulsorily forced to be adopted, encouraged to be adopted, and/or incentivized to be adopted (Varma 2019). In the context of Big Data and BDA, Malaysia has introduced encouraging policies for companies that are within the scope of Big Data and BDA for tax-exempt business operation years, and similarly, funds and grants are even created to support organizations within certain industries to transfer, assimilate, and adopt big data into operations (Hamzah et al. 2020; MDeC official website). These kinds of government supports have proven to enhance the adoption rate of any technology in an organization.

Accordingly, we proposed the hypotheses as follows:

H1: There are significant effects of technological factors on the adoption of BDA among Malaysian SMEs in the manufacturing industry.
H2: There are significant effects of organizational factors on the adoption of BDA among Malaysian SMEs in the manufacturing industry.
H3: There are significant effects of environmental factors on the adoption of BDA among Malaysian SMEs in the manufacturing industry.
H4: There are significant effects of technological factors on the adoption of BDA among Malaysian SMEs in the construction industry.

H5: There are significant effects of organizational factors on the adoption of BDA among Malaysian SMEs in the construction industry.
H6: There are significant effects of environmental factors on the adoption of BDA among Malaysian SMEs in the construction industry.

8.3 Research Method

In order to establish a framework, this study used a content analysis based on several papers and previous studies. Several cited papers support the aforementioned concept's derivation and proposal.

8.4 Conclusion

The study began with a pertinent question with regards to BDA adoption in industry-related SMEs. In order to accomplish the objectives and answer the research question formulated for this study, an exploration was conducted, and a further investigation would be conducted to populate the responses with regards to the TOE framework, whereby the parameters were treated as independent variables on the level of BDA adoption. It is understood that big data and BDA will be the footprint of our advancing Industry 4.0 and the impending Industry 5.0. It is anticipated that the study was able to elucidate and assist on future implementations of BDA infrastructure within the vicinity of the construction and manufacturing industries, and we expect further research to be conducted across industries as a preamble for an optimized BDA adoption across all industries.

References

Al-Balushi FM (2016) Technology, organizational and environmental (TOE) factors influencing Enterprise Application Integration (EAI) implementation in Omani Government Organizations. Indian J Sci Technol 9(1):1–5. https://doi.org/10.17485/ijst/2016/v9i45/107114

Alalawneh AA, Alkhatib SF (2020) The barriers to big data adoption in developing economies. Electron J Inf Syst Dev Ctries 87(1). https://doi.org/10.1002/isd2.12151

Ali H (2021) Perceived applicability of educational management information system [EMIS] in secondary schools using the toe framework. Pak Soc Sci Rev 5(I):581–596. https://doi.org/10.35484/pssr.2021(5-i)44

Ali O, Soar J (2018) Technology innovation adoption theories. Technol Adopt Soc Issues, 821–860.https://doi.org/10.4018/978-1-5225-5201-7.ch037

Aslan E (2021) How supply chain management will change in the Industry 4.0 Era? In: Research anthology on cross-industry challenges of Industry 4.0, pp 1015–1035. https://doi.org/10.4018/978-1-7998-8548-1.ch051

Atal M, Mike M (2018) Creating value from big data. Big Data Manag 30–54.https://doi.org/10.4324/9780429489679-3

Baird A, Davidson E, Mathiassen L (2017) Reflective technology assimilation: facilitating electronic health record assimilation in small physician practices. J Manag Inf Syst 34(3):664–694. https://doi.org/10.1080/07421222.2017.1373003

Baker J (2011) The technology–organization–environment framework. Inf Syst Theory, 231–245.https://doi.org/10.1007/978-1-4419-6108-2_12

Bala P (2022) Introduction of big data with analytics of big data. In: Research anthology on big data analytics, architectures, and applications, pp 54–66.https://doi.org/10.4018/978-1-6684-3662-2.ch003

Becker D, King TD, McMullen B (2015) Big data, big data quality problem. In: 2015 IEEE international conference on big data (Big Data). https://doi.org/10.1109/bigdata.2015.7364064

Belhadi A, Zkik K, Cherrafi A, Yusof SM, El Fezazi S (2019) Understanding big data analytics for manufacturing processes: insights from literature review and multiple case studies. Comput Ind Eng 137:106099.https://doi.org/10.1016/j.cie.2019.106099

Benbernou S, Ouziri M (2017) Enhancing data quality by cleaning inconsistent big RDF data. In: 2017 IEEE international conference on big data (Big Data). https://doi.org/10.1109/bigdata.2017.8257913

Berti-Equille L, Ba ML (2016) Veracity of big data. J Data Inf Qual 7(3):1–3. https://doi.org/10.1145/2935753

Bhuvaneswari V (2021) Role of big data analytics in industrial revolution 4.0. Big Data Appl Ind 4.0, 85–105. https://doi.org/10.1201/9781003175889-5

Bieh-Zimmert O, Felden C (2015) Shaping data: visualization under construction. In: 2015 IEEE international conference on big data (Big Data). https://doi.org/10.1109/bigdata.2015.7364039

Bohr, M. (2008). Using innovation to drive moore's law. 2008 9th International Conference on Solid-State and Integrated-Circuit Technology. https://doi.org/10.1109/icsict.2008.4735115

Burk S, Miner GD (2020) Statistics, causation, and prescriptive analytics. It's All Analytics!, 225–254. https://doi.org/10.4324/9780429343988-10

Catherine C, Abdurachman E (2018) ERP system adoption analysis using TOE framework in Permata Hijau Group (PHG) Medan. Int J Enterp Inf Syst 14(3):91–105. https://doi.org/10.4018/ijeis.2018070105

Dautav R, Distefano S (2017) Quantifying volume, velocity, and variety to support (BIG) data-intensive application development. In: 2017 IEEE international conference on big data (Big Data). https://doi.org/10.1109/bigdata.2017.8258252

Deshpande PS, Sharma SC, Peddoju SK (2019) Predictive and prescriptive analytics in big-data era. Stud Big Data, 71–81.https://doi.org/10.1007/978-981-13-6089-3_5

Edu SA, Agozie DQ (2022) Exploring factors influencing big data and analytics adoption in healthcare management. In: Research anthology on big data analytics, architectures, and applications, pp 1433–1449.https://doi.org/10.4018/978-1-6684-3662-2.ch069

Elitzur R, Krass D, Zimlichman E (2020) Developing a pre-testing diagnostic tool for COVID-19 using big data predictive analytics. SSRN Electron J.https://doi.org/10.2139/ssrn.3634884

Estrada R, Ruiz I (2016) Big data, big challenges. Big Data SMACK, 3–7.https://doi.org/10.1007/978-1-4842-2175-4_1

Falk TH, Sejdić E (2018) Signal processing in the era of biomedical big data. In: Signal processing and machine learning for biomedical big data, pp 3–7.https://doi.org/10.1201/9781351061223-1

Fang L (2016) Entry barriers, competition, and technology adoption. Econ Inq 55(2):794–805. https://doi.org/10.1111/ecin.12391

Finlay S (2014) The predictive analytics process. In: Predictive analytics, data mining and big data, pp 134–156.https://doi.org/10.1057/9781137379283_7

Fortune 500: Eastman Kodak (n.d.) https://archive.fortune.com/magazines/fortune/fortune500_archive/snapshots/196 4/455.html. Accessed 22 Feb 2022

Fowler P (1996) The challenge of information technology transfer and diffusion. Diffus Adopt Inf Technol, 8–11.https://doi.org/10.1007/978-0-387-34982-4_2

Gupta GU, Gupta A (2015) Vision: a missing key dimension in the 5v big data framework. J Int Bus Res Mark 1(3):40–47. https://doi.org/10.18775/jibrm.1849-8558.2015.13.3005

Ganapathi A, Chen Y (2016) Data quality: experiences and lessons from operationalizing big data. In: 2016 IEEE international conference on big data (Big Data). https://doi.org/10.1109/bigdata.2016.7840769

Ganguly KK (2022) Understanding the challenges of the adoption of blockchain technology in the logistics sector: the TOE framework. Technol Anal Strat Manag, 1–15.https://doi.org/10.1080/09537325.2022.2036333

Girasa R (2020) Ai as a disruptive technology. Artif Intell Disruptive Technol, 3–21.https://doi.org/10.1007/978-3-030-35975-1_1

Gonçalves A (2017) Prescriptive analytics. Soc Media Anal Strat, 289–295.https://doi.org/10.1007/978-1-4842-3102-9_18

Gwadabe UM, Arumugam N (2021) Adoption of big data in Agripreneurship. Entrep Big Data, 71–82.https://doi.org/10.1201/9781003097945-6

Haldorai A, Ramu A (2022) The impact of big data analytics and challenges to cyber security. In: Research anthology on big data analytics, architectures, and applications, pp 1216–1230.https://doi.org/10.4018/978-1-6684-3662-2.ch058

Hall B, Khan B (2003) Adoption of new technology.https://doi.org/10.3386/w9730

Hamzah MA, Mat Yatin SF, Yusof M, Rashid TS, Shuhaimi H, Suleiman AB, Mansor AN, Taib KM (2020) Big data implementation in Malaysian public sector: a review. Int J Acad Res Bus Soc Sci 10(11). https://doi.org/10.6007/ijarbss/v10-i11/9072

Hartung T (2018) Making big sense from big data. Front Big Data, 1.https://doi.org/10.3389/fdata.2018.00005

Hempell T, Zwick T (2008) New technology, work organisation, and innovation. Econ Innov New Technol 17(4):331–354. https://doi.org/10.1080/10438590701279649

Ho JC, Chen H (2018) Managing the disruptive and sustaining the disrupted: the case of Kodak and Fujifilm in the face of digital disruption. Rev Policy Res 35(3):352–371. https://doi.org/10.1111/ropr.12278

Hodeghatta UR, Nayak U (2017) Introduction to descriptive analytics. Bus Anal Using R-Pract Approach, 59–89.https://doi.org/10.1007/978-1-4842-2514-1_4

Holmes DE (2017) 6. Big data, big business. Big Data Very Short Introd, 75–89. https://doi.org/10.1093/actrade/9780198779575.003.0006

Hussain A, Roy A (2016) The emerging era of big data analytics. Big Data Anal 1(1). https://doi.org/10.1186/s41044-016-0004-2

Kalaian SA, Kasim RM, Kasim NR (2016) Descriptive and predictive analytical methods for big data. In: Advances in business information systems and analytics, pp 66–83.https://doi.org/10.4018/978-1-5225-0293-7.ch005

Kockum F, Dacre N (2021) Project management volume, velocity, variety: a big data dynamics approach. SSRN Electron J. https://doi.org/10.2139/ssrn.3813838

Kordic G, Grgurevic I, Husnjak S (2017) Identification of factors relevant for the estimation of smartphone life cycle. In: 2017 25th telecommunication forum (TELFOR). https://doi.org/10.1109/telfor.2017.8249276

Kumaraguru PV, Chakravarthy VJ (2017) A study of big data definition, layered architecture and challenges of big data analytics. Indian J Forensic Med Toxicol 11(2):635. https://doi.org/10.5958/0973-9130.2017.00201.8

Leo Handoko B, Edward Riantono I, Wigna Sunarto F (2021) Determinants affecting intention of use of big data analytics on remote audits: toe framework approach. In: 2021 7th international conference on education and training technologies. https://doi.org/10.1145/3463531.3463539

Li C, Chen Y, Shang Y (2022) A review of industrial big data for decision making in intelligent manufacturing. Eng Sci Technol Int J 29:101021. https://doi.org/10.1016/j.jestch.2021.06.001

Li E (2015) deliberate learning, dynamic capability and organizational schema: a case study of kodak. Acad Manag Proc 2015(1):14715. https://doi.org/10.5465/ambpp.2015.14715abstract

Loshin D (2013a) Business problems suited to big data analytics. Big Data Anal, 11–19.https://doi.org/10.1016/b978-0-12-417319-4.00002-8

Loshin D (2013b) Data governance for big data analytics. Big Data Anal, 39–48.https://doi.org/10.1016/b978-0-12-417319-4.00005-3

Malik S, Chadhar M, Vatanasakdakul S, Chetty M (2021) Factors affecting the organizational adoption of blockchain technology: extending the technology–organization–environment (TOE) framework in the Australian context. Sustainability 13(16):9404. https://doi.org/10.3390/su13169404

Margara A, Rabl T (2018) Definition of data streams. Encycl Big Data Technol, 1–4.https://doi.org/10.1007/978-3-319-63962-8_188-1

Mohamad Shafi RH, Misman FN (2019) The perceived benefit and risk framework of E-wallet adoption among the early adopters in Malaysia. In: Proceedings of the 2nd economics and business international conference. https://doi.org/10.5220/0009326905790584

Moktadir MA, Ali SM, Paul SK, Shukla N (2019) Barriers to big data analytics in manufacturing supply chains: a case study from Bangladesh. Comput Ind Eng 128:1063–1075. https://doi.org/10.1016/j.cie.2018.04.013

Nagy J, Oláh J, Erdei E, Máté D, Popp J (2018) The role and impact of Industry 4.0 and the internet of things on the business strategy of the value chain—the case of Hungary. Sustainability 10(10):3491. https://doi.org/10.3390/su10103491

Nair MK, Rao A, Patel M (2017) Big data predictive modeling and analytics. Big Data Anal, 117–150.https://doi.org/10.1201/b21822-6

Nobanee H (2021) A bibliometric review of big data in Finance. Big Data 9(2):73–78. https://doi.org/10.1089/big.2021.29044.edi

Nyamful C, Agrawal R (2017) Big variety data. Encycl Big Data, 1–4.https://doi.org/10.1007/978-3-319-32001-4_23-1

Olszak CM, Mach-Król M (2018) Conceptual framework for assessing organization's readiness to big data adoption. https://doi.org/10.20944/preprints201808.0335.v1

Pendyala V (2018) The big data phenomenon. Veracity Big Data, 1–15.https://doi.org/10.1007/978-1-4842-3633-8_1

Prause M (2019) Challenges of Industry 4.0 technology adoption for SMEs: the case of Japan. Sustainability 11(20):5807. https://doi.org/10.3390/su11205807

Przechlewski T, Strzała K (2009) Determinants of open source software adoption–an application of TOE framework. Inf Syst Dev, 461–469.https://doi.org/10.1007/b137171_48

Ram J, Afridi NK, Khan KA (2019) Adoption of big data analytics in construction: development of a conceptual model. Built Environ Proj Asset Manag 9(4):564–579. https://doi.org/10.1108/bepam-05-2018-0077

Riahi Y, Riahi S (2018) Big data and big data analytics: concepts, types and technologies. Int J Res Eng 5(9):524–528. https://doi.org/10.21276/ijre.2018.5.9.5

Rivera Vargas MI (2018) Technology transfer and assimilation. Technol Transf Univ-Ind Relat-Ship, 77–131.https://doi.org/10.4324/9781315829395 5

Rouse WB (2019) Technology adoption. Comput Possible Futur, 73–90.https://doi.org/10.1093/oso/9780198846420.003.0005

Sayginer C, Ercan T (2020) Understanding determinants of cloud computing adoption using an integrated diffusion of innovation (doi)-technological, organizational and environmental (TOE) model. HumIties Soc Sci Rev 8(1):91–102. https://doi.org/10.18510/hssr.2020.8115

Shi Y (2022) Big data and big data analytics. Adv Big Data Anal, 3–21.https://doi.org/10.1007/978-981-16-3607-3_1

Shirin Abbasi (2016) Big data; definition and challenges. J Energy Power Eng 10(7). https://doi.org/10.17265/1934-8975/2016.07.004

Shrafat FD (2018) Examining the factors influencing knowledge management system (KMS) adoption in small and medium enterprises smes. Bus Process Manag J 24(1):234–265. https://doi.org/10.1108/bpmj-10-2016-0221

Sima V, Gheorghe IG, Subić J, Nancu D (2020) Influences of the industry 4.0 revolution on the human capital development and consumer behavior: a systematic review. Sustainability 12(10):4035. https://doi.org/10.3390/su12104035

SME Business Digitalisation Grant. MDEC (13 Dec 2021) https://mdec.my/digital-economy-initiatives/for-the-industry/sme-digitalisation-grant/. Accessed 22 Feb 2022

Stadler JG, Donlon K, Siewert JD, Franken T, Lewis NE (2016) Improving the efficiency and ease of healthcare analysis through use of data visualization dashboards. Big Data 4(2):129–135. https://doi.org/10.1089/big.2015.0059

Stojanovic N, Dinic M, Stojanovic L (2015) Big data process analytics for continuous process improvement in manufacturing. In: 2015 IEEE international conference on big data (Big Data). https://doi.org/10.1109/bigdata.2015.7363900

Surya Saputra R (2011) Understanding the impact of internet to industry structure and value system. In: 1st business strategy and organizational behaviour (BizStrategy 2011). https://doi.org/10.5176/978-981-08-8227-3_bizstrategy28

Varma A (2019) Fintech adoption choices of small businesses: a technology organization environment (TOE) framework study. Account Financ Res 8(2):86. https://doi.org/10.5430/afr.v8n2p86

Venkatram K, Geetha MA (2017) Review on big data & analytics–concepts, philosophy, process and applications. Cybern Inf Technol 17(2):3–27. https://doi.org/10.1515/cait-2017-0013

Verma S, Chaurasia S (2022) Understanding the determinants of big data analytics adoption. In: Research anthology on big data analytics, architectures, and applications, pp 1549–1577.https://doi.org/10.4018/978-1-6684-3662-2.ch075

Walker R (2015) Benefits of scale and velocity in big data. From Big Data Big Profits, 35–60.https://doi.org/10.1093/acprof:oso/9780199378326.003.0002

Yadegaridehkordi E, Hourmand M, Nilashi M, Shuib L, Ahani A, Ibrahim O (2018) Influence of big data adoption on manufacturing companies' performance: an integrated DEMATEL-ANFIS approach. Technol Forecast Soc Chang 137:199–210. https://doi.org/10.1016/j.techfore.2018.07.043

Yuzawa T (2018) The digital revolution and business behaviour: the case of Kodak versus Fujifilm. Entrep Hist 90(1):37. https://doi.org/10.3917/eh.090.0037

Zhang X, Xiang S (2015) Data quality, analytics, and privacy in big data. Stud Big Data, 393–418.https://doi.org/10.1007/978-3-319-11056-1_14

9

Unpacking the Potential of Crowdsourcing via Social Media to Foster New Product Development Among Small and Medium-Sized Enterprises

Muhammad Dharma Tuah Putra Nasution, Pipit Buana Sari, Henry Aspan, and Yossie Rossanty

Abstract

The focal point of this chapter is to explore the potential of social media-enabled crowdsourcing capabilities for new product development in small and medium-sized enterprises (SMEs). The proposed research methodology utilizes a content analysis approach to analyze the relevant literature. The study anticipates to establish the empirical relevance of crowdsourcing in new product development by leveraging the capability of social media platforms. In addition, the study seeks to explore the relationship between crowdsourcing capability and new product development while considering the moderating effect of company size and type of business. Through this comprehensive exploration, we intend to make a significant contribution to the existing knowledge on crowdsourcing in marketing from a theoretical perspective. The study is expected to present an initial framework for future research in this area.

Keywords

Crowdsource • Crowdseeker • Crowdsolver • New product development • SMEs

M. D. T. P. Nasution (✉) · P. B. Sari · H. Aspan · Y. Rossanty
Universitas Pembangunan Panca Budi, Medan, North Sumatera, Indonesia
e-mail: dharmatuah@gmail.com

A. Rafiki et al. (eds.), *Open Innovation in Small Business*, Contributions to Environmental Sciences & Innovative Business Technology,
https://doi.org/10.1007/978-981-99-5142-0_9

9.1 Introduction

Social media platforms have become increasingly popular for both personal and business purposes (Kaplan and Haenlein 2010). This is evident from the significant growth in the number of active social media accounts, which is expected to reach 3.43 billion, surpassing the 2.95 billion accounts recorded in 2019 (Statistica 2020).

Businesses have integrated social media into various aspects of their operations, as highlighted in several studies (Appel et al. 2020; Karkkainen et al. 2010; Kaplan and Haenlein 2010; Brabham 2011; Kietzmann et al. 2011; Lee et al. 2015). One of the significant benefits of social media for businesses is its ability to engage directly with consumers, business partners, and brand communities at a lower cost than traditional media (Karkkainen et al. 2010; Scott 2010; Kietzmann et al. 2011).

Moreover, social media has emerged as a powerful channel through which businesses can reach a vast number of individuals (Kaplan and Haenlein 2010). Additionally, social networking platforms have enabled businesses to access diverse backgrounds, skills, and experiences (Lee et al. 2015).

Finally, social media has become an essential tool for organizations to collect and leverage ideas and knowledge from a seemingly endless pool of individuals (Brabham 2008, 2011). This capability provides businesses with the opportunity to gather information from a vast number of sources, potentially leading to valuable insights and innovations.

Crowdsourcing involves organizations using the collective knowledge and expertise of users on popular social media platforms to discover innovative business ideas and products/services (Simula and Ahola 2014; Yang et al. 2020). Evidence indicates the effectiveness of crowdsourcing in providing low-cost problem-solving solutions, generating innovative ideas for organizations, and reducing uncertainty (Yang et al. 2020). While crowdsourcing may be more critical for small companies than for larger enterprises, every business must contend with limitations on time, money, and resources (Van de Vrande et al. 2009).

Small and medium-sized enterprises (SMEs) face more significant resource constraints than larger organizations. As noted by Vossen (1998), SMEs have difficulty dispersing risk due to a lack of financial resources, limited opportunities to recruit qualified workers, and a smaller portfolio of innovation. Therefore, the challenge for SMEs is to find solutions to issues that appear to be resource-constrained (Van de Vrande et al. 2009). Consequently, SMEs often rely on internal search patterns within their businesses and eventually encounter organizational limitations (Laursen 2012).

However, crowdsourcing allows SMEs to leverage their networks to identify untapped resources. Social media facilitate resource and time efficiency, and crowdsourcing may enable SMEs to overcome their limitations (Karkkainen et al. 2010). Therefore, SMEs can use crowdsourcing to tap into a vast pool of ideas and resources that they would not have access to through internal search patterns.

The limited number of empirical studies on crowdsourcing in SMEs, particularly those focusing on social media-enabled crowdsourcing for new product development, has prompted the need for further research. Specifically, this study sought to address the following research questions:

- What is the nature of crowdsourcing and its potential benefits for SMEs in new product development, particularly when enabled by social media platforms?
- How do firm size and business type moderate the relationship between crowdsourcing capability and new product development in SMEs?

To answer these questions, a thorough review of the existing literature on crowdsourcing, new product development, and SMEs will be conducted. The resource-based theoretical perspective is also employed to explore the key capabilities required by SMEs to effectively leverage crowdsourcing for new product development. The results of this study contribute to the existing literature on crowdsourcing in SMEs and provide practical insights for SME owners and managers. The model with principles linked to social media-enabled crowdsourcing capabilities in new product development will offer SMEs a clear roadmap for enhancing their crowdsourcing capabilities. The results of this study may also be useful to policymakers seeking to promote innovation and growth in the SME sector.

9.2 Literature Review

9.2.1 Social Media-Enabled Crowdsourcing Capability

According to Brabham (2011), companies can manage the crowdsourcing process and control the initiative's outcome. Crowdsourcing involves outsourcing work to a large number of unspecified individuals (Yang et al. 2020; Maiolini and Naggi 2011) and has evolved from a practice previously performed by employees to open calling for contributions (Howe 2006). Crowdsourcing can be managed by contacting individuals, assigning tasks, and evaluating work using specific applications (Grier 2013).

In this process, the individuals who complete the tasks assigned by the company are referred to as "crowdsolvers," and the company assigning the tasks is known as the "crowdseeker" (Blohm et al. 2013). Crowdsolvers can be individuals, formal or informal groups, or other companies that receive incentives after fulfilling the crowdseeker's requests. In situations in which the crowdsourcing process requires specific knowledge, skills, or ideas that crowdsolvers can provide, crowdseekers and crowdsolvers share roles (Aitamurto 2015). Thus, crowdsourcing allows companies to outsource work to a large number of individuals, and the process can be managed by companies that assign tasks to crowdsolvers. Crowdseekers and crowdsolvers share roles in situations in which specific knowledge, skills, or ideas are required for the crowdsourcing process to produce expected outcomes.

Crowdsourcing has become a popular tool for various marketing activities, including promotions and new product development (Whitla 2009). For instance, collaborative design, research and development, and branding can be achieved through crowdsourcing in the context of new product development (Mladenow et al. 2014). Digital platforms have made crowdsourcing more beneficial for companies, as they can increase consumer engagement and drive marketing communications (Füller et al. 2013). Typically, companies handle most of their communications through their main websites, email marketing, social media, and search engine optimization efforts (Taiminen and Karjaluoto 2015).

Crowdsourcing has become an attractive option for companies to offer various marketing communication initiatives to customers (Whitla 2009). For instance, in content marketing, crowdsourcing can be used to engage brand enthusiasts to create brand content, thereby saving money and improving marketing by outsourcing content production to the public (Marsden 2009). However, although crowdsourcing can provide advantages in the creation or development of new products or content, involvement, and quality assurance must be prioritized.

Marketing initiatives that rely on crowdsourcing require project management capabilities, and the SME business owner-manager must be capable of managing the entire process to achieve expected results (Füller et al. 2013). As such, an effective crowdsourcing system must have quality control and crowdseekers must choose the most qualified crowdsolvers to carry out their initiatives to achieve certain expected goals (Lakhani 2013).

Similar to launching a new product, quality assurance is critical to any crowdsourcing system and the final product must meet the expectations of crowdseekers (Dowson and Bynghal 2011). Therefore, crowdseekers must identify the extent to which crowdsolvers are competent, knowledgeable, creative, and valuable to their companies (Whitla 2009).

When a company engages in crowdsourcing, it asks a group of individuals or the public to help complete a task; in exchange, rewards are given to crowdsolvers (Grier 2013). Crowdsourcing can be beneficial to companies if it can reduce the costs of completing tasks compared to internal methods and produce superior outcomes that align better with consumer needs (Whitla 2009). Therefore, the success of crowdsourcing initiatives depends on a balance between cost savings and output quality.

9.2.2 New Product Development

The importance of new product development in achieving long-term sustainability and competitive advantage for innovative organizations is well recognized in contemporary business (Ganesan et al. 2005; Akgün et al. 2006; Morgan et al. 2019). However, the efficient execution of new product development initiatives in a highly competitive environment is challenging. The literature emphasizes the importance of the speed and uniqueness of new product development compared to competitors (Dahan and Hauser, 2001; Rundquist and Halila 2010; Ma et al. 2012; Wu et al. 2020). Ma et al. (2012) suggested that balancing resource exploitation and exploration is critical for successful new product development. Innovation and speed are two key factors contributing to the success of new product development (Sheng et al. 2013; Cheng and Yang 2019; Cooper 2019). Therefore, companies must synergize resource exploitation and exploration to quickly develop new products and respond to customer needs, or differentiate themselves from competitors through unique product offerings.

9.2.3 The Relationship Between the Capability to Leverage Crowdsourcing and the Success of New Product Development

This study predicts that social media-enabled crowdsourcing has a significant impact on new product development. Füller et al. (2014) argued that crowdsourcing enables SMEs to leverage the specialized knowledge of external sources when they are unable to develop new products internally. To address these challenges, SMEs can use crowdsourcing facilitated by social media, which involves broadcasting the challenge of open calls online. Social media platforms have made it easier to reach problem solvers who are capable of providing high-quality solutions. Crowdsourcing can also help companies overcome problems encountered during new product development by accessing individuals outside the boundaries of the organization with specific skills or expertise. Therefore, it is necessary to follow the principles of crowdsourcing to explain the essence of this concept (Füller et al. 2014).

Afuah and Tucci (2012) concluded that crowdsourcing can outperform internal problem-solving efforts and help companies generate customized solutions by leveraging the expertise of crowdsolvers. To effectively use crowdsourcing, companies must meet four key criteria. First, companies must be able to clearly define the problem at hand, as self-chosen crowdsolvers are best equipped to solve problems that they feel confident about tackling. Second, companies must carefully select crowdsolvers with relevant experience and expertise to effectively solve problems. The relevance of the generated solution is largely dependent on the level of knowledge and expertise of the crowdsolver (Füller et al. 2014). Third, companies must persuade crowdsolvers to participate and remain engaged throughout the crowdsourcing process. Finally, a company must be able to effectively evaluate and choose the best solution from options generated through crowdsourcing efforts (Afuah & Tucci, 2012).

To achieve more productive and profitable new product development, it is important to synchronize internal and external sources (Tzabbar et al. 2013). Scholars have suggested that rethinking outdated ideas and utilizing new methods can act as key catalysts for new product development (Liu 2021; Zhan et al. 2020; Qin et al. 2016; Kamp 2014). Therefore, it is crowdsourcing capabilities of SMEs expected to enhance their ability to develop new products (Morgan et al. 2019). Companies can benefit from a variety of external knowledge sources, remain updated on changing customer demands and expectations, and respond quickly to emerging opportunities (Morgan et al. 2019). Caridi-Zahavi et al. (2016) also emphasize that new products often stem from a combination of outside knowledge sources and that companies can bring new ideas to the market more quickly by synergizing their internal knowledge with external sources. Thus, this study investigates how crowdsourcing capabilities significantly influence new product and contributes to research in this area.

SMEs are expected to benefit from the crowdsourcing capabilities on social media to aid in the development of new products. Crowdsourcing depends on the availability of knowledge resources (Leiponen and Helfat 2010) and the specific quantity of resources that can be used (Kamp 2014). Companies that utilize social media to gather ideas and creativity are more likely to improve the quality of the knowledge received (Kamp 2013). To simplify the conceptual model, a few simple characteristics that are directly related to product development are actionable (Chirico and Salvato 2008). The explanatory power of the overall model can be increased by considering the crowdsourcing and NPD capabilities.

This study proposes three hypotheses to examine the relationship between crowdsourcing capabilities and new product development, moderated by firm size and business type.

Hypothesis 1 (H1) suggests a direct relationship between SMEs' crowdsourcing capabilities and new product development. This hypothesis assumes that SMEs can leverage crowdsolvers' knowledge and expertise to develop new and innovative products.

Hypothesis 2 (H2) proposes that the relationship between crowdsourcing capability and new product development is moderated by firm size. This hypothesis assumes that the impact of crowdsourcing on new product development varies depending on SME size.

Hypothesis 3 (H3) suggests that the relationship between crowdsourcing capability and new product development is moderated by business type. This hypothesis assumes that the impact of crowdsourcing on new product development varies depending on the type of business that an SME operates.

Overall, the study aims to provide insights into the impact of crowdsourcing capabilities on new product development while considering the moderating effects of firm size and business type.

9.3 Research Method

The framework of this study was established through content analysis of various papers and previous studies that support the proposed concept. The selected papers were cited based on their relevance to the proposed concept and potential to support its derivation.

9.4 Discussion and Conclusion

The primary objective of this chapter is to explore the relationship between social media-enabled crowdsourcing capabilities and new product development. Business owners and managers use social media to support marketing activities. The proposed model anticipates that crowdsourcing capabilities will play a significant role in the development of new products. Specifically, crowdsourcing capability is closely related to a company's ability to define problems and understand the characteristics and knowledge required by crowdsolvers. This implies that SME owners and managers need to have the skill to invite and engage external crowdsolvers.

Furthermore, the capability to define the actual problem that will be presented to crowdsolvers to generate a solution will depend on how much in-depth and specialized knowledge matches the crowdsolver's area of expertise to solve the problem. Previous studies by Afuah and Tucci (2012) and Füller et al. (2014) found that crowdsolvers are more likely to participate in activities relevant to their qualifications if the problem is precisely defined, which in turn results in a higher rate of qualified solutions.

It is important to note that crowdsolvers are knowledgeable and capable crowd-marketers who choose their own defined problem-solving activities. The ability of SME owner-managers to precisely identify problems and announce them to the masses makes it possible for SMEs to attract the most competent crowdsolvers. In brief, this study highlights the crucial role of social media-enabled crowdsourcing capabilities in SMEs' NPD and emphasizes the importance of accurate problem definition and involvement of qualified crowdsolvers for successful outcomes.

The effectiveness of SMEs in using crowdsourcing is closely linked to the ability of their owners and managers to motivate and inspire crowdsolvers both before and during the crowdsourcing process. This principle is consistent with Afuah and Tucci's (2012) argument that the ability to motivate crowdsolvers is a key factor for achieving optimal results in crowdsourcing processes. The ability to evaluate the ideas submitted by crowdsolvers is another important aspect of crowdsourcing. Once the crowdsolver has contributed enough to complete the task, the company must then analyze and choose the most suitable solution from among the proposed alternatives. As crowdsolvers may offer several alternative solutions, it is essential for a company to identify the optimal solution.

Crowdsourcing is correlated with SMEs' capability to develop new products. The underlying reasoning behind this is that, through crowdsourcing, SMEs can expand their pool of relevant resources to tap into and leverage. The more resources they have access to, the greater the likelihood of finding a solution that can lead to the development of new products that meet consumer demand. A second key factor contributing to this explanation is that crowdsourcing often generates ideas and solutions that are more practical and applicable because crowdsolvers can explore and exploit various knowledge sources. Social media have been found to be a valuable external source of knowledge for identifying useful ideas related to specific issues, as noted by Füller et al. (2014). Consequently, the benefits of crowdsourcing are more pronounced when crowdsolvers propose suitable solutions that can benefit SME.

The development of new products in SMEs is positively linked to their effective use of crowdsourcing. This correlation can be logically explained by the fact that crowdsourcing provides SMEs with access to a larger pool of relevant resources, thereby increasing their chances of finding a suitable solution to meet consumer needs. Another reason for this result is that crowdsourcing can generate more practical and applicable ideas and solutions owing to crowdsolvers' ability to explore and exploit various knowledge sources. Füller et al. (2014) suggested that social media can be particularly useful for finding valuable ideas related to specific issues. Therefore, when crowdsolvers propose a suitable solution, the practical advantages of crowdsourcing are amplified, resulting in greater benefits for the SME.

We predict that the relationship between SME crowdsourcing capabilities and new product development can be strengthened by firm size. The underlying reason for this assumption is that the resources available to SMEs to produce new products that meet consumer expectations are implicitly linked to their crowdsourcing capabilities, which depend on the size of the company they manage. Similarly, we predict that the relationship between crowdsourcing capabilities and new product development can be strengthened by the type of business that SMEs operate. The logic behind this outcome shows that the ability to crowdsource solutions depends on the type of business in which they manage to develop new products that cater to consumers' expectations. SMEs can strategically leverage the digital ecosystem to find solutions, and the approach taken depends on the type and size of their business, resulting in faster, more creative, and fulfilling new product development for consumers.

9.5 Implications

The evaluation of a model has resulted in the presentation of evidence demonstrating a relationship between crowdsourcing capabilities and new product development. These predictions hold significance for scholars and academics, who can incorporate this relationship as an integral component of the crowdsourcing model utilized by small and medium enterprises. This study may have overlooked some relevant concepts, including the importance of social media for supporting task completion through crowdsourcing. Although the concept of crowdsourcing has been well documented (Füller et al., 2014), there is still room for the development of crowdseekers and crowdsolvers in implementing collaborative problem-solving, which is essential for successful new product development. This is an interesting research avenue for both scholars and academics.

An important implication of this study for business owners and managers is the identification of the key aspects of crowdsourcing capabilities that SMEs must focus on to develop new products that meet customer expectations for attractive products. Hence, business owners and managers must be able to define a problem, identify appropriate crowdsolvers, motivate individuals involved in the crowdsourcing process, and evaluate the quality of the generated ideas and solutions.

9.6 Funding Acknowledgment

This work was supported by the Indonesian Ministry of Education, Culture, and Technology Research in 2022: Research Grant Scheme "PDUPT" No. 124/LL1/LT/K/2022.

References

Afuah A, Tucci C (2012) Crowdsourcing as a solution to distant search. Acad Manag Rev 37(3):355–375

Aitamurto T (2015) The role of crowdfunding as a business model in journalism: a five-layered model of value creation. Crowdfunding Futur:189–205

Akgün AE, Lynn GS, Byrne JC (2006) Antecedents and consequences of unlearning in new product development teams. J Prod Innov Manag 23(1):73–88

Appel G, Grewal L, Hadi R, Stephen AT (2020) The future of social media in marketing. J Acad Mark Sci 48(1):79–95

Blohm I, Leimeister JM, Krcmar H (2013) Crowdsourcing: how to benefit from (too) many great ideas. MIS Q Exec 4(12):199–211

Brabham D (2011) Crowdsourcing: a model for leveraging online communities

Brabham DC (2008) Crowdsourcing as a model for problem solving an introduction and cases. Converg: Int J Res Mew Media Technol 14(1):75–90

Caridi-Zahavi O, Carmeli A, Arazy O (2016) The influence of CEOs' visionary innovation leadership on the performance of high-technology ventures: the mediating roles of connectivity and knowledge integration. J Prod Innov Manag 33(3):356–376

Cheng C, Yang M (2019) Creative process engagement and new product performance: the role of new product development speed and leadership encouragement of creativity. J Bus Res 99:215–225

Chirico F, Salvato C (2008) Knowledge integration and dynamic organizational adaptation in family firms. Fam Bus Rev 21(2):169–181

Cooper RG (2019) The drivers of success in new-product development. Ind Mark Manage 76:36–47

Dahan, Hauser (2001) Product development: managing a dispersed process. Handb Mark.179–222

Dowson R, Bynghal S (2011) Getting results from crowds: the definitive guide to using crowdsourcing to grow your business. Advanced Human Technologies Inc., San Francisco

Füller J, Bilgram V, Koch G, Rapp M (2013) The potential of crowdsourcing for co-marketing: how consumers may be turned into brand ambassadors. Transf Werbeforsch Prax 59(4):42–48

Füller J, Lemmer S, Hutter K (2014) Crowdsourcing: how social media and the wisdom of the crowd change future companies. In: Management of the fuzzy front end of innovation. Springer, Cham, pp 243–249

Ganesan S, Malter AJ, Rindfleisch A (2005) Does distance still matter? Geographic proximity and new product development. J Mark 69(4):44–60

Grier DA (2013) Crowdsourcing for dummies. John Wiley & Sons

Howe J (2006) The rise of crowdsourcing. Wired Mag 14(6):1–4

Kamp P (2013) Literature review: utilizing social media to facilitate innovation in SMEs. Technical University, Eindhoven

Kamp PJC (2014) Effects of crowdsourcing, facilitated by social media, on knowledge integration and product advantage during NPD in SMEs
Kaplan A, Haenlein M (2010) Users of the world, unite! The challenges and opportunities of Social Media. Bus Horiz 53(1):59–68
Karkkainen H, Jussila J, Väisänen J (2010) Social media use and potential in business-to- business companies' innovation. In: 14th international academic mindtrek conference: envisioning future media environments. ACM, Tampere, Finland, pp 228–236
Kietzmann J, Hermkens K, McCarthy I, Silvestre B (2011) Social media? Get serious! Understanding the functional building blocks of social media. Bus Horiz 54(3):241–251
Lakhani KR (2013) Using the crowd as an innovation partner. Harv Bus Rev 91(4):60–69
Laursen K (2012) Keep searching and you'll find: What do we know about variety creation through firms' search activities for innovation? Ind Corp Chang 21(5):1181–1220
Lee CKM, Chan CY, Ho S, Choy KL, Ip WH (2015) Explore the feasibility of adopting crowdsourcing for innovative problem solving. Ind Manag Data Syst115(5):803–832
Leiponen A, Helfat C (2010) Innovation Objectives, Knowledge Sources, and the Benefits of Breadth. Strateg Manag J 31(2):224–236
Liu B (2021) Matching external search strategies with radical and incremental innovation and the role of knowledge integration capability. Balt J Manag 16(5):765–784
Ma C, Yang Z, Yao Z, Fisher G, Fang EE (2012) The effect of strategic alliance resource accumulation and process characteristics on new product success: Exploration of international high-tech strategic alliances in China. Ind Mark Manage 41(3):469–480
Maiolini R, Naggi R (2011) Crowdsourcing and SMEs: opportunities and challenges. Inf Technol Innov Trends Organ, 399–406
Marsden P (2009) Crowdsourcing. Contagious Magazine 18:24–28
Mladenow A, Bauer C, Strauss C (2014) Social crowd integration in new product development: crowdsourcing communities nourish the open innovation paradigm. Glob J Flex Syst Manag 15(1):77–86
Morgan T, Anokhin SA, Song C, Chistyakova N (2019) The role of customer participation in building new product development speed capabilities in turbulent environments. Int Entrep Manag J 15(1):119–133
Qin S, Van der Velde D, Chatzakis E, McStea T, Smith N (2016) Exploring barriers and opportunities in adopting crowdsourcing based new product development in manufacturing SMEs. Chin J Mech Eng 29(6):1052–1066
Rundquist J, Halila F (2010) Outsourcing of NPD activities: a best practice approach. Eur J Innov Manag 13(1):5–23
Scott M (2010) The new rules of marketing and PR: How to use social media, blogs, news releases, online video, and viral marketing to reach buyers directly. John Wiley & Sons Inc., New Jersey
Sheng S, Zhou KZ, Lessassy L (2013) NPD speed vs. innovativeness: The contingent impact of institutional and market environments. J Bus Res 66(11):2355–2362
Simula H, Ahola T (2014) A network perspective on idea and innovation crowdsourcing in industrial firms. Ind Mark Manage 43(3):400–408
Statistica (2020) Number of social network users worldwide from 2010 to 2023. https://www.statista.com/statistics/278414/number-of-worldwide-social-network-users/. Accessed on 10 July 2022
Taiminen HM, Karjaluoto H (2015) The usage of digital marketing channels in SMEs. J Small Bus Enterp Dev 22(4):633–651
Tzabbar D, Aharonson BS, Amburgey TL (2013) When does tapping external sources of knowledge result in knowledge integration? Res Policy 42(2):481–494
Van de Vrande V, de Jong J, Vanhaverbeke W, de Rochemont M (2009) Open innovation in SME's: trends and management challenges. Technovation 6(7):423–437
Vossen RW (1998) Relative strengths and weaknesses of small firms in innovation. Int Small Bus J 16(3):88–94
Whitla P (2009) Crowdsourcing and its application in marketing activities. Contemp Manag Res 5(1):15–28

Wu L, Liu H, Su K (2020) Exploring the dual effect of effectuation on new product development speed and quality. J Bus Res 106:82–93

Yang Y, Dong C, Yao X, Lee PK, Cheng TCE (2020) Improving the effectiveness of social media-based crowdsourcing innovations: roles of assurance mechanism and innovator's behaviour. Ind Manag Data Syst 121(2):478–497

Zhan Y, Tan KH, Chung L, Chen L, Xing X (2020) Leveraging social media in new product development: Organisational learning processes, mechanisms and evidence from China. Int J Oper Prod Manag 40(5):671–695

10 How the Innovation Climate and Open Innovation Practices Contribute to Firm Innovativeness in Small- and Medium-Scale Enterprises? An Empirical Analysis

Onan Marakali Siregar and Nicholas Marpaung

Abstract

Open innovation has emerged as a crucial strategic element for enhancing the efficiency and commercialization of innovation among large companies. However, the potential applicability of this concept to small- and medium-scale enterprises (SMEs) in Indonesia, particularly those based on digital technology, remains relevant. Although research on open innovation for SMEs has yielded important insights, it is relatively limited in scope, with a focus on inbound innovation activities and little consideration for medium- or small-scale enterprises, even though they are often seen as radical innovators in their fields. To address this gap, it is necessary to examine the potential of SMEs in the process of business innovation through digital platforms, with special attention given to their better specialization capabilities than larger companies. This specialization is more profitable when the market is open to innovative activities and when innovation is supported by SME owners–managers who create an innovation climate. Empirical analysis is required to answer the key research questions. The first question pertains to the extent of SME innovativeness, particularly in terms of their capacity to utilize internal, external, and hybrid knowledge sources in open innovation. The second question pertains to the extent to which the innovation climate supports SME owner–manager efforts to find creative ways to exploit and explore innovation sources. The ultimate goal of this study is to develop an extensive framework that elucidates the link between the innovation climate, open innovation, and firm innovativeness in the context of SMEs.

O. M. Siregar (✉) · N. Marpaung
Faculty of Social and Political Science, Universitas Sumatera Utara, Medan, Indonesia
e-mail: onan@usu.ac.id

A. Rafiki et al. (eds.), *Open Innovation in Small Business*, Contributions to Environmental Sciences & Innovative Business Technology,
https://doi.org/10.1007/978-981-99-5142-0_10

Keywords

Innovation climate • SMEs • Open Innovation • Inbound • Outbound • Firm innovativeness

10.1 Introduction

Globalization is now significantly more complex and rapid; however, its interconnectedness may be a pathway for business sector growth. The digitization of business, shorter product life cycles, and greater consumer connectedness are some of the hallmarks of globalization. In a world that is growing more interconnected, globalization has posed new challenges on organizations to innovate in order to achieve competitive advantages and ensure their long-term survival (Tidd and Bessant 2020). As a consequence of this rivalry, there have been suggestions and practices in the field of innovation management for a change from closed innovation toward an open innovation paradigm (Chesbrough et al. 2014; Chesbrough 2012).

In a dynamic environment influenced by social change and technological evolution, innovation assumes relevance in determining economic growth, creating competitive advantage, and organizational sustainability. Innovation can be understood as a learning process that results in the exploration of new ideas for the future, which can be incorporated into new products or services, processes, or methods to improve company performance. The current paradigm shift demands new ways to organize internal processes and trigger the emergence of open and collaborative models of innovation. The evolution of new communication technologies, the growing mobility of highly skilled workers, and the need to save costs have consequences for the optimization of internal operations and integration of external activities.

The concept of open innovation (OI) in this context proposes that firms, individuals, and the general public collaborate, which encourages the creation of new products and services. The innovation process has been facilitated by the rapid development and deployment of several web-based ICT, which have simplified interactions among stakeholder groups and led to the rise of new forms of network collaboration (Hutter et al. 2013). According to a 2008 McKinsey report, 60–70% of large, established companies have integrated customers and external expertise into their enterprise innovation processes. Most scholarly investigations into OI have exclusively concentrated on large-scale enterprises, primarily in high-tech industries (Hutter et al. 2013). However, it has been suggested that an OI approach might have both benefits and advantages for SMEs, particularly those enabled by web-based technologies (Hossain 2015; Bianchi et al. 2010).

Although crowdsourcing, co-creation, and user innovation communities have been discussed in the context of OI in SMEs, previous research has neglected to consider the benefits of Web 2.0, technologies, or platforms (Parida et al. 2012; Lasagni 2012; Su et al. 2015). However, despite these significant contributions to management theory and practice, further research is needed to fill the gaps in the current knowledge of OI models, as they apply to SMEs (Colombo et al. 2012).

The challenge of OI is relevant to increasing market intrusion into business innovation, where SME have better specialization abilities than large companies, and this specialization is more profitable when the market is more open to innovative activity processes. Likewise, SME owner-manager initiatives that promote an innovation climate (IC) are more profitable with the continuation of OI practices. Thus, this study focuses on the main issues, the first of which is related to FI in SMEs, particularly in terms of utilizing internal, external, and hybrid knowledge sources for OI. Second, to what extent does IC apply to SME company owners' efforts to find creative ways to exploit and explore sources of innovation? Thus, this study investigates the expansion of the framework linking the innovation climate (IC), open innovation (OI), and firm innovativeness (FI) among SMEs.

10.2 Literature Review

10.2.1 Open Innovation (OI)

Current challenges in business include increasing the costs of technology development, shorter product life cycles, and customers who are more connected than ever before owing to the fast-paced flow of information (Chesbrough 2006, 2017). As a result, businesses have shifted their focus toward OI, which involves utilizing both internal and external sources of knowledge to enhance the innovation process and expand the market for external innovation (Chesbrough 2017).

OI can be broadly classified into two categories: the exploration and exploitation of innovation opportunities. Outbound innovation refers to the application of technology through external commercial channels, while inbound innovation refers to the process of identifying and mobilizing external sources of knowledge for business growth (Chesbrough et al. 2014).

Overall, businesses must embrace OI to keep up with the rapidly evolving market landscape. They should seek to leverage external knowledge sources to improve internal innovation and take advantage of external opportunities to enhance their market positions. By adopting an OI strategy, businesses can better navigate the challenges of the current economy and remain competitive in the long run.

Small- and medium-scale enterprises can greatly benefit from adopting the OI paradigm, which has been shown to lead to enhanced innovation profitability, expansion or enhancement of the knowledge base, and increased customer satisfaction. In particular, the use of web-based apps and other types of information and communication technology (ICT) has become increasingly important in the OI process (Bianchi et al. 2010).

The rapid evolution of ICT provides SMEs with a range of interactive, low-cost options to facilitate access to external ideas and to identify new market prospects for current ideas. This technology can be used in two ways: first, to help SMEs access external ideas and knowledge and second, to help identify new market opportunities for their current ideas. The specific use of ICT depends on the business context, including factors such as the use of social networks and virtual communities to support a company's OI process (Hutter et al. 2013).

In summary, the OI paradigm can provide numerous benefits for SMEs, including increased innovation profitability, a wider knowledge base, and greater customer satisfaction. The use of ICT, including web-based apps and social networks, can help facilitate the OI process by providing SMEs with access to external ideas and knowledge, as well as identifying new market opportunities.

10.2.2 Innovation Climate (IC)

In the current highly competitive global business environment, it is crucial for organizations to encourage innovation, differentiate themselves from competitors, and provide value to customers. One effective way to promote innovation is to cultivate an internal innovative work climate (Newman et al. 2020).

According to Najar and Dhaouadi (2020), CEOs can act as agents of change to overcome organizational resistance and foster an innovative climate. They argue that positive attitudes toward OI from leaders are essential in reducing employee resistance, and that both factors are equally important in OI practice. Popa et al. (2017) identified three main determinants of an innovation climate (IC), including commitment-based human resources, connectedness between departments, and centralization of decision-making. Hoang et al. (2019) highlighted the importance of leaders in creating a climate for innovation, by mobilizing organizational resources and motivating individuals to work toward creative outcomes.

Najar and Dhaouadi (2020) further describe IC as a working climate characterized by a clear vision, shared concern for excellence in task performance, an interpersonal atmosphere that provides opportunities to participate, and support for innovation characterized by an open work environment and support.

Overall, developing an internal innovative work climate is essential for organizations to remain competitive and drive innovation. CEOs can act as change agents, and positive attitudes toward OI are crucial for reducing resistance and fostering an innovative culture. IC can be further enhanced by factors such as commitment-based human resources, connectedness between departments, and centralization of decision-making, as well as by leaders who motivate and mobilize organizational resources..

Najar and Dhaouadi (2020) have highlighted that bold, proactive, and risky decisions are necessary for organizations to adopt OI and create an IC. While individuals can generate innovative ideas, their willingness to do so depends on the climate within the organization (Hoang et al. 2019). To achieve an IC at the corporate level, firms must create an atmosphere that promotes knowledge sharing and collaboration across departments (Najar and Dhaouadi 2020). Innovative companies are often characterized by a climate that supports OI, which sets them apart from their competitors (Hoang et al. 2019). However, the adoption of OI requires a new mindset and strong aspirations from owners and managers (Najar and Dhaouadi 2020).

Popa et al. (2017) suggest that firms with supportive internal IC are more likely to encourage lateral thinking, calculated risks, and exploration of external knowledge sources. Such firms may explore beyond their borders and expand their expertise, especially when operating within an open inbound innovation ecosystem. Therefore, it is reasonable to assume that the climate of innovation is significantly correlated with OI.

10.2.3 Innovation and SMEs

The adoption of OI by large companies to develop new products and services using web platforms has been well established, with at least 70% implementing such initiatives, as reported by McKinsey in 2008. However, it is also worthwhile to explore the transferability of these ideas to small- and medium-scale enterprises, as they tend to innovate differently from large firms and can benefit from OI processes to achieve innovation outcomes (Parida et al. 2012).

The subject of innovation in SMEs is of increasing interest due to its significant contribution to economic growth and employment opportunities. SMEs are a major source of employment, foster an entrepreneurial spirit, and possess the ability to innovate, thereby enhancing business competitiveness. SMEs with customer-centricity, adaptability, and capability to promptly identify innovation opportunities determine their performance, setting them apart from large enterprises and enabling them to adjust more rapidly to changes in market circumstances and customer demands (Bigliardi and Galati 2018).

SMEs have a unique advantage in fostering an innovative culture due to their flat hierarchies, efficient communication, and minimal bureaucracy (Bocconcelli et al. 2018). Small and medium-scale enterprises typically have expertise in a particular field or product line and are closely connected to the local community due to their limited service area, allowing them to tailor their offerings to meet the specific needs of their market while also streamlining the innovation process (Davis and Bendickson 2021; Bianchi et al. 2010).

However, SMEs often encounter difficulties in managing the entire innovation process, particularly in terms of commercialization, as they lack the necessary complementary assets for marketing and promoting innovation effectively (Purchase and Volery 2020). Although SMEs are proficient in generating innovative ideas, they frequently lack the resources and know-how to bring them to the market successfully (Bianchi et al. 2010).

10.2.4 Open Innovation in Small and Medium-Scale Enterprise

The success rate of innovations that have increased the chances of survival of SMEs ranges from only 22%, highlighting the need for SMEs to overcome the constraints that limit their ability to innovate effectively (Golovko and Valentini 2011; Keupp and Grassmann 2009). OI offers a viable option for SMEs to enhance their innovation performance and profitability, thereby ensuring their competitiveness and survival (Hutter et al. 2013). However, SMEs face greater challenges in adopting OI compared to larger firms, such as being less involved in strategic alliances with other companies, resulting in a lower propensity to collaborate with external partners of all kinds (Ebersberger et al. 2010, 2012).

According to scholarly literature, the size of a company is a crucial factor to consider when implementing OI approaches (Adam and Alarifi 2021; Keupp and Grassmann 2009). Despite the size of the company and a lack of internal commitment, obstacles to applying OI practices primarily relate to cultural and organizational issues (Hutter et al. 2013). SMEs face particular challenges in implementing OI due to a lack of knowledge and awareness among managers or business owners (Parida et al. 2012), as well as a limited ability to disseminate risk, which leads them to be hesitant to experiment with OI activities. Ramirez-Portilla et al. (2017) outline how SME companies often implement OI practices through three main types of activities: inbound, outbound, and coupled. Despite implementing inbound or outbound practices, companies are not precluded from establishing internal innovations (Ramirez-Portilla et al. 2017). Based on these insights, we propose the following hypotheses:

H1: The innovation climate has a significant correlation with firm innovativeness.

H2: The innovation climate has a significant correlation with open innovation.

H3: Open innovation has a significant correlation with firm innovativeness.

H4: The innovation climate indirectly correlates with firm innovativeness through open innovation.

10.3 Research Method

The present study adopts a descriptive and empirical research design, with a focus on exploring the variables of IC, OI practices, and FI. Through a descriptive approach, the study seeks to provide a comprehensive account of these variables, while an explanatory approach is employed to test a hypothesis related to the relationship among them. Causality is a key consideration in this study, as it aims to establish a causal link between the variables under investigation. A cross-sectional time horizon is employed, whereby data is collected directly from business owners and managers of SMEs in Medan, North Sumatera, Indonesia, to examine the phenomenon over a specific period. The unit of analysis is the business owners and managers of SMEs. To measure the variables, the study adopts and adapts four measurement instruments for IC from the study by Popa et al. (2017), and eight instruments from their studies to measure OI practices. Additionally, the measurement instrument used to assess FI is adapted from the study by Ramirez-Portilla et al. (2017).

Population and Sample

The population in research refers to a group of individuals, events, or objects that researchers are interested in investigating. The sample, on the other hand, is a subset of the population that is selected to represent the population as a whole. The unit of analysis in this study was small and medium-scale enterprises registered with the Cooperatives and SMEs Office of Medan City. The sample size was determined based on analytical methods, such as structural equation modeling (SEM), to test the hypotheses. Following the rule of thumb for the sample size in SEM, a minimum ratio of 5 to 15 respondents for each parameter in the study is required. Therefore, with 18 parameters, the minimum sample size for this study was 216. The study added approximately 30% to account for potential sample dropouts and response rates, resulting in a total sample size of 281. Data were collected using a structured questionnaire consisting of statements adopted and modified from previous literature and studies. The questionnaire limited respondents' responses to a set of predetermined answers.

10.4 Data Analysis

Descriptive and Inferential Statistics

A descriptive analysis was conducted for each study variable to provide an overview of the unit of analysis. It is essential to define the scope of research analysis units, specifically SMEs, and to describe their contributions and roles. The average score for each variable was computed and interpreted according to this classification.

Statistical analysis was performed using structural equation modeling (SEM), specifically the Smart Partial Least Squares (PLS) program. SEM-PLS was used because it allows a more in-depth investigation of complex statistical models. It does not make any distributional assumptions when predicting causation or estimating the parameters. Therefore, there was no need for significant testing procedures. The SEM-PLS was used to assess both the outer and inner models. The outer model is a measurement model that predicts the relationship between the estimated indicators or parameters and their latent variables, while the inner model predicts the relationship between the latent variables and which one causes the other.

In research models that use multidimensional constructs in the form of constructs formed from dimensions and indicators that form dimensional latent constructs, analysis involves two phases. The first phase is the analysis of first-order constructs or lower-order constructs, which are dimensional latent constructs that are reflected or shaped by their indicators. The second phase is the analysis of second-order or higher order constructs, which are constructs that are reflected by or formed by dimensional latent constructs.

Model Evaluation

Measurement and Structural Model
In order to test a causal relationship prediction model, it is necessary to subject the research model to the purification stage of the measurement model. To assess construct validity and instrument reliability, a measurement model was employed. Two approaches, convergent and discriminant validity, were used to evaluate construct validity. Measuring reliability can be done using either Cronbach's alpha or composite reliability. While Cronbach's alpha measures the lower limit of a construct's reliability value, composite reliability estimates the construct's actual reliability value and is ideal for assessing internal consistency. However, even when construct validity is achieved, the assessment of internal consistency is not absolute, as a construct's reliability is not always indicative of its validity.

Inner model analysis was used to evaluate the path coefficients or t-values that are linked to the study variables. The path coefficient measures the correlation between exogenous and endogenous variables, and the size of the standardized path coefficient is indicated by the latent variable relationship arrows. A standardized path coefficient of less than 0.1 indicates that the influence of exogenous variables on endogenous variables is statistically significant. The t-statistic value can also be used to evaluate the path coefficient score. To test the hypothesis at a significance level of 5%, the result must exceed 1.96 for the two-tailed hypothesis and 1.64 for the one-tailed hypothesis (Hair et al. 2014).

Bootstrap or non-parametric methods were utilized to determine the final results of testing the structural model and the significance of the model, as PLS does not assume normal distribution of the data (Hair et al. 2014). The significance test results were used to test the hypotheses. The coefficient of determination (R^2) was used to determine the size of the influence and relationship between the latent variables in the study.

Coefficient of determination (R^2)

The coefficient of determination is an indicator of the effectiveness of the model or independent variables in explaining the dependent variable. The coefficient of determination is evaluated using criteria such as 0.75, 0.5, or 0.25, which represent model assessments that can explain the data well, moderately, or weakly, respectively. When assessing exogenous variables with different measurements or an unequal number of observations, adjusted R^2 is used instead of standard R^2 (Hair et al. 2014).

10.5 Results

Subsequently, the data collected from small and medium-scale enterprises (SMEs) owners–managers in Medan, North Sumatera were sorted according to respondent profiles. The distribution of 281 questionnaires was based on predetermined target respondents. Of these, 217 were successfully completed, resulting in a response rate of 77.22%.

10.5.1 Respondent Profile

Table 10.1 presents the profile of the respondents as follows:

According to the findings presented in Table 10.1, it can be observed that 99 out of 217 SME owner-managers who participated in the study were women, accounting for 45.62% of the respondents. Men, on the other hand, constituted the majority, with 118 respondents, accounting for 54.37% of the sample. The distribution of respondent profiles by business type was as follows: 57 of the respondents were owners or managers of cafes and restaurants, accounting for 26.23% of the sample, while 57 individuals were in retail, representing 23.96%. The fashion business had 47 participants (21.66%) and 42 respondents were in the furniture business, accounting for 19.35% of the sample. The remaining 19 individuals were involved in other types of enterprise, accounting for 8.76% of the sample. In terms of educational attainment, 82 respondents had high school education, representing

Table 10.1 Profile of respondents

	Frequency	Proportion (%)
Gender		
Woman	99	45.62
Man	118	54.37
Type of business		
Furniture	42	19.35
Fashion	47	21.66
Cafes and restaurants	57	26.27
Retail	52	23.96
Other	19	8.76
Education		
High school or equivalent	82	37.79
Diploma	37	17.05
Bachelor	72	33.18
Postgraduate	26	11.98
Operating revenues		
Below IDR 500 million	99	45.62
IDR 501 million to IDR 1.5 billion	74	34.10
IDR 1.51 billion to IDR 2.5 Billion	41	18.89
Above IDR 2.5 billion	3	1.38

37.79% of the sample, while 37 individuals had a diploma (17.05%), 72 individuals had a bachelor's degree (33.18%), and 26 individuals had a postgraduate degree (11.92%). Regarding business income, 99 SME owner–managers had a business income below IDR 500 million (45.62%), 74 individuals had a business income between IDR 501 million and IDR 1.5 billion (34.1%), 41 individuals had a business income between IDR 1.51 billion and IDR 2.5 billion (18.89%), and 3 individuals had a business income above IDR 2.5 billion (1.38%).

10.5.2 Descriptive Statistical Analysis

Descriptive statistics offer a concise overview of the respondents' answers to each instrument used to measure the study variables of IC, OI practices, and FI. The instruments employed a five-point Likert scale and consisted of specific statements that assessed each variable using a questionnaire. Participants indicated their level of agreement or disagreement with the statements. The focus of descriptive statistics is on the description of participants' responses rather than making general inferences. Table 10.2 presents the participants' responses to the aforementioned variables.

Table 10.2 presents an overview of the results of the study, which measured the IC, OI practices, and FI among SMEs. The findings reveal that SMEs have a positive overall IC with an average score of 3.67 out of 5.00. Additionally, SMEs showed a favorable attitude toward OI practices, with an average score of 3.73 out of 5.00. The data further indicate that FI among SMEs is rated as good with a score of 3.70 out of 5.00.

10.5.3 Inferential Statistical Analysis

This study's inferential statistical analysis was conducted utilizing the structural equation method. The partial least squares (PLS) approach is used to complete structural equations due to the complexity of the research model, the usage of second-order constructs, the non-normal distribution of the data, and the exploratory nature of the effects between variables. The PLS analysis is conducted by examining the consistency of the constructs and validating the indicators of each construct using outer model analysis. If the research model fits the requirements for a good outer model, the study continues by investigating the relationship between variables using the inner model analysis. The data analysis was facilitated by the statistical application SmartPLS version 3.0.

10.5.4 Measurement Model Analysis

This study was conducted to identify a fit model for each indicator on each construct, such that the estimate of the inner model is not biased. Analyzing the

Table 10.2 Descriptive statistics for each variable

	Total	Means	Classification	
	F	%		
Innovation climate				
IC. 1	217	100	3.67	Good
IC. 2	217	100	3.66	Good
IC. 3	217	100	3.64	Good
IC. 4	217	100	3.69	Good
Innovation climate			3.67	Good
OI Practices				
OI. 1	217	100	3.73	Good
OI. 2	217	100	3.75	Good
OI. 3	217	100	3.73	Good
OI. 4	217	100	3.72	Good
OI. 5	217	100	3.76	Good
OI. 6	217	100	3.68	Good
OI. 7	217	100	3.71	Good
OI. 8	217	100	3.76	Good
OI practices			3.73	Good
Firm innovativeness				
FI. 1	217	100	3.76	Good
FI. 2	217	100	3.67	Good
FI. 3	217	100	3.65	Good
FI.4	217	100	3.70	Good
FI. 5	217	100	3.68	Good
FI. 6	217	100	3.77	Good
Firm innovativeness			3.70	Good

reliability of each indicator on each construct and the validity of all indicators in assessing each construct constitutes the outer model analysis. To assess the constructs' reliability, composite reliability was used. The indicators' validity was evaluated using convergent and discriminant validity. Referring to the research method, indicator criteria and constructs are considered reliable and valid; if reliability and validity criteria are not fulfilled, the model is adjusted by eliminating problematic indicators from the construct. If the reliability criteria are not met, the analysis will not proceed to the outer model's measurement validity step.

Convergent validity is a component of the measurement model, which is also called the "outer model" in SEM-PLS. Two criteria must be met to determine whether the measurement model meets the convergent validity requirements for the reflective construct: (1) the loading must be greater than 0.7, and (2) the p-value must be significant (<0.05). However, in some cases, frequently, standards for loading more than 0.70 are not met, especially for newly developed surveys. Therefore, indicators with loadings below 0.40 should be removed from the model. Indicators with loadings less than 0.40 should thus be eliminated from the model. It is ideal to investigate the influence of deleting indicators with loadings between 0.40 and 0.70 on AVE and composite reliability.

Indicators with a loading between 0.40 and 0.70 may be deleted if they raise AVE and composite reliability above their respective thresholds. The AVE limit value is 0.50, and composite reliability is 0.7. Another factor to consider when removing indicators is the effect on the construct's content validity. Low loading indicators can be kept because they contribute to the construct content validity. Table 10.3 presents the tabulation of the measurement model results.

Table 10.3 presents the validity test of factor loading, which shows that all factor loading values are greater than 0.7. This means that in the second order, the validity measure based on the loading value has met the validity requirements. Likewise with the validity requirements based on AVE in the second order, indicating that all AVE values are greater than 0.5. This confirms that the overall AVE value meets the validity requirements. Furthermore, the reliability measure is based on CR at the second-order stage, which shows that the overall CR value is greater than 0.7. This result means that it meets the reliability requirements based on CR.

Tabel 10.3 The measurement model results

Constructs	Item	Item-loading	AVE	CR rho _a	CR rho _c	CA	Discriminant Validity	R^2
Innovation climate	IC1	0.856	0.674	0.840	0.892	0.838	0.699	
	IC2	0.809						
	IC3	0.807						
	IC4	0.809						
Open innovation	OI1	0.742	0.561	0.888	0.911	0.888	0.744	0.365
	OI2	0.753						
	OI3	0.763						
	OI4	0.791						
	OI5	0.730						
	OI6	0.759						
	OI7	0.736						
	OI8	0.714						
Firm innovativeness	FI1	0.778	0.581	0.855	0.892	0.855	0.720	0.493
	FI2	0.780						
	FI3	0.797						
	FI4	0.765						
	FI5	0.717						
	FI6	0.731						

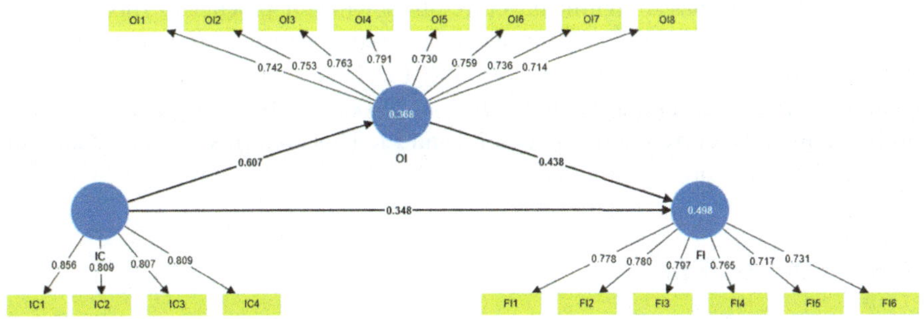

Fig. 10.1 Validity testing based on second-order factor loading. *Source* Author's research output

Similarly, the calculation of Cronbach's alpha on the second order shows that all CA values are greater than 0.7. This result can be interpreted as meaning that the results of reliability testing based on Cronbach's alpha have fulfilled the reliability requirements. Testing the discriminant validity using the Fornell-Larcker approach for the second order shows that the correlation value between a latent variable and other latent variables is less than the square root of the AVE for that latent variable. These results lead to the conclusion that the calculation of the test meets the requirements of discriminant validity. Furthermore, the R-Square presented in Table 10.3 shows that the R-Square value of FI is 0.493. The results of this acquisition explain that the IC and OI practices can influence FI by 49.3%. The R-squared value of OI practices is 0.365. This means that the IC can influence OI practices by 36.5% (Fig. 10.1).

10.5.5 Structural Model Analysis

The significance of the influence test and the path coefficient are summarized in the results of the measurement model in Table 10.4.

Table 10.4 shows that the results of each hypothesis's significance test are as follows: With p-values of $(0.000 < 0.05)$, IC has a significant correlation with FI. This confirms that the proposed hypothesis (H1) is accepted. With p-values of

Table 10.4 Structural model results

	Correlation	Original	Mean	Std. deviation	t-value	p-value	Support
H1	IC → FI	0.613	0.616	0.069	8.897	0.002	Supported ***
H2	IC → OI	0.607	0.610	0.074	8.239	0.000	Supported ***
H3	OI → FI	0.438	0.444	0.109	4.011	0.000	Supported ***
H4	IC–FI (indirect effect)	0.266	0.271	0.078	3.400	0.001	Supported ***

***p-value < 0.01

(0.000 < 0.05), IC has a significant correlation with OI practices. This confirms that the proposed hypothesis (H2) is accepted. With *p*-values of (0.000 < 0.05), OI practices have a significant correlation with FI. This confirms that the proposed hypothesis (H3) is accepted. With *p*-values of (0.000 < 0.05), OI practices significantly mediate the correlation between IC and FI. This confirms that the mediation hypothesis (H4) is proven and accepted.

10.6 Discussion

The significance of the correlation between IC and FI

The present study yields empirical evidence that highlights the significance of the IC in relation to FI. The analysis revealed a strong correlation between the two variables, with a path coefficient of 0.613 and a significance level of 0.000. A favorable IC is characterized by the active participation of employees who allocate their resources and time to generate innovative solutions, which may include knowledge sharing or exchange. Consequently, the contributions of such employees can enhance the company's ability to produce more innovative products or processes. Notably, innovative firms typically exhibit traits such as reduced costs associated with new product development, time, and cost efficiency in reaching target markets. The current findings reinforce prior research, particularly the arguments posited by Popa et al. (2017), who suggest that an innovative organizational climate encourages a culture of innovation. This culture is shaped by shared values, beliefs, and assumptions that facilitate the innovation process (Martín-de Castro et al. 2013; Popa et al. 2017). Our study's results further strengthen Popa et al.'s (2017) empirical findings, revealing a positive relationship between the IC and OI in SMEs. For instance, SMEs benefit from an organizational climate that promotes innovation as it enhances their problem-solving capabilities, fosters a greater willingness to take risks, and improves their ability to leverage external knowledge. In light of these findings, it is evident that SMEs with a supportive IC are more likely to expand beyond their current boundaries and enrich their knowledge base.

The significance of the correlation between IC and OI practices

The present study has revealed a statistically significant correlation between IC and OI practices, as attested by a path coefficient of 0.607 and a significance level of 0.000. The findings have demonstrated that organizations endowed with a favorable IC and the capacity to guide their personnel toward collaborative work are more likely to embrace OI. This empirical evidence corroborates the previous work of Popa et al. (2017), who have established that SMEs exhibiting a robust IC are more prone to harness inbound innovation to augment their collective knowledge. The results of this study have further validated the importance of IC in determining a company's commitment to outbound innovation, as manifested in the management system for the exploitation of technical knowledge through patents or intellectual property rights. Therefore, nurturing a conducive IC is of utmost significance.

The significance of the correlation between OI practices and FI
The study has established a significant correlation between OI practices and FI, with a path coefficient of 0.438 and a significance level of 0.000. This finding bolsters the existing body of research emphasizing the importance of enhancing SMEs innovation by adopting OI practices (Ramirez-Portilla et al. 2017). The study also supports prior findings indicating that not all innovation practices are equally impactful on innovation outcomes, with inbound innovation (such as purchasing technical services, conducting technology prospecting, and external personnel training) having a greater effect on OI practices than outbound innovation. Additionally, selling technical or scientific services based on company experience is deemed more important than open (outbound) innovation. These results further corroborate that SMEs utilize search, acquisition, and knowledge sourcing activities as well as knowledge sharing practices to augment their level of innovation (Ramirez-Portilla et al. 2017).

The significance of indirect relationship between IC and FI
This study has provided empirical evidence that OI practices mediate the relationship between the IC and FI, with a path coefficient of 0.266 and a significance level of 0.001. This finding underscores the significance of an innovative corporate climate, characterized by skilled and competent employees connected by OI practices in promoting innovative firms. Specifically, SMEs tend to rely heavily on the contributions of external partners by promoting open (inbound) innovation, which involves direct engagement with external partners to generate innovation. This is exemplified by utilizing research and development results from external partners for new and innovative products, systems, or services. Conversely, companies adopting open (outbound) innovation tend to derive greater benefits from their innovation results by seeking to sell trading licenses to external parties such as a particular brand, recipe, or process. The study's findings further corroborate prior research positing that a firm's IC is a critical determinant of its innovation performance (Popa et al. 2017). Therefore, the stronger a company's focus on innovation, the more significant the impact of IC improvement, with OI practices bridging the relationship between the two.

10.7 Conclusion

This study has significant implications for both academic and practical circles, as small- and medium-scale enterprises (SMEs) cannot generalize the results of previous empirical research on OI practices as they pertain to large firms (Lee et al. 2010; Spithoven et al. 2013). Instead, SMEs may promote innovation by using both inbound and outbound OI practices, supported by the firm's IC. This study contributes to the existing empirical findings on OI in SMEs and empirically tests a research extension model to evaluate the impact of IC on FI as well as the mediating role of inbound and outbound OI practices.

The study results indicate that SME owners and managers must be aware of the significance of creating a conducive IC to enhance the company's emphasis on innovation, complemented by OI practices. It is also important for SMEs to be aware that OI strategies are becoming more prevalent, and adapting to them is necessary to remain competitive. However, this study has some limitations that must be addressed in future research. First, the results cannot be generalized as they are based on a sample of SMEs from only one city and province. Future studies should include larger sample sizes and different regions, either within a country or across nations. Second, this study only provides a cross-sectional overview of the causes and outcomes of OI practices, and timing changes cannot be explained. Therefore, a longitudinal investigation may be necessary to corroborate these results. Including these recommendations in future research could improve the accuracy and generalizability of the results.

References

Adam NA, Alarifi G (2021) Innovation practices for survival of small and medium enterprises (SMEs) in the COVID-19 times: the role of external support. Journal of Innovation and Entrepreneurship 10(1):1–22

Bianchi M, Campodall'orto S, Frattini F, Vercesi P (2010) Enabling open innovation in small- and medium- sized enterprises: how to find alternative applications for your technologies. R&D Management 40(4):414–431

Bigliardi, B., & Galati, F. (2018). An open innovation model for SMEs. In *Researching open innovation in SMEs* (pp. 71–113).

Bocconcelli R, Cioppi M, Fortezza F, Francioni B, Pagano A, Savelli E, Splendiani S (2018) SMEs and marketing: a systematic literature review. Int J Manag Rev 20(2):227–254

Chesbrough H (2006) Open Innovation: A New Paradigm for Understanding Industrial Innovation, *Open Innovation: Researching a New Paradigm*. Oxford University Press, Oxford

Chesbrough H (2012) Open innovation: Where we've been and where we're going. ResearchTechnology Management 55(4):20–27

Chesbrough H (2017) The future of open innovation: The future of open innovation is more extensive, more collaborative, and more engaged with a wider variety of participants. ResearchTechnology Management 60(1):35–38

Chesbrough H, Vanhaverbeke W, West J (eds) (2014) New frontiers in open innovation. Oup Oxford

Colombo MG, Laursen K, Magnusson M, Rossi-Lamastra C (2012) Introduction: small business and networked innovation: organizational and managerial challenges. J Small Bus Manage 50(2):181–190

Davis PE, Bendickson JS (2021) Strategic antecedents of innovation: variance between small and large firms. J Small Bus Manage 59(1):47–72

Ebersberger B, Marsili O, Reichstein T, Salter A (2010) Into thin air: using a quantile regression approach to explore the relationship between R&D and innovation. Int Rev Appl Econ 24(1):95–102

Ebersberger B, Bloch C, Herstad SJ, Van De Velde ELS (2012) Open innovation practices and their effect on innovation performance. Int J Innov Technol Manag 9(06):1250040

Golovko E, Valentini G (2011) Exploring the complementarity between innovation and export for SMEs' growth. J Int Bus Stud 42(3):362–380

Hair JF Jr, Sarstedt M, Hopkins L, Kuppelwieser VG (2014) Partial least squares structural equation modeling (PLS-SEM): An emerging tool in business research. Eur Bus Rev 26(2):106–121

Hoang G, Wilson-Evered E, Lockstone-Binney L (2019) Leading innovation among tourism small and medium enterprises: Examining the mediating role of climate for innovation. Leadersh Organ Dev J 40(5):647–666

Hossain M (2015) A review of literature on open innovation in small and medium-sized enterprises. J Glob Entrep Res 5(1):1–12

Hutter K, Hautz J, Repke K, Matzler K (2013) Open innovation in small and micro enterprises. ProblPerspect Manag (11, Iss. 1):12–22

Keupp MM, Grassmann O (2009) Determinants and archetype users of open innovation. R&D Manag 39(4):331–341

Lasagni A (2012) How can external relationships enhance innovation in SMEs? New evidence for Europe. J Small Bus Manage 50(2):310–339

Lee S, Park G, Yoon B, Park J (2010) Open innovation in SMEs—an intermediated network model. Res Policy 39(2):290–300

Martín-de Castro G, Delgado-Verde M, Navas-López JE, Cruz-González J (2013) The moderating role of innovation culture in the relationship between knowledge assets and product innovation. Technol Forecast Soc Chang 80(2):351–363

Mckinsey (2008) Building the web 2.0 enterprise

Najar T, Dhaouadi K (2020) Chief Executive Officer's traits and open innovation in small and medium enterprises: The mediating role of innovation climate. J Small Bus Enterp Dev 27(4):607–631

Parida V, Westerberg M, Frishammar J (2012) Inbound open innovation activities in high-tech smes: the impact on innovation performance. J Small Bus Manage 50(2):283–309

Popa S, Soto-Acosta P, Martinez-Conesa I (2017) Antecedents, moderators, and outcomes of innovation climate and open innovation: an empirical study in SMEs. Technol Forecast Soc Chang 118:134–142

Purchase S, Volery T (2020) Marketing innovation: a systematic review. J Mark Manag 36(9–10):763–793

Ramirez-Portilla A, Cagno E, Brown TE (2017) Open innovation in specialized SMEs: the case of supercars. Bus Process Manag J 23(6):1167–1195

Spithoven A, Vanhaverbeke W, Roijakkers N (2013) Open innovation practices in SMEs and large enterprises. Small Bus Econ 41(3):537–562

Su CY, Lin BW, Chen CJ (2015) Technological knowledge co-creation strategies in the world of open innovation. Innovation 17(4):485–507

Tidd J, Bessant JR (2020) Managing innovation: integrating technological, market and organizational change. John Wiley & Sons

Characteristics of Small and Medium-Scale Enterprises: Orientation, Innovation, Performance, and Competitiveness

11

Onan Marakali Siregar and Selwendri

Abstract

The aim of this study is to explore the link and distinction between market orientation, product innovation capability, and marketing performance in determining competitive advantage among small and medium-scale enterprises (SMEs). This study employs a quantitative methodology and involves business owners and managers in three categories. A Bayesian regression model is used to test the proposed research model. The study's results show a positive correlation between all proposed constructs and competitive advantage. Although these constructs have led firms to become more competitive, there are notable differences between small and medium-scale enterprises in terms of their characteristics related to market orientation, product innovation capability, marketing performance, and competitive advantage, which demand attention. This study attempts to make a valuable contribution to the expansion of knowledge in the marketing field by investigating and analyzing the relationships between constructs that are essential for defining and establishing a competitive edge. The ramifications of this study are envisaged to assist entrepreneurs and managers in making appropriate decisions. Firms must capitalize on their unique strengths if they remain competitive in an ever-changing market. SMEs are viewed as promising sources for creating jobs and driving economic growth in the emerging markets.

Keywords

Market orientation • Product innovation capability • Marketing performance • Competitive advantage • SMEs

O. M. Siregar (✉) · Selwendri
Faculty of Social and Political Science, Universitas Sumatera Utara, Medan, Indonesia
e-mail: onan@usu.ac.id

A. Rafiki et al. (eds.), *Open Innovation in Small Business*, Contributions to Environmental Sciences & Innovative Business Technology,
https://doi.org/10.1007/978-981-99-5142-0_11

11.1 Introduction

The importance of SMEs in promoting economic growth has been widely acknowledged across both emerging and developed countries, with various research studies (Barbero et al. 2011; Gilmore et al. 2013; O'Cass and Sok 2014; Wolf et al. 2012) supporting this assertion. According to the OECD (2017), these SMEs account for almost 99% of all business activities and have been recognized as crucial drivers of employment generation and overall organizational performance, as evidenced by several studies (Afriyie et al. 2019; Gruber-Muecke and Hofer 2015).

Although SMEs are recognized as contributors to the growth of a country and organizations, many fail to survive in a dynamic market environment beyond three years, as highlighted by several research studies (e.g., Sirmon et al. 2007; Lafuente et al. 2020; Storey and Greene 2010). The reason why some SMEs thrive while others stagnate has been a subject of inquiry in various studies (Anderson and Eshima 2013; Hansen and Hamilton 2011). Furthermore, SMEs performance is crucial for facilitating their transition toward greater diversification. Ironically, SMEs in many emerging countries lack knowledge of how to survive while neglecting the importance of product and marketing innovation (O'Cass and Sok 2014) or lack managerial competence, adequate resources, market-related experience, and the capacity for continuous innovation (Ali et al. 2020; Gruber et al. 2010; Afriyie et al. 2019), which involves transforming ideas into new products, methods, and systems continually (Saunila 2020).

As SMEs strive to compete with larger and better-resourced companies, their ability to innovate has become a crucial challenge. Research studies suggest that combining sufficient resources enhances a firm's innovation capability (such as O'Cass and Sok 2014; Sanzo et al. 2011; Gruber et al. 2010). Innovation capability refers to the ability to create new and valuable products or knowledge (Saunila and Ukko 2014; Zheng et al. 2010). In SMEs, innovation capability encompasses a set of processes and outcomes. The processes involve evaluating the organization's level of innovation, while the outcomes pertain to the impact of the organization's innovation efforts (O'Cass and Sok 2014). Additionally, when a firm prioritizes meeting customer demands and expectations with innovative and valuable products, it adopts a market-oriented approach (Jaworski and Kohli 1993). As outlined in previous studies, firm'market orientation comprises various resources, flexible processes, and responsiveness (Raju et al. 2011; Buli 2017).

In the current challenging environment, it is crucial to address the challenges faced by SMEs in terms of organizational market orientation (MO), product innovation capability (PIC), and marketing performance (MP). To enhance competitiveness, it is necessary to use these factors more effectively. This study examines the degree to which the correlation and difference between small and medium-scale enterprises in terms of MO, PIC, and MP are relevant in determining the level of competitiveness. This paper is divided into several sections. The first section provides an overview of the theoretical framework that supports the proposed hypotheses. The next section outlines the methodology used to conduct the investigation, followed by a presentation of the empirical results. The conclusion of the paper features a discussion that addresses both the limitations and the practical implications of the study.

11.2 Literature Review

11.2.1 Market Orientation (MO)

The concept of MO was initially introduced by Narver and Slater (1990) and Kohli and Jaworski (1990) to refer to a set of cultural dimensions, behaviors, and processes that aim to generate customer value. Nowadays, MO is considered as a group of cross-functional activities and processes aimed at retaining and satisfying customers by continually assessing their demands and expectations, as described in contemporary literature (De Toni et al. 2021; Chakravarty et al. 2014).

MO refers to a company's cognitive, behavioral, and cultural characteristics that prioritize the creation of greater customer value (Shehu and Mahmood 2014; Jogaratnam 2017; Montiel-Campos 2018; D'souza et al. 2021). It is considered an essential component of a company's strategic approach to anticipate and meet evolving market demands (Jansson et al. 2017; Iyer et al. 2019). Market-oriented firms tend to have resources that can be utilized to produce valuable outputs for customers, and those with a high level of MO tend to offer greater value to customers (Alnawas and Hemsley-Brown 2019; Sahi et al. 2018; Yeh 2016; Gotteland et al. 2020).

Gebhardt (2012) posits that market-oriented firms possess the ability to gather and share information concerning existing and potential customers through collaborations. Additionally, such firms must aim to meet or surpass customer expectations as this can provide them with various CA (De Toni et al. 2021). Essentially, market-oriented firms must organize, coordinate, and exchange customer-related data across departments to comprehend customer needs and offer enhanced value to their products (Fang et al. 2014; Sahi et al. 2018; Yeh 2016). MO is considered a significant source of advantage and is associated with high-level marketing capabilities (Kirca et al. 2005; Iyer et al. 2019; Pérez-Luño et al. 2016). It is also connected with innovation, firm performance, MP, and CA (D'souza et al. 2021; Alhakimi and Mahmoud 2020; Fernandes et al. 2020). Similarly, organizational capabilities stemming from a market-focused approach contribute to a firm's competitive edge. The term "capability" refers to a firm's ability to differentiate its products from those of its competitors, which is crucial in today's dynamic business environment. A profound understanding of MO is crucial for businesses seeking to generate CA. Consequently, we propose the following hypothesis:

H1: Among small and medium-scale enterprises, market orientation holds substantial influence over competitive advantage.

11.2.2 Product Innovation Capability (PIC)

Innovation, which involves the integration of novel ideas and generation of fresh knowledge into an organization's operations, has been discussed in several studies (Kahn 2018; Rajapathirana and Hui 2018; Müller et al. 2021). Lukovszki et al. (2021) categorized innovation into three primary types: product, process, and market innovation. The introduction of a new product into the market or the improvement of an existing product's features or quality falls under product innovation. It also encompasses ways to make products more useful to customers, both functionally and otherwise (Najafi-Tavani et al. 2018; Zaefarian et al. 2017).

Vega-Jurado et al. (2009) assert that product innovation does not depend solely on a company's internal capabilities, such as their proficiency in managing design functions. Instead, it is a method of enhancing value that can assist firms in achieving a competitive edge and maintaining their market share (Bowonder et al. 2010). Therefore, innovative products are used to increase a business's success and provide it with a competitive edge. A resource-based approach to innovation is predicated on the idea that organizational resources and capabilities can both underlie and determine a firm's innovative potential. Tangible and intangible organizational resources are used to create inputs that are then transformed and integrated into new forms of CA through capabilities (Lukovszki et al. 2021). Thus, the potential for innovation is contingent on a firm's capabilities to generate beneficial value for customers, which may ultimately lead to a CA.

Innovation capability is an important factor for SMEs to enhance firm performance as it involves various activities aimed at generating novel products, processes, and operating systems by leveraging diverse sources of knowledge, experience, and ideas (Keskin 2006; Lawson and Samson 2001; Saunila 20202014; Saunila and Ukko 2014; Taques et al. 2021; Zhang and Hartley 2018). Specifically, the concept of product innovation capability refers to a set of interrelated processes used to develop new products or upgrade existing ones (Mafimisebi et al. 2020; O'Cass and Sok 2014).

CA is achieved by firms when their resources cannot be easily imitated by current or future rivals, or it is too costly to do so (Falahat et al. 2020). By adopting a resource-based view approach, firms must leverage their valuable resources to exploit opportunities and mitigate risks while also ensuring that their resources are not perfectly replicable (Elia et al. 2021; Alonso and Austin 2016). Product innovation can impact a firm's CA by utilizing its resources and capabilities to create higher product or service value, making resources and capabilities increasingly critical for innovation to occur (Falahat et al. 2020). As the level of PIC increases, the potential for a firm's CA to emerge also increases (Ferreira and Coelho 2020). Thus, the following hypothesis is proposed:

H2: The higher the level of product innovation capability, the greater the potential for a firm's competitive edge.

11.2.3 Marketing Performance (MP)

The topic of performance has gained increasing attention from scholars, academics, and practitioners, as evident from various studies (Clark et al. 2005; Lamprinopoulou and Tregear 2011; Lamberti and Noci 2010; Nwokah 2009; O'Sullivan et al. 2009; O'Sullivan and Abela 2007; Pimenta da Gama 2011). These studies highlight that performance is a multidimensional construct that requires comparisons across time, place, or planned versus actual outcomes. Therefore, performance measurement is a continuous process that can be evaluated in various ways, and there is no one-size-fits-all performance measure.

Pimenta da Gama (2011) proposes three conceptual frameworks for assessing firm performance: the goals approach, resource systems approach, and process approach. According to Pimenta da Gama (2011), marketing can be viewed as a management philosophy that explains how and why enterprises must adapt and impact markets. The marketing subsystem is responsible for the development of a set of marketing-related activities. Marketing practices are frequently criticized for their inefficiency and inadequacy in assessing the relationship between activities and outcomes (Pimenta da Gama 2011). As a discipline, marketing tends to focus more on outcomes than on processes and system support (Grewal et al. 2009; O'Sullivan et al. 2009).

O'Sullivan and Abela (2007) propose three perspectives to categorize the measurement of MP: marketing productivity measurement, identification of metrics used, and brand equity measurement. Measuring marketing productivity can indicate the connection between marketing efforts and a company's performance. However, the difficulty in measuring marketing productivity arises from the management's inability to account for marketing initiatives. To assess marketing's contribution to a company's performance, the measurement of marketing productivity requires both management attention and adequate resources.

Marketing plays a critical role in ensuring long-term business performance given that an organization's existence is contingent upon its capacity to create value. As such, measuring MP is the primary responsibility of management. Eusebio et al. (2006) argue that management is charged with the responsibility of developing acceptable measures in order to respond to the changing marketing paradigm. However, due to the lack of firms that pay attention to a rigorous evaluation of marketing results, including the design of an appropriate MP measurement system (Lamberti and Noci 2010), the measurement of satisfaction represents the firm's assessment of MP measurement. Morgan et al. (2002) suggested measuring user satisfaction as a key outcome indicator to measure how well it worked.

SMEs typically do not focus on discussing their market share. Instead, they concentrate on identifying the specific niche market in which they operate and their main priority is to identify the most suitable customers for their products or services (Carson et al. 1995). This approach involves offering high-quality products and personalized customer service, which is the most effective way to encourage consumers to recommend a company's products or services to others (O'Donnell et al. 2002). Thus, SMEs can measure the effectiveness of their marketing efforts based on the number and type of referrals received from their customers.

In addition to the outcomes of marketing factors, MP in small firms differs from that in large firms. This is due to the unique characteristics of small firms, which make them face greater obstacles than larger firms. Therefore, examining the impact of MP on achieving superior competitiveness becomes imperative. As a result, we put forth the following hypothesis:

H3: Among small and medium-scale enterprises (SMEs), marketing performance holds significant influence over competitive advantage

11.2.4 Competitive Advantage (CA)

Despite numerous attempts, there is no widely accepted definition of CA that is clear and unambiguous (Sigalas and Pekka Economou 2013). For instance, Teece et al. (1997) argue that a firm's CA arises from its ability to consistently enhance the organizational capabilities underlying its products and services. Another feature of CA is the ability to create value that existing or potential rivals cannot replicate (Barney 1991; O'Donnell et al. 2002).

Sigalas and Pekka Economou (2013) implicitly propose a two-fold categorization of CA. The first stream is primarily based on performance-based explanations, whereas the second is mainly based on resource-based explanations.

Literature suggests that a business's success depends on its ability to manage distinctive resources (Singh et al. 2020). These resources must be adaptable to the changing environmental conditions and the overall strategy of the firm (Akter et al. 2020; Schilke et al. 2018; Teece 2014). In the vein, Barney (2001a, 2001b) emphasizes the importance of identifying relevant, novel, and difficult-to-replicate sources of CA.

According to Correia et al. (2020), a firm's CA may originate from its unique and dynamic assets and capabilities. To gain access to superior resources, a firm must acquire scarce, valuable, irreplaceable, and difficult-to-replicate resources and capabilities, giving it an edge over its competitors as per Barney's (1991) view. Several aspects of an organization may affect its CA, including PIC, MO, and MP, which are assumed to be important in this study. MO refers to a business culture that encourages consistent delivery of more value to consumers. On the other hand, PIC refers to a firm's ability to create products more effectively, making it difficult for competitors to replicate. A firm's CA is determined by its MO and PIC as well as its utilization of resources to aid in the development of MP, which is dependent on the firm's size. Based on this, the study posits the following hypotheses:

H4: There exist significant differences between small and medium-scale enterprises in terms of market orientation.

H5: There exist significant differences between small and medium-scale enterprises in terms of product innovation capability.

H6: There exist significant differences between small and medium-scale enterprises in terms of marketing performance.

H7: There exist significant differences between small and medium-scale enterprises in terms of their level of competitiveness.

11.3 Research Methods

This study utilized both descriptive and inferential analyses (Hair et al. 2010) in order to test the proposed framework with a Bayesian regression model, using SPSS version 25. A cross-sectional approach was used to collect quantitative data using a structured questionnaire. The sample frame consisted of Indonesian SMEs listed in the SMEs directory, specifically bakeries and coffee shops, restaurants, lodging, and accommodations. The study population comprised SMEs from four cities in North Sumatera Province and was selected for their representation of typical SMEs and the presence of SME centers, as well as their rapid growth compared to other areas. The sample size was determined using the estimate interval formula with a random sample selected from the population of the relevant sector to reflect the two sizes of firms and subsectors in Indonesia. Data collection began in 2019 and continued until February, 2020.

The definition of "SMEs" by the Indonesian Central Statistics Agency is based on employment criteria, whereby a small enterprise is defined as having between five and 19 employees, while a medium-scale enterprise has between nineteen to ninety-nine employees. In order to conduct this study, the researchers determined the appropriate sample size and chose to include individuals who were typically responsible for overseeing the daily operations of the organization, such as owners or managers. To ensure that the sample consisted of firms with fit criteria, SMEs with at least two years of commercial activity were evaluated. At least 60 business owners and managers who met the aforementioned criteria were identified in each of the four locations, resulting in 160 survey responses or 66.7% of the overall sample size. Five responses from non-SMEs were excluded, leaving 155 responses for analysis. The questionnaire utilized in this study was designed in a straightforward manner to facilitate respondents' easy understanding of the questions and to enable them to provide their perceptions on the matter. The questions were derived from prior research and only underwent minor adjustments, if deemed necessary. The study provides a comprehensive overview of each section, detailing the content and sources of the questions. A five-point Likert scale was employed to measure both independent and dependent variables, with scores ranging from strongly disagree to strongly agree.

To measure the CA, a construct developed by Falahat et al. (2020) and Kaleka and Morgan (2017) was adopted, consisting of ten items with minor adjustments to suit the research objectives of this study. MO, on the other hand, refers to SMEs' inclination to obtain, disseminate, and react to market information. Ten items adapted from Deshpandé and Farley (1998) were used to measure this construct. In terms of PIC, eight items derived from previous research studies, such as Pham et al. (2017) and Salavou and Avlonitis (2008), were utilized, with slight adaptations. Finally, to measure MP, four items were adopted from studies by Kirca et al. (2005), Hooley et al. (2000), and Jalilvand (2017), and were slightly modified to align with the objectives of this research.

To collect the necessary data, the authors collaborated with enumerators to evaluate the responses of business owners and managers. A data collection approach that used written closed questions was employed. Direct question-and-answer sessions were held with business owners and managers at four locations in North Sumatera Province, Indonesia. The constructed items were used in the investigation, and the validity of the questionnaire responses was verified. This study aimed to establish a basis for future research and to compare it with current approaches. The approach considered significant topics such as MO, PIC, MP, and CA. By providing empirical evidence based on a theoretical framework, this study contributes to the existing literature.

11.4 Data Analysis

11.4.1 Respondents Profiling

According to Table 11.1, the majority of the respondents were either business owners (81 individuals, 52.3%) or managers (74 individuals, 47.7%). The data also indicate that most managers work for medium-scale enterprises, whereas most owners work for small enterprises. The survey showed that 49 owner-managers (31.6%) had between 12 and 16 years of experience, followed by 42 (27.1%) with 7 to 11 years of experience, 40 (25.7%) with more than 16 years of experience, and 24 (15.5%) with 2 to 6 years of experience. This suggests that SME owner-managers are well informed about their business environment. Furthermore, the survey revealed that 83 owner-managers of small-scale enterprises (53.5%) employed between five and 19 workers, while 72 owner-managers of medium-scale enterprises (46.5%) had hired between 19 and 99 employees. In terms of industry, 63 SME owner-managers operate restaurants (40.6%), 59 SME owner-managers operate bakeries and coffee shops (38.1%), and 33 SME owner-managers run lodging and accommodation enterprises (21.3%).

Table 11.1 Specific characteristics of participants

Classification	Freq	%
Status:		
Owner	81	52.3
Manager	74	47.7
Experience		
2–6 years	24	15.5
7–11 years	42	27.1
12–16 years	49	31.6
More than 16 years	40	25.8
Years of operation		
Under 5 years	12	7.7
5–10 years	53	34.2
11–15 years	47	30.3
More than 15 years	43	27.7
Number of employee		
5–19 employees	83	53.5
More than 19 employees	72	46.5
Type of business		
Bakery and coffee shop	59	38.1
Restaurant	63	40,6
Lodging accomodation	33	21.3

11.4.2 Factor Analysis

This section presents an EFA performed on the available data using PCA with varimax rotation to extract the hypothesized components. To determine if a factor loading value is high, it must be above 0.4 (Hair et al. 2010). The results of the factor loading that passed the rotation process using the varimax rotation approach are presented in Table 11.2.

To determine the suitability of the EFA method, the Kaiser–Meyer–Olkin MSA value was utilized. If the KMO MSA value exceeds 0.5, EFA can be executed, while values below 0.5 indicate that EFA cannot be performed. The KMO MSA value in this study is 0.809, which is deemed "meritorious." Additionally, Bartlett's test of sphericity yielded a significant result of 0.000, indicating that the matrix was factorable and that the conditions for the factor analysis were met, as presented in Table 11.3.

An eigenvalue-based method was employed to determine the appropriate number of factors to be included in the study. The analysis resulted in the extraction of four factors based on the criterion that the eigenvalues should be greater than one, as presented in Table 11.4. The eigenvalues of the four factors were 6.039,

Table 11.2 Factor loadings rotated factor matrix

Item	Factor 1	Factor 2	Factor 3	Factor 4
Market orientation				
MO-1	0.586			
MO-2	0.637			
MO-3	0.612			
MO-4	0.600			
MO-5	0.549			
MO-6	0.615			
MO-7	0.554			
MO-8	0.561			
MO-9	0.579			
MO-10	0.586			
Product innovation capability				
PIC-1		0.720		
PIC-2		0.605		
PIC-3		0.528		
PIC-4		0.549		
PIC-5		0.504		
PIC-6		0.593		
PIC-7		0.589		
PIC-8		0.662		
Marketing performance				
MP-1			0.639	
MP-2			0.534	
MP-3			0.520	
MP-4			0.569	
Competitive advantage				
CA-1				0.611
CA-2				0.511
CA-3				0.498
CA-4				0.464
CA-5				0.490
CA-6				0.498
CA-7				0.597
CA-8				0.676
CA-9				0.595
CA-10				0.668

Table 11.3 KMO test and Bartlett's test

KMO and Bartlett's test		
KMO measure of sampling adequacy		0.809
Bartlett's test of sphericity	Approximate χ^2	1504.061
	df	496
	Significance	0.000

Table 11.4 Total variance explained

Factor	Initial eigenvalues			Extraction sums of square loadings			Rotation sums of square loadings		
	Total	Var	%	Total	Var	%	Total	Var	%
1	6.039	18.870	18.870	5.411	16.908	16.908	3.749	11.714	11.714
2	3.780	11.813	30.683	3.173	9.917	26.825	3.525	11.015	22.730
3	2.597	8.114	38.797	1.970	6.157	32.982	3.017	9.428	32.157
4	1.909	5.965	44.762	1.273	3.979	36.962	1.537	4.804	36.962

Note Extraction Method: Maximum Likelihood

3.780, 2.597, and 1.909, respectively, all exceeding one, which implies that the four factors can be further examined.

11.4.3 Hypotheses Testing

A regression analytic technique based on a Bayesian approach was utilized to first investigate the linkage among constructs and, second, the difference between the small and medium-scale enterprises in terms of MO, PIC, MP, and CA (Table 11.5).

Table 11.5 Bayesian regression

Bayesian estimates of coefficients					
Parameter	Posterior			95% credible interval	
	Mode	Mean	Variance	LB	UB
(Intercept)	1.135	1.135	0.181	0.299	1.971
MO	0.318	0.318	0.005	0.178	0.457
PIC	0.206	0.206	0.005	0.066	0.346
MP	0.183	0.183	0.004	0.054	0.312

Note Dependent variable: CA; assume standard reference priors. LB (Lower Bound); UB (Upper Bound)

According to the results of the regression analysis in the empirical model, a positive relationship exists between MO and CA. The regression coefficient for this relationship was 0.318 and was statistically significant at a credible interval ranging from 0.178 to 0.457. Hence, the first hypothesis (H1) was accepted.

Similarly, the analysis finds that PIC has a direct positive impact on CA. The regression coefficient for this relationship is 0.206 and the credible interval is between 0.066 and 0.346, which does not include 0. Therefore, the second hypothesis (H2) is accepted.

Additionally, the results show a positive relationship between MP and CA. The regression coefficient for this relationship was 0.183, and its significance was confirmed by a credible interval ranging from 0.054 to 0.312. Thus, the third hypothesis (H3) is supported. Please refer to Table 11.6 for further detail.

In addition, an independent sample t-test was used to examine the hypotheses regarding the distinctions between small and medium enterprises in terms of MO, PIC, MP, and CA.

The empirical findings of this study suggest that small-scale enterprises have a significantly higher level of MO (3.7867) than medium-scale enterprises (3.5500), with a *p*-value of 0.039. This indicates that there is a difference in MO between small and medium enterprises, supporting H4. However, the average PIC of medium-scale enterprises (3.8594) was found to be significantly higher than that of small-scale enterprises (3.6160), with a *p*-value of 0.029, indicating that SMEs

Table 11.6 Hypotheses result

Hypotheses	Regression coefficient	95% credible interval		Result
		LB	UB	
MO → CA	0.318	0.178	0.457	Supported
PIC → CA	0.206	0.066	0.346	Supported
MP → CA	0.183	0.054	0.312	Supported

Note There is no null value in the 95% credible interval (lower and upper bounds)

Table 11.7 Independent sample t-test

Factor	Firm scale	N	Mean	SD	*p*-value	Result
MO	Small	83	3.7867	0.56585	0.039**	Supported
	Medium	72	3.5500	0.80596		
PIC	Small	83	3.6160	0.75728	0.029**	Supported
	Medium	72	3.8594	0.58703		
MP	Small	83	3.5663	0.76816	0.001***	Supported
	Medium	72	3.9479	0.68810		
CA	Small	83	3.6554	0.66429	0.046**	Supported
	Medium	72	3.8722	0.67786		

Note(s) ($p < 0.10$)*; ($p < 0.05$)**; ($p < 0.01$)***

have distinct PIC, supporting H5. Moreover, the study found that medium-scale enterprises have significantly better MP (3.9479) than small-scale enterprises (3.5663), with a *p*-value of 0.001, suggesting differences in MP between small and medium enterprises, supporting H6. Finally, with a *p*-value of 0.046, the study concluded that medium-scale enterprises have a significantly higher level of CA (3.8722) than small-scale enterprises (3.6554), indicating that there is a difference in CA of SMEs, supporting H7.

11.5 Discussion and Conclusion

This study focuses on how small and medium-scale enterprises in developing countries such as Indonesia can gain CA by implementing MO, PIC, and MP. First, this study investigates the relationship between MO and CA among SMEs. The results show that MO has a significant impact on CA, as demonstrated by the credible interval, which does not include the null value (with a lower bound of 0.178 and an upper bound of 0.457). This implies that MO helps SMEs build strong customer relationships and leverage their strengths and opportunities to effectively respond to competitive challenges.

MO has the potential to improve SMEs' ability to offer valuable products that meet customer expectations and preferences by adding new attributes or by creating innovative products. Moreover, the ability to gather customer information and develop strategic plans based on this information to meet needs and expectations is another significant advantage of MO, leading to increased competition. This finding aligns with earlier research indicating that market-oriented firms can enhance their competitiveness (Barney 2014a, b; Day 2014). However, there are differences between small and medium-scale enterprises in terms of MO. On an average, small-scale enterprises are more market-oriented than medium-scale enterprises because they have a better understanding of their customers, including managing and promoting their products to increase customer value.

The second research question examines the impact of PIC on CA. The empirical findings show a significant relationship between PIC and CA among SMEs, with a credible interval that does not include a null value (a lower bound of 0.066 and an upper bound of 0.346), thus supporting H2. This indicates that SMEs can gain CA by developing innovative products that meet their customers' needs and expectations. This finding is consistent with previous research (e.g., Lukovszki et al. 2021; Pham et al. 2017; Srivastava et al. 2013), which suggests that firms can improve their competitive positions by enhancing their PICs.

Furthermore, the empirical data also demonstrate differences in PIC between small and medium enterprises, with medium-scale enterprises having, on average, greater PIC. This difference can be attributed to the resources available to each type of firm, as medium enterprises are better equipped to use tangible and intangible resources to create competitive and innovative products.

The third research question examined the relationship between MP and CA in SMEs. The results indicate that MP has a significant impact on CA with a credible interval that does not include the null value (a lower bound of 0.054 and an upper bound of 0.312), thus supporting H3. MP focuses on the results achieved, rather than on the processes and systems used. This study found that marketing outcomes are linked to CA.

Additionally, the empirical data reveal differences in MP between small and medium- enterprises. Medium-scale enterprises tend to have greater MP, on average, than small enterprises. This could be because smaller enterprises face more challenges such as time and financial constraints and may have limited strategic planning and marketing expertise. The study also finds that medium enterprises have a higher average CA than small enterprises, which could be attributed to their ability to deal with these challenges better and have stronger market expertise.

11.5.1 Implications

The present study is based on relevant prior research and aims to establish a research framework by examining the influence of MO, PIC, MP, and CA on SMEs. The authors conclude that further research is required in SMEs, particularly in three categories of businesses: coffee shops, restaurants, and lodging accommodation. Consequently, this study's significant contribution is to empirically test the differences between small and medium enterprises in terms of their MO, PIC, and MP in relation to CA.

This study's findings have several practical implications for business owners and managers. They can effectively utilize research outcomes to address SME-related business challenges. Businesses should also consider sustainable development goals to take a new direction in business by collaborating openly and enhancing their ability to integrate internal and external resources. Furthermore, business owners, managers, and stakeholders should prioritize their goals in a more market-oriented way, as this approach can generate substantial value and improve their survival prospects. Thus, businesses should focus on identifying lucrative niche market opportunities by strengthening their PICs.

In an ever-changing business landscape, SMEs must exhibit agility and adaptability to attain optimal MP and achieve their goals. Business owners and managers can leverage their unique skills and proximity to customers to gain CA. It is imperative that they comprehend the process by which MO, PICs, and MP can be developed simultaneously, as this will significantly contribute to CA.

However, MO alone is insufficient for enhancing competitiveness. Other capabilities, such as the ability to develop innovative products, can complement and contribute to SMEs' competitiveness. Thus, business owners and managers must regularly integrate internal and external resources to enhance production efficiency, create new processes, and commercialize new products. SMEs can achieve optimal MP and increase their competitiveness by creating new products or features and developing innovative products.

11.5.2 Limitations

There are several limitations to this study that require attention in future research. First, it only examined the relationship between MO, PIC, MP, and CA among SMEs in Indonesia. Therefore, future research should investigate other factors that may influence SME competitiveness. Second, the study was limited to three business categories and one country; therefore, future research should include a variety of industries and countries to ensure that the findings are generalizable. Third, this study relied on subjective assessments of MP by SME owners and managers. Future research should incorporate objective measures of MP. Finally, the study's data were cross-sectional, and the long-term effects of MO, PIC, and MP on SME competitiveness are unknown. More research is needed to explore the longitudinal effects of these constructs on CA among SMEs.

References

Afriyie S, Du J, Ibn Musah AA (2019) Innovation and marketing performance of SME in an emerging economy: the moderating effect of transformational leadership". J Glob Entrep Res 9(1):1–25

Akter S, Gunasekaran A, Wamba SF, Babu MM, Hani U (2020) Reshaping competitive advantages with analytics capabilities in service systems". Technol Forecast Soc Chang 159:120180

Al Mamun A, Fazal SA, Mohiuddin M, Su Z (2019) Strategic orientations, the mediating effect of absorptive capacity and innovation: a study among Malaysian manufacturing SMEs". Int J Innov Manag 23(2):1–25

Alhakimi W, Mahmoud M (2020) The impact of MO on innovativeness: evidence from Yemeni SMEs. Asia Pacific Journal of Innovation and Entrepreneurship 14(1):47–59

Ali GA, Hilman H, Gorondutse AH (2020) Effect of entrepreneurial orientation, market orientation and total quality management on performance. Benchmarking: Int J 27(4):1503–1531

Alnawas I, Hemsley-Brown J (2019) Market orientation and hotel performance: investigating the role of high-order marketing capabilities. Int J Contemp Hosp Manag 31(4):1885–1905

Anderson BS and Eshima Y (2011) The influence of firm age and intangible resources on the relationship between entrepreneurial orientation and firm growth among Japanese SMEs. J Bus Ventur 28(3):413–429

Barbero JL, Casillas JC, Feldman HD (2011) Managerial capabilities and paths to growth as determinants of high-growth small and medium-sized enterprises. Int Small Bus J 29(6):671–694

Barney J (1991) Firm resources and sustained competitive advantage. J Manag 17(1):99–120

Barney JB (2001a) Is the resource-based 'view' a useful perspective for strategic management research? Yes. Acad Manag Rev 26(1):41–56

Barney JB (2001b) Resource-based theories of competitive advantage: a ten-year retrospective on the resource-based view. J Manag 27(6):643–650

Barney JB (2014a) How marketing scholars might help address issues in resource-based theory. J Acad Mark Sci 42(1):24–26

Barney JB (2014b) Gaining and sustaining competitive advantage. Pearson Higher Ed

Bowonder B, Dambal A, Kumar S, Shirodkar A (2010) Innovation strategies for creating competitive advantage. Res Technol Manag 53(3):19–32

Buli BM (2017) Entrepreneurial orientation, market orientation and performance of SMEs in the manufacturing industry: evidence from Ethiopian enterprises. Manag Res Rev 40(3):292–309

Carson D, Cromie S, McGowan P, Hill J (1995) Marketing and Entrepreneurship in SMEs. Prentice-Hall, London

Chakravarty A, Kumar A, Grewal R (2014) Customer orientation structure for internet-based business-to-business platform firms. J Mark 78(5):1–23
Clark BH, Abela AV, Ambler T (2005) Organizational motivation, opportunity and ability to measure marketing performance. J Strateg Mark 13(4):241–259
Correia RJ, Dias JG, Teixeira MS (2021) Dynamic capabilities and competitive advantages as mediator variables between market orientation and business performance. J Strateg Manag 14(2):187–206
Day GS (2014) An outside-in approach to resource-based theory. J Acad Mark Sci 42(1):27–28
De Toni D, Reche RA, Milan GS (2021) Effects of market orientation, innovation strategies and value-based pricing on market performance. Int J Prod Perform Manag, Vol. ahead-of-print No. ahead-of-print
Deshpandé R, Farley JU (1998) Measuring market orientation: generalization and synthesis. J Mark-Focus Manag 2(3):213–232
D'souza C, Nanere M, Marimuthu M, Arwani M, Nguyen N (2021) Market orientation, performance and the mediating role of innovation in Indonesian SMEs. Asia Pac J Mark Logist. Vol. ahead-of-print No. ahead-of-print
Elia S, Giuffrida M, Mariani MM, Bresciani S (2021) Resources and digital export: an RBV perspective on the role of digital technologies and capabilities in cross-border e-commerce. J Bus Res 132:158–169
Eusebio R, Andreu JL, Belbeze MPL (2006) 'Measure of marketing performance; a comparative study from Spain.' Int J Contemp Hosp Manag 18(2):145–155
Falahat M, Ramayah T, Soto-Acosta P, Lee YY (2020) SMEs internationalization: the role of product innovation, market intelligence, pricing and marketing communication capabilities as drivers of SMEs' international performance. Technol Forecast Soc Chang 152:119908
Fang SR, Chang E, Ou CC, Chou CH (2014) Internal market orientation, market capabilities and learning orientation. Eur J Mark 48(1/2):170–192
Fernandes CI, Ferreira JJ, Lobo CA, Raposo M (2020) The impact of market orientation on the internationalisation of SMEs. Rev Int Bus Strat 30(1):123–143
Ferreira J, Coelho A (2020) Dynamic capabilities, innovation and branding capabilities and their impact on competitive advantage and SME's performance in Portugal: the moderating effects of entrepreneurial orientation. International Journal of Innovation Science 12(3):255–286
Gebhardt GF (2012) Marketing orientation. In: Shankar V, Carpenter GS (eds) Handbook of Marketing Strategy. Edward Elgar Publishing, Cheltenham, pp 28–38
Gilmore A, McAuley A, Gallagher D, Massiera P, Gamble J (2013) Researching sme entrepreneurial research: a study of journal of research in marketing and entrepreneurship (jrme) 2000–2011. J Res Mark Entrep 15(2):87–100
Gotteland D, Shock J, Sarin S (2020) Strategic orientations, marketing proactivity and firm market performance. Ind Mark Manag 91:610–620
Grewal D, Iyer G, Kamakura W, Mehrotra A, Sharma A (2009) Evaluation of marketing subsidiary performance: combining process and outcome performance metrics. J Acad Mark Sci 37(2)
Gruber M, Heinemann F, Brettel M, et al (2010) Configurations of resources and capabilities and their performance implications: An exploratory study on technology ventures. Strat Manag J 30(12):1337–1356
Gruber-Muecke T, Hofer KM (2015) Market orientation, entrepreneurial orientation and performance in emerging markets. Int J Emerg Mark 10(3):560–571
Hair JF, Ortinau DJ, Harrison DE (2010) Essentials of marketing research, vol 2. McGraw-Hill/Irwin, New York, NY
Hansen B, Hamilton RT (2011) Factors distinguishing small firm growers and non-growers. Int Small Bus J 29(3):278–294
Hooley G, Cox T, Fahy J, Shipley D, Beracs J, Fonfara K, Snoj B (2000) Market orientation in the transition economies of central Europe: tests of the narver and slater market orientation scales. J Bus Res 50(3):273–285
Hutcheson GD, Sofroniou N (1999) The multivariate social scientist: introductory statistics using generalized linear models. Sage.

Iyer P, Davari A, Zolfagharian M, Paswan A (2019) Market orientation, positioning strategy and brand performance. Ind Mark Manag 81:16–29
Jalilvand MR (2017) The effect of innovativeness and customer-oriented systems on performance in the hotel industry of Iran. Journal of Science and Technology Policy Management 8(1):43–61
Jansson J, Nilsson J, Modig F, Hed Vall G (2017) Commitment to sustainability in small and medium-sized enterprises: The influence of strategic orientations and management values. Bus Strat Environ 26(1):69–83
Jaworski BJ, Kohli AK (1993) Market orientation: antecedents and consequences. J Mark 57(3):53–70
Jogaratnam G (2017) The effect of market orientation, entrepreneurial orientation and human capital on positional advantage: evidence from the restaurant industry. Int J Hosp Manag 60:104–113
Kahn KB (2018) Understanding innovation. Bus Horiz 61(3):453–460
Kaleka A, Morgan NA (2017) Which competitive advantage(s)? Competitive advantage–market performance relationships in international markets. J Int Mark 25(4):25–49
Keskin H (2006) Market orientation, learning orientation, and innovation capabilities in SMEs: An extended model. Eur J Innov Manag 9(4):396–417
Kirca AH, Jayachandran S, Bearden WO (2005) Market orientation: a meta-analytic review and assessment of its antecedents and impact on performance. J Mark 69(2):24–41
Kohli AK, Jaworski BJ (1990) Market orientation: the construct, research propositions, and managerial implications. J Mark 54:1–18
Lafuente E, Szerb L, Rideg A (2020) A system dynamics approach for assessing SMEs' competitiveness. J Small Bus Enterp Dev 27(4):555–578
Lamprinopoulou C, Tregear A (2011) Inter-firm relations in SME clusters and the link to marketing performance. J Bus Ind Mark 26(6):421–429
Lamberti L, Noci G (2010) Marketing strategy and marketing performance measurement system: Exploring the relationship. Eur Manag J 28(2):139–152
Lawson B, Samson D (2001) Developing innovation capability in organisations: a dynamic capabilities approach. Int J Innov Manag 5(03):377–400
Lukovszki L, Rideg A, Sipos N (2021) Resource-based view of innovation activity in SMEs: an empirical analysis based on the global competitiveness project. Compet Rev 31(3):513–541
Mafimisebi OP, Obembe D, Aluko O (2020) Organization and product design pairings: a review of product innovation capabilities, conceptualization, and future directions. Strat Chang 29(1):13–24
Malhotra N, Birks D (2006) Marketing research: An applied perspective. Prentice Hall, Harlow
Montiel-Campos H (2018) Entrepreneurial orientation and market orientation: systematic literature review and future research. J Res Mark Entrep 20(2):292–322
Morgan NA, Clark BH, Gooner R (2002) Marketing productivity, marketing audits, and systems for marketing performance assessment: integrating multiple perspectives. J Bus Res 55(5):363–375
Müller JM, Buliga O, Voigt KI (2021) The role of absorptive capacity and innovation strategy in the design of industry 4.0 business Models-A comparison between SMEs and large enterprises. Eur Manag J 39(3):333–343
Najafi-Tavani S, Najafi-Tavani Z, Naudé P, Oghazi P, Zeynaloo E (2018) How collaborative innovation networks affect new product performance: Product innovation capability, process innovation capability, and absorptive capacity. Ind Mark Manag 73:193–205
Narver J, Slater S (1990) The effect of a market orientation on business profitability. J Mark 54:20–35
Nwokah NG (2009) Customer focus, competitor focus and marketing performance. Meas Bus Excell 13(3):20–28
O'Cass A, Sok P (2014) The role of intellectual resources, product innovation capability, reputational resources and marketing capability combinations in firm growth. Int Small Bus J 32(8):996–1018

O'Sullivan D, Abela A, Hutchinson M (2009) Marketing performance measurement and firm performance: evidence from the European high-technology sector. Eur J Mark 43(5/6):843–862
O'Donnell A, Gilmore A, Carson D, Cummins D (2002) Competitive advantage in small to medium-sized enterprises. J Strateg Mark 10(3):205–223
OECD (2017) Enhancing the contributions of SMEs in a global and digitalised economy. Paris
O'Sullivan D, Abela AV (2007) Marketing performance measurement ability and firm performance. J Mark 71(2):79–93
Pérez-Luño A, Saparito P, Gopalakrishnan S (2016) Small and medium-sized enterprise's entrepreneurial versus market orientation and the creation of tacit knowledge. J Small Bus Manage 54(1):262–278
Pham TSH, Monkhouse LL, Barnes BR (2017) The influence of relational capability and marketing capabilities on the export performance of emerging market firms. Int Mark Rev 34(5):606–628
Pimenta da Gama A (2011) An expanded model of marketing performance. Mark Intell Plan 29(7):643–661
Rajapathirana RJ, Hui Y (2018) Relationship between innovation capability, innovation type, and firm performance. J Innov Knowl 3(1):44–55
Raju PS, Lonial SC, Crum MD (2011) Market orientation in the context of SMEs: a conceptual framework. J Bus Res 64(12):1320–1326
Sahi GK, Gupta MC, Lonial SC (2018) Relating strategic market orientation and market performance: Role of customer value types. J Strat Mark 26(4):318–338
Salavou H, Avlonitis G (2008) Product innovativeness and performance: a focus on SMEs. Manag Decis 46(7):969–985
Sanzo MJ, Santos ML, Garcia N, et al (2011) Trust as a moderator of the relationship between organizational learning and marketing capabilities: Evidence from Spanish SMEs. Int Small Bus J 30(6):700–726
Saunila M (2020) Innovation capability in SMEs: a systematic review of the literature. J Innov Knowl 5(4):260–265
Saunila M, Ukko J (2014) Intangible aspects of innovation capability in SMEs: impacts of size and industry. J Eng Technol Manag 33:32–46
Schilke O, Hu S, Helfat CE (2018) Quo vadis, dynamic capabilities? A content-analytic review of the current state of knowledge and recommendations for future research. Acad Manag Ann 12(1):390–439
Shehu AM, Mahmood R (2014) The relationship between market orientation and business performance of Nigerian SMEs: the role of organizational culture. Int J Bus Soc Sci 9(1):159–168
Sigalas C, Pekka Economou V (2013) Revisiting the concept of competitive advantage: problems and fallacies arising from its conceptualization. J Strateg Manag 6(1):61–80
Sigalas C, Pekka Economou V, Georgopoulos BN (2013) Developing a measure of competitive advantage. J Strateg Manag 6(4):320–342
Singh S, Akbani I, Dhir S (2020) Service innovation implementation: a systematic review and research agenda. Serv Ind J 40(7/8):491–517
Singh S, Paul J, Dhir S (2021) Innovation implementation in Asia-Pacific countries: a review and research agenda. Asia Pac Bus Rev 27(2):180–208
Sirmon DG, Hitt MA, Ireland RD (2007) Managing firm resources in dynamic environments to create value: looking inside the black box. Acad Manag Rev 32(1):273–292
Slater FS, Narver JC (1995) Market orientation and the learning organization. J Mark 59:63–74
Slater SF, Narver JC (1998) Customer-led and market-oriented: Let's not confuse the two. Strat Manag J 19:1001–1006
Slater SF, Narver JC (2000) The positive effect of a market orientation on business profitability: A balanced replication. J Bus Res 48(1):69–73
Srivastava M, Franklin A, Martinette L (2013) Building a sustainable competitive advantage. J Technol Manag Innov 8(2):47–60
Storey DJ and Greene FJ (2010) Small Business and Entrepreneurship. Harlow: Pearson Education

Taques FH, López MG, Basso LF, Areal N (2021) Indicators used to measure service innovation and manufacturing innovation. J Innov Knowl 6(1):11–26

Teece DJ (2014) The foundations of enterprise performance: dynamic and ordinary capabilities in an (economic) theory of firms. Acad Manag Perspect 28(4):328–352

Teece DJ, Pisano G, Shuen A (1997) Dynamic capabilities and strategic management. Strateg Manag J 18(7):509–533

Vega-Jurado J, Gutierrez-Gracia A, Fernandez-de-Lucio I (2009) Does external knowledge sourcing matter for innovation? Evidence from the Spanish manufacturing industry. Ind Corp Chang 18(4):637–670

Wolf P, Kaudela-Baum S, Meissner JO (2012) Exploring innovating cultures in small and medium-sized enterprises: Findings from Central Switzerland. Int Small Bus J 30(3):242–274

Yeh Y-P (2016) Market orientation and service innovation on customer perceived value: The case of supermarket retailers. Manag Res Rev 39(4):449–467

Zaefarian G, Forkmann S, Mitręga M, Henneberg SC (2017) A capability perspective on relationship ending and its impact on product innovation success and firm performance. Long Range Plan 50(2):184–199

Zhang M, Hartley JL (2018) Guanxi, IT systems, and innovation capability: The moderating role of proactiveness. J Bus Res 90:75–86

Zheng Y, Liu J, George G (2010) The dynamic impact of innovative capability and inter-firm network on firm valuation: A longitudinal study of biotechnology start-ups. J Bus Ventur 25(6):593–609

A Review on Innovation Audits

12

Rana Fathinah Ananda, Sari Nuzullina Rahmadhani,
Fauziah Rahman, and Sucitra Dewi

Abstract

This chapter aims to elaborate and discuss on audit and innovation through research database (Scopus). The study adopts content analysis technique where various sources are referred including reports, official documents and research articles. The chapter also explains on the auditor innovations during a pandemic as well as the artificial intelligence based audit integration. The research on audit and innovation has been done and became the concern by researchers around the globe particularly from the developed countries since 1973. Other opportunities and challenges on audit innovation are revealed and should be known by practitioners, academicians and other stakeholders, which to be used as valuable information for strategic planning and initiatives.

Keywords

Auditing • Audit • Small business • Content analysis

12.1 Introduction

The internal audit function has gradually adopted the latest developments in technology and methodology, and has directly contributed to expanding its services in areas such as assessing cultural values and ways of handling talent which can provide greater benefits to an enterprise.

Basically, the internal audit function around the world varies, depending on what is the main focus of a company in a country. As much as possible the internal audit department continues to innovate to provide added value to the company. Of

R. F. Ananda (✉) · S. N. Rahmadhani · F. Rahman · S. Dewi
Faculty of Economics and Business, Universitas Medan Area, Medan, Indonesia
e-mail: rana@staff.uma.ac.id

A. Rafiki et al. (eds.), *Open Innovation in Small Business*, Contributions to Environmental Sciences & Innovative Business Technology,
https://doi.org/10.1007/978-981-99-5142-0_12

course, the company in this case continues to follow the various changes that occur and strives to understand and apply the best approaches, such as using the latest and most qualified technologies. In addition to using more various analytics and other types of innovation, the head of internal audit or better known as the Chief Audit Executive (CAE) prioritizes their main strategy by continuing to strengthen what is the strength of the internal audit department itself and establishing close relationships with various departments which of course become their partners.

Innovations carried out by the internal audit department are associated with stronger impacts and influences in general, surveys show that internal audit groups that have adopted innovative approaches tend to have greater influence than those that have not taken innovative actions. Manufacturing companies' approaches for managing innovation are changing as a result of openness, servitization, and digitalization. Managers that ignore these patterns run the danger of making two different mistakes. The first kind of error is failing to take advantage of chances. The competitive positions of managers' companies may be compromised if they are unable to adapt to changes in their environment as a result of openness, servitization, and digitalization. The second mistake is underestimating the difficulties. In other words, managers could change their innovation management technique hastily and without thoroughly evaluating the difficulties they entail. In any case, failing to evaluate innovation in light of these trends may result in a loss of competitive advantage (Frishammar et al. 2019; Verhaeghe and Kfir 2002).

The events of the last two years that have had an impact on the global economy due to the pandemic have had a significant impact on the adaptation of the business world. Most businesses rely on innovation and transformation in the face of the turbulence of a pandemic to survive and compete. This change then needs to be adopted by internal audit in terms of the innovation and transformation mindset.

12.2 Literature Review

Innovations carried out by the internal audit department are associated with stronger impacts and influences in general (Cherney et al. 2023; Hallgren 2009). Some surveys show that internal audit groups that have adopted innovative approaches tend to have greater influence than those that have not taken innovative actions.

Every internal audit department can innovate, and innovations can be initiated in small ways. For example, the internal audit team could start by identifying inefficiencies in its processes and activities. This can generate various ideas for the use of technology, various activities carried out, such as data extraction training, to be able to reduce these inefficiencies. While some innovation involves technology and larger budgets to be sure, the most obvious innovation required in internal auditing usually involves a change in mindset. The desired mindset is how existing teams can be more proactive, forward-looking, more engaged with stakeholders, more focused on delivering business value and insights rather than the traditional internal audit mindset.

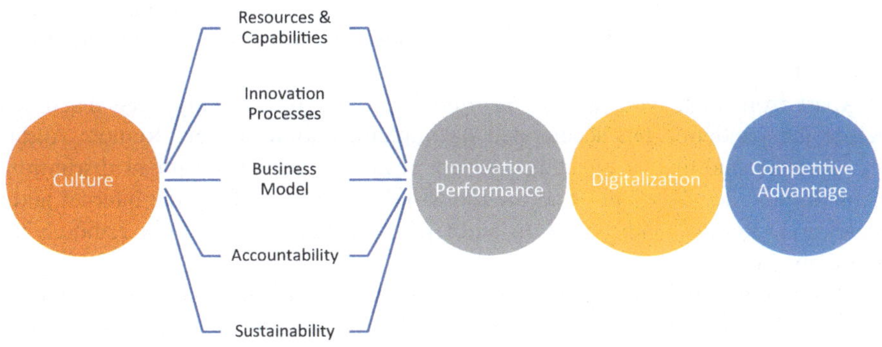

Fig. 12.1 Key dimension of prior innovation audit frameworks. *Source* Authors' adapted model from Frishammar et al. (2019)

An analytical methodology called an "innovation audit" enables comparing a company's present levels of innovation to previous or desired performance levels (Burgelman, 2009). Both empirical and theoretical research has been done on innovation auditing (e.g. Björkdahl and Holmén 2016; Chiesa et al. 1996; Cormican and O'Sullivan 2004; Radnor and Noke 2002; Frishammar et al. 2019) (Fig. 12.1).

Managers can find innovation's advantages and disadvantages by using audits of innovation. Also, an innovation audit enables businesses to develop and maintain a competitive edge through the development of innovative capabilities (Björkdahl and Holmén 2016). A future-focused innovation audit goes beyond simple quantitative measurement. It focuses on performance gaps between present and desired performance, combines quantitative data with qualitative insights, and enables managers to create action plans (Chiesa et al. 1996). The majority of currently used frameworks are the result of technology advancement or the creation of new products (Hallgren 2009). Yet, recent audit frameworks recognize that innovation goes beyond simply creating new physical objects (Rao and Weintraub 2013; Richtnér et al. 2017).

12.2.1 Auditor Innovations During a Pandemic

In a quick survey conducted of auditors, when asked what auditors were doing in response to the pandemic, more than half of the auditors answered that they had stopped the scope of the audit assignment being carried out. Some even only continued the audit assignments requested by the regulators. It didn't stop there, apart from stopping the ongoing audit, several respondents also canceled the planned audit assignment.

Even though there were audit assignments that were cancelled, some respondents instead added new assignments, especially those related to Covid-19 or new risks that emerged. In line with this spirit, there were also respondents who expanded or added to the scope of several relevant assignments, such as becoming advisors. Innovation for agile or adapting to the conditions of this pandemic will make auditors still needed by companies (Gaosong and Leping 2021; Geng 2019; Tysiac 2017).

Apart from that, there is also an innovation that has actually been discussed before the pandemic but accelerated its implementation, namely Remote Auditing. Remote auditing is an audit concept resulting from widespread disruption, technological developments, and the industrial revolution 4.0 by the internal audit profession. Remote auditing is an audit process that is carried out remotely or indirectly (remote).

This remote auditing is very appropriate to be held in a pandemic situation that does not encourage crowds. Moreover, many corporate systems have adopted information technology, such as using ERP (enterprise resource planning) where all data is integrated in digital form. ERP systems are software packages that combine different parts of an organization's information system using relational database technology. For any business, ERP systems offer a variety of independent yet interconnected modules that can be installed as a bundle (Scapens and Jazayeri 1998). With the presence of this system, the auditor should be able to understand the state of the company without having to go directly to the field (Barr-Pulliam et al. 2022; Frishammar et al. 2019; Humphrey et al. 2021).

The bulk of significant clients that audit companies serve seem to employ ERP systems. As a result, although audit firm clients utilize a range of systems to handle accounting transactions, ERP systems are the dominant system environment for auditors serving public clients (Brazel 2005). The change brought on by the ERP system must be effectively handled by auditing professionals, according to Yen et al. (2018) and Coppers and Lybrand (2002).

Throughout the past ten years, there have been major changes to the conventional audit approach. Due to challenges from the market, such as price competition, market saturation, and increasing training and technology, the financial audit's focus has changed (Eilifsen et al. 2001). The audit has a risk management focus, and information technology specialists are increasingly being included on audit engagement teams (Winograd et al. 2000). Moreover, the field has begun to offer more systems assurance services. For instance, assurance services have undergone a paradigm change that places more emphasis on internal control systems throughout the life cycle of an information system (Arnold et al. 2000; O'Donnell and Schultz 2005).

The remote auditing stage is actually the same as the conventional auditing stage, namely the initial meeting, document analysis, observation, interview, and closing audit meeting. All stages in the implementation of the audit can be done with remote auditing. For example, initial meetings can be held using various meeting platforms in the network, as has been done by many groups.

Even to make observations, auditors do not have to go directly to the field. Auditors can ask the company to tour the company's location, live-stream and even use CCTV footage for several days which is controlled centrally to test the production implementation in the company. The interview stage can be done by video conferencing in order to capture the client's non-verbal movements.

These innovations such as agility and remote auditing are just a few, of course there are many other innovations that must be cultivated by inspection service industry players such as auditors. This has been accelerated by the presence of the pandemic and digitalization, so auditors should play a strategic role in assisting entity leaders in making the right decisions.

12.2.2 Artificial Intelligence Based Audit Integration

One of the objectives of the Technology Audit is to improve (improvement) and not to see what is right. One of the tools for assessing is by benchmarking similar businesses or state of the art from similar businesses both locally and globally. Benchmarks can refer to technical standards or industry standards or national standards or from reference standards (Abdel-Razek and Alsanad 2014; Curtis et al. 2016). If there is a gap, then to what extent can this gap be managed to increase technological capability.

Technology audit is a systematic activity that aims to match or compare technology assets (humanware, infoware, technoware, and orgaware) with predetermined criteria/standards. In terms of time, technology assessment is carried out before the technology is implemented, while technology audits can be carried out before and after the technology is implemented, which aims to protect the public interest and avoid technological disasters.

Auditors in carrying out the stages of audit procedures require a long time, starting from preparing audit plans to presenting audit reports that provide opinions to increase assurance to users of the company's financial statements (Cormican and O'Sullivan, 2004; Chiesa et al. 1996). Every year, the auditor will mobilize more funds, time and energy to prepare a transparent and accountable audit process according to the client's background, the data obtained, to explore every aspect skeptically. The purpose of carrying out a financial audit is to collect findings supported by competent and reliable evidence in order to form an opinion that supports the quality of the audited financial statements. However, with the times, it is increasingly possible for auditors to move flexibly in carrying out audits to clients with the help of technology.

Today, the acceleration of innovation has delivered rapid growth in life that integrates technology-based human activities. Progress is very visible in the era of the industrial revolution 4.0 which presents sophisticated technology to support industrial activities to move faster so that they are able to be competitive and meet increasing market demands, where this affects auditors who are starting to synergize with the use of technology in order to increase effectiveness and efficiency in managing time and minimizing audit risk (Widuri et al. 2016; Chan and Vasarhelyi 2018).

This has been seen in the research of Munoko et al. (2020) that implementing AI in auditing can provide several advantages such as speeding up the data analysis process, increasing the accuracy and validation of big data, forming agile auditing, to guaranteeing service to clients through the quality of existing information data.

The Big 4 Public Accounting Firms (PwC, Deloitte, EY, and KPMG) have also implemented technology to assist reporting which improves audit planning on an ongoing basis through focused innovation in risk assessment, substantive testing, to preparing audit working papers. Investigating further, the application of AI also affects the code of ethics in auditing and companies gradually begin to integrate with the times (Munoko et al. 2020; Johne and Snelson 1988).

In the report by Bizarro et al. (2019) entitled "The Intelligent Audit" reveals that AI can be a game changer for auditors and how this can be improved through computer training of large data sources to recognize patterns and identify development trends as expected. For its application, it can be assisted by branches of Artificial Intelligence, namely Machine Learning and Natural Language Processing to assist auditors in tracing and evaluating all large data sets in a fast and efficient manner, by producing audit information in an accurate and accountable manner. Through the application of AI, it can increase the auditor's flexibility to explore insights and expand brilliant innovations that can be oriented to audit plans, where at this stage the auditor will never be replaced by technology, considering that human strengths are showing feelings and giving professional judgment (Bizarro et al. 2019; Kafetzopoulos and Skalkos 2019).

AI implementation can be adapted in detail in audit procedures which consist of audit planning and approach, intenal control, substantive tests, to the presentation of audit working papers which contain opinions and descriptions of audit reports. Following are attached some details developed by Bizarro et al. (2019) related to implementing and without implementing AI in the audit process.

AI can assist risk managers or internal auditors as a tool to make objective and more accurate estimates of emerging risks in calculating the subsequent impact on business, for example:

- Estimate when a business risk will result in a large loss, or the business will fail
- Predicting cyber-attacks based on monitoring messages and external chats, or identifying intentions to make fraudulent financial trades
- Determine when a project will fall behind schedule or substantially over budget, or when a new product will likely fail to deliver the promised benefits

In the short term, risk managers and auditors are required to have significant skill changes (Björkdahl and Holmén 2016). They must be proficient in data science, statistical modeling and technology and apply them in the context of risk. It is necessary to build the same enthusiasm to innovate by adopting AI. Risk managers and internal auditors can play a key role in helping organizations stay in control through the adoption of new technologies.

In conclusion, it can be seen that the role of technology is very helpful for the continuity of the organization in various aspects, if it is applied appropriately according to needs. Through technology, it can support the auditor's performance prospects to move more quickly and effectively, where one of the advantages of using technology can open new opportunities, such as knowing trend developments, finding anomalies or data discrepancies that have the potential for fraud, to transforming professions oriented to the realm of technology (Hull et al. 2000; Jeppesen 2007). As stated by Bizarro et al. (2019) that auditors will never be replaced, because even though AI has the ability to think for itself like humans, it does not have the capability to provide assessments and deliver communications to stakeholders.

12.3 Methods

This study adopts a content analysis with additional few items on bibliometric analysis. Bibliometric analysis manifests interconnections among the articles whereby the data was retrieved from Scopus, a reputable and comprehensive database. The primary reason for choosing Scopus is its availability, nearly 60 percent larger than the Web of Science (WoS) database (Zhao and Strotmann 2015).

12.4 Discussion

There are 2,148 documents sourced from reputable database called Scopus discussed on auditing and innovation which have been explored by researchers around the world. Below is the summary of elements related to research on auditing and innovation (Table 12.1).

It is also mentioned that, the highest hit on the number of documents is in the year 2019 which reaches to 152 documents. It is a long period of research history on auditing and innovation which starts in 1973. There is a dynamic process of auditing and innovation which could bring an impact to any accounting matters includes the financial statement. The financial institutions considered the innovation in auditing is crucial which safe costs, faster process and more accountable.

Meanwhile, there are three authors that actively produced documents on halal supply chain (Crabtree, B.F., Michie, S., Yam, R.C.M.). The highest one is with 6 documents. There are many researchers equally concern on auditing and innovation in various countries.

Moreover, the University of Manchester as an institution that actively conducted researches on auditing and innovation, followed by the University College London and the University of Toronto. Most of the institutions located in the Western countries which consider as developed countries. It means, those developed countries are more concern on the innovation in audit process.

Furthemore, United Kingdom known as a country that mostly engaged in research on audit/auditing and innovation which has 465 documents compare to other 9 countries. This followed by countries such as United States, Australia and China in the 2nd, 3rd and 4th place. Based on several articles, it is proven that the

Table 12.1 A Summary of top ten research output, affiliation and document types

No	Year/ number document	Author/number of documents	Country/number of documents	Affiliation/ number of documents	Document type/ number of documents
1	2022 (149)	Crabtree, B.F. (6)	United Kingdom (456)	University of Manchester (20)	Article (1366)
2	2021 (152)	Michie, S. (6)	United States (391)	University College London (18)	Conference paper (328)
3	2020 (132)	Yam, R.C.M. (5)	Australia (162)	University of Toronto (17)	Review (198)
4	2019 (117)	Grimshaw, J.M. (4)	China (145)	University of Sheffield (15)	Conference review (70)
5	2018 (98)	Hanson, K.L. (4)	Canada (93)	King's College London (15)	Book chapter (68)
6	2017 (95)	Jordan, A. (4)	India (90)	Monash University (15)	Note (25)
7	2016 (84)	Matos, F. (4)	Germany (55)	University of Queensland (14)	Book (21)
8	2015 (75)	Miller, W.L. (4)	Russian Federation (46)	University of Ottawa (13)	Short survey (18)
9	2014 (77)	Stange, K.C. (4)	Italy (45)	University of Sydney (12)	Editorial (15)
10	2013 (74)	Tang, E.P.Y. (4)	France (40)	University of Leicester (12)	Letter (7)

innovation audits became a potential topic to be explored as the need of accountability and good governance is increasing (Le Anh et al. 2021; Lhuillery et al. 2023).

12.5 Conclusion

Auditors always face new challenges and scenarios due to the rapidly changing nature of business, culture, and technology. Auditors must now deal with big clients who conduct millions of transactions each month as a result of globalization. The sampling approach and application of a materiality limit are utilized since the auditor cannot analyze that much data due to human limitations. However, this procedure still has numerous drawbacks when carried out manually.

In order to remain relevant with the times, auditors must begin to look at the risks associated with changes in technology, organizational reputation, and changes in business models due to technology. Being adaptive to change is one important factor. Prepare yourself by always updating your knowledge and improving your competence, and be open to collaboration with various parties.

The real-time economy has rendered the conventional audit paradigm obsolete. To provide real-time assurance, the traditional auditing procedure must be improved. Continuous auditing is being studied by academics and practitioners as a possible replacement for the conventional audit paradigm. Continuous auditing approach boosts the effectiveness and efficiency of the audit process to support real-time assurance using technology and automation.

References

Abdel-Razek R, Alsanad DS (2014) Auditing and comparing innovation management in organizations. Glob J Bus Res 8(2):49–56

Arnold V, Collier PA, Leech SA, Sutton SG (2000) The effect of experience and complexity on order and recency bias in decision making by professional accountants. Account Financ 40(2):109–134

Barr-Pulliam D, Brown-Liburd HL, Munoko I (2022) The effects of person-specific, task, and environmental factors on digital transformation and innovation in auditing: A review of the literature. J Int Financ Manag Acc 33(2):337–374

Bizarro W, Sucari W, Quispe-Coaquira A (2019) Evaluación formativa en el marco del enfoque por competencias. Rev Innova Educ 1(3):374–390

Björkdahl J, Holmén M (2016) Innovation audits by means of formulating problems. R&D Manag 46(5):842–856

Brazel JF (2005) A measure of perceived auditor ERP systems expertise. Dev, Assess Uses, Manag Audit J 20(6):619–631

Burgelman RA (2009) Combining grounded theorizing and historical methods: a proposal to strengthen the power of qualitative research. Stanford University, Graduate School of Business

Cherney OT, Smirnova ZV, Romanovskaya EV, Garin AP, Sidorenko YA (2023) The main components of the innovation audit of the economy. In Technological trends in the AI economy: international review and ways of adaptation. Springer Nature Singapore, Singapore, pp 111–117

Cormican K, O'Sullivan D (2004) Auditing best practice for effective product innovation management. Technovation 24(10):819–829

Coppers C, Lybrand LLP (2002) Security, audit and control features SAP R/3: a technical and risk management reference guide. IT Governance Institute, Illinois

Curtis E, Humphrey C, Turley WS (2016) Standards of innovation in auditing. Audit: J Pract Theory 35(3):75–98

Chiesa V, Coughlan P, Voss CA (1996) Development of a technical innovation audit. J Prod Innov Manag: Int Publ Prod Dev & Manag Assoc 13(2):105–136

Chan DY, Vasarhelyi MA (2018) Innovation and practice of continuous auditing. In: Chan DY, Chiu V, Vasarhelyi MA (eds) Continuous auditing (rutgers studies in accounting analytics), Bingley, pp 271–283. https://doi.org/10.1108/978-1-78743-413-420181013

Eilifsen A, Knechel WR, Wallage P (2001) Application of the business risk audit model: a field study. Account Horiz 15:193–207

Frishammar J, Richtnér A, Brattström A, Magnusson M, Björk J (2019) Opportunities and challenges in the new innovation landscape: implications for innovation auditing and innovation management. Eur Manag J 37(2):151–164

Gaosong Q, Leping Y (2021) Measurement of internal audit effectiveness: construction of index system and empirical analysis. Microprocess Microsyst, 104046

Geng X (2019) Technological innovation audit. In: The routledge companion to innovation management. Routledge, pp 516–538

Hallgren EW (2009) How to use an innovation audit as a learning tool: a case study of enhancing high-involvement innovation. Creat Innov Manag 18(1):48–58

Humphrey C, Sonnerfeldt A, Komori N, Curtis E (2021) Audit and the pursuit of dynamic repair. Eur Account Rev 30(3):445–471

Hull R, Coombs R, Peltu M (2000) Knowledge management practices for innovation: an audit tool for improvement. Int J Technol Manage 20(5–8):633–656

Jeppesen KK (2007) Organizational risk in large audit firms. Manag Audit J 22(6):590–603. https://doi.org/10.1108/02686900710759398

Johne A, Snelson P (1988) Auditing product innovation activities in manufacturing firms. R&D Manag 18(3):227–233

Kafetzopoulos D, Skalkos D (2019) An audit of innovation drivers: some empirical findings in Greek agri-food firms. Eur J Innov Manag 22(2):361–382

Le Anh T, Nguyen T, Tran L (2021) Relationships between innovation, its antecedents, and organisational performance: evidences from auditing service industry. Knowl Manag Res & Pract, 1–15

Lhuillery S, Tellechea M, Thiéry S (2023) Innovation in lieu of compliance: internal audit departments' standardized and non-standardized knowledge sources. Technovation 123:102720

Munoko I, Brown-Liburd HL, Vasarhelyi M (2020) The ethical implications of using artificial intelligence in auditing. J Bus Ethics 167:209–234

O'Donnell E, Schultz JJ Jr (2005) The halo effect in business risk audits: can strategic risk assessment bias auditor judgment about accounting details? Account Rev 80(3):921–939

Radnor ZJ, Noke H (2002) Innovation compass: a self–audit tool for the new product development process. Creat Innov Manag 11(2):122–132

Rao J, Weintraub J (2013). How innovative is your company's culture? MIT Sloan Manag Rev

Richtnér A, Brattström A, Frishammar J, Björk J, Magnusson M (2017) Creating better innovation measurement practices. MIT Sloan Manag Rev 59(1):45

Scapens R, Mostafa J (1998) SAP: Integrated information systems and the implications for management accountants. Management Accounting, London

Tysiac K (2017) How to enable audit innovation. J Account 223(4):33

Verhaeghe A, Kfir R (2002) Managing innovation in a knowledge intensive technology organisation (KITO). R&D Manag 32(5):409–417

Widuri R, O'Connell B, Yapa PWS (2016) Adopting generalized audit software: an Indonesian perspective. Manag Audit J 31(8/9):821–847. https://doi.org/10.1108/MAJ-10-2015-1247

Winograd BN, Gerson JS, Berlin BL (2000) Audit practices of pricewaterhousecoopers. Audit: J Pract Theory 19:175–182

Yen CC, Huang SM, Li CL, Hsiah YC (2006) Application. Influ Impact Sarbanes-Oxley Act, Comput Audit J 15:1–11

Yen JC, Lim JH, Wang T, Hsu C (2018) The impact of audit firms' characteristics on audit fees following information security breaches. J Account Public Policy 37(6):489–507

Zhao D, Strotmann A (2015) Analysis and visualization of citation networks. Synth Lect Inf Concepts, Retr, Serv 7(1):1–207

A Study of Muslim Women Entrepreneurs' SMEs Challenges and Motivation in the Asia Pacific Region

13

Ilhaamie Abdul Ghani Azmi, Rosmawani Che Hashim, Hasan Al-Banna Mohamed, and Norashikin Sahol Hamid

Abstract

Almost half of the population of those who are employed are women. Nonetheless, they encounter several issues and difficulties juggling their family and job obligations. In Asian Pacific nations, men are expected to be the primary caregiver for their families, and women must be their subservient. Being in charge of their own business is one way for women to have more freedom with their family time. As a result, they have a very high level of drive to work for their business. A mixed-method quantitative and qualitative technique is used to gather research data for this study. Using a quantitative research method, IBM SPSS is used to do a descriptive analysis of the survey data. Most respondents in Vietnam, Thailand, and Pakistan are inspired to start their businesses by personal, family, social, market, networking, and legal considerations. Parallel to this, a quantitative study in Vietnam and Thailand identifies similar factors like personal, family, social, political, and economic factors. In addition to these elements, Muslim women in these two neighbouring nations confront personal, economic, political, and governance obstacles.

Keywords

Women • Muslim • Motivation • Challenges • Entrepreneurship • Asia–Pacific

I. Abdul Ghani Azmi · R. Che Hashim
University of Malaya, Kuala Lumpur, Malaysia

H. A.-B. Mohamed (✉) · N. Sahol Hamid
National Defence University of Malaysia, Kuala Lumpur, Malaysia
e-mail: hasanalbanna@upnm.edu.my

A. Rafiki et al. (eds.), *Open Innovation in Small Business*, Contributions to Environmental Sciences & Innovative Business Technology,
https://doi.org/10.1007/978-981-99-5142-0_13

13.1 Introduction

13.1.1 Definition of Women Entrepreneurs, Motivation and Challenges

Female entrepreneurs actively manage their businesses, own at least 50% of them and have been in operation for over a year. They utilise their abilities and resources to grow or create new business prospects (Anwar and Rashid 2011). In writing about entrepreneurship, the terms female and women are often used interchangeably, and business innovation is often associated with entrepreneurship. Nonetheless, entrepreneurship and business have often been used synonymously, particularly in SMEs with a lack of or restriction on innovations (Ilhaamie et al. 2015). A psychological force known as "motivation" motivates people to act. It relates to a person's desire, willingness, and behavioural intention (Shaikh Ali et al. 2004); (Mat and Mansor 2010). The most crucial aspect of starting a business is motivation, which is the business's objective, aim, or goal. It relates to how motivated women are to pursue a career in business (Kalyani and Kumar 2011). Most importantly, research shows a link between motivation and business success (Lerner et al. 1997; Osman et al. 2009).

Restrictions, problems, obstructions, disagreements, and issues are all examples of "challenges." Ahmad (2011) argued that challenges include social, economic, and political factors and impediments, challenges, hardships, and limits that businesswomen face. Women entrepreneurs face personal, cultural, and institutional challenges (Shmailan 2014). Business challenges arise throughout the establishment, operation, and growth stages (Sadi and Al-Ghazali 2010). Al-Saidi et al. (2011) classified businesswomen's problems into seven categories: infrastructural, professional, educational and training, social and cultural, legal, and behavioural roles. According to Meyer and Mostert (2016), the biggest barriers inhibiting female entrepreneurs from establishing successful businesses are a lack of financial support and an inability to identify clear business goals.

Similarly, financial and resource constraints, a lack of capital support, talent and education constraints, institutional and cultural constraints, and gender constraints are all considered entrepreneurial challenges (Mohsin and Lei 2020); (Holmen et al. 2011).

13.1.2 Entrepreneurship in Islam

Women are not prohibited from running businesses in Islam. The fact that the Prophet's wife is a successful businesswoman demonstrates this. Allah Subhana Wa Taala (S.W.T) said:

> *Whoever does righteousness, whether male or female, while he is a believer–We will surely cause him to a good life, and We will surely give them their reward [in the Hereafter] according to the best of what they used to* (*Surah al-Nahl*: 97)

Businesswomen must follow Islamic norms to achieve al-Falah, or success in this world and the Hereafter (Surah al-Nahl: 97) (Ullah et al. 2013). Free mingling between genders leads to backbiting, close contact, and infidelity (Yusof 2011).

The Prophet PBUH said: "*Whenever a man is alone with a woman, the Devil makes a third.*" (*Sahih Bukhari*, Book 25, Number 5403, Narrated Abdullah ibn Amr ibn al-'As). Allah S.W.T. also said in Holy Quran:

> *O wives of the Prophet, you are not like anyone among women. If you fear Allah, then do not soft in speech [to men], lest he in whose heart is disease should covet, but speak inappropriate speech* (*Surah al-Ahzab*: 32).

This rule protects women's dignity so they may be respected. Because Allah S.W.T. said:

> *O Prophet, tell your wives and your daughters and the women of the believers to bring down over themselves (part) of their outer garments. That is more suitable that they will be known and not be abused. And ever is Allah Forgiving and Merciful* (*Surah al-Ahzab*: 59).

The Quran and Hadith instruct Muslim women in business (Alina 2013). Before married women could work, Islam had two further requirements. First and foremost, neither the husband's approval nor employment should break the family (Yusof 2011). The Holy Quran states this requirement:

> *The believer, men and women, are protectors of one another: they enjoin what is just and forbid what is evil: they observe regular prayers, practice regular charity and obey Allah and His messenger. On them will Allah pour His mercy; for Allah is Exalted in power, Wise* (*Surah at-Taubah*: 71).

Consequently, Muslim women may work as long as they follow Islam's laws. Married women must obtain consent from their spouses and guarantee their employment will not impact their marriage (Salfiya Ummah et al. 2021).

13.1.3 Vietnam SMEs

Since 2009, the definition of a Vietnam SME has been revised by Decree 56/2009/ND-CP. The decree defined small and medium-sized enterprises are "companies that have registered their business under the law and are divided into three levels: micro, small, and medium based on the sizes of their capital employed (amounting to the total assets identified in an enterprise's accounting records) or the average annual number of labourers (total capital is the priority criterion)" (Le et al. 2011).

The World Bank classified Vietnam as a low-middle-income country (Avin and Kinney 2014). Their SMEs employ 50.1% of the workforce and provide more than 40% of GDP (The Vietnam Women Entrepreneurs Council 2007a, 2007b). In 1991, women accounted for 60% of labour and 52% of the population (Bunck 1997).

Research by the Vietnam Chambers of Commerce and Industry found that women in Vietnam are involved in enterprises of all sizes (VCCI). Women-owned and operated 21% of the firms. Women-led firms account for 26% of all microenterprises and 17% of medium-and larger-sized companies. Women were estimated to own 24% of the 113,352 incorporated enterprises in December 2005. Moreover, according to The Vietnam Women Entrepreneurs Council (VWEC), women were considered to own 27% of the 3 million domestic enterprises in 2007. Over 97% of women-led firms are private, while the remaining 3% are state-owned or involve foreign participation (Le et al. 2011).

According to Nguyen et al. (2016), 16.1% of these women managed existing businesses, 13.9% were early-stage entrepreneurs, and women-led enterprises had lower average wages and revenue per worker. Nevertheless, many more women are working in women-owned enterprises, and this might be because these female business owners recruit more female staff (Le et al. 2011). Women are more likely than males to own restaurants, hotels, wholesale, and retail companies (The Vietnam Women Entrepreneurs Council 2007a, 2007b). Female business owners often operate in businesses that rely on traditional weaving, embroidery, and craft skills. Other sectors with a substantial number of female entrepreneurs include education-related services (56%) and hotels and restaurants (47%). (The Vietnam Women Entrepreneurs Council 2007a, 2007b).

Vietnamese women run micro and small businesses under financial restrictions. They cannot sponsor marketing, computer literacy, leadership, general management, human resource management, and communication training programmes. Despite the government's incentive credit programmes, only 5 to 10% of SMEs in Vietnam have access to capital. Banks' high-interest rates make matters worse. Small company loan requests are often rejected due to the lack of a standardised accounting system and yearly financial statement audits. Capital loan collateral is scarce for less creditworthy businesses. Consequently, approximately 80% of Vietnamese SMEs obtained loans from family and friends and non-profit organisations (The Vietnam Women Entrepreneurs Council 2007a, 2007b).

13.1.3.1 Motivation of Women Entrepreneurs in Vietnam SMEs

Three common factors motivating women in Vietnam to become entrepreneurs are providing family jobs, good skills, and traditional family business (Scheela and Van Hoa 2004; Benzing et al. 2005).

Scheela and Van Hoa (2004) interviewed six Vietnamese women entrepreneurs. The motivation for starting a business is a market need, independence, making money and getting rich, and providing family jobs. Benzing et al. (2005) discovered that participants were motivated to assist their families by earning. Acquiring public recognition and displaying one's ability to run a business efficiently are additional motivators. According to the Vietnam Women Entrepreneurs Council (VWEC), September (2007a), (2007b); Poon et al. (2012), 72% of female entrepreneurs thought that having excellent skills and familial support from social capital provided them with strong possibilities for starting a business. In addition, several recommendations for female entrepreneurs have recently been implemented in Vietnam, including considerable requirements for developing human resources, counselling help, and financial services (Gooty and Narekapudi 2022).

According to Hampel-Milagrosa et al. (2010), traditional family enterprises inspired 28% of respondents to become entrepreneurs. Avin and Kenney (2014) identified two reasons for establishing a company in Vietnam in 2007: "opportunity" owing to high work skills (72%), and "necessity" due to unemployment/low income/traditional family business/others urged me/employment for my children (28%). In another research, Perri and Chu (2012) found that the demand for employment stability motivates Vietnamese women entrepreneurs.

In 2013, the primary drivers of early-stage entrepreneurial activity (businesses operating for less than three and a half years) in Vietnam were opportunity (74.6%) and needs (25.4%). Although Le and Raven (2015) observed that most respondents said that their top incentive for becoming businesswomen was to be their boss, their secondary objective was safeguarding their freedom. This finding supports Zhu et al. (2015)'s contention that the two primary motives for owning a business are increasing income and autonomy.

13.1.3.2 Challenges of Women Entrepreneurs in Vietnam SMEs

Women entrepreneurs in Vietnam face hurdles. Vietnamese women entrepreneurs must consult their families before making major business choices. They work long hours due to money, family, and a dearth of social services in Vietnam (Bunck 1997). Therefore, time constraint is a major obstacle in business pioneered by women (Hampel-Milagrosa et al. 2010).

Due to time constraints, Le et al. (2011) concurred that women could not network with bankers and other entrepreneurs, and it inhibits loan applications and other commercial ventures. Vietnamese businesswomen struggle with low-skilled and unproductive workers. This barrier has harmed Vietnamese businesses due to the high rate of staff job transfers (via re-hiring and re-training). Several female entrepreneurs admit that managing male employees may be difficult (Hampel-Milagrosa et al. 2010).

In addition, financial institutions and the government's policies' rigidity caused Vietnamese women to suffer (Nguyen 2020). They struggle to get monetary assistance due to lengthy loan application processes, complex lending terms and conditions, low asset valuations, and stringent collateral requirements (Hampel-Milagrosa et al. 2010). Moreover, since married women do not own title to real estate property, such as a home or land, they are not eligible to benefit from the credit market (Nguyen 2006). Their lack of education worsens things. Management, communication, and decision-making training are unavailable due to time, networking, and family duties. Because of this, individuals can only acquire a small loan with a hefty interest rate. Therefore, many women-owned enterprises are uncompetitive due to capital constraints, small operating sizes, technology, information, management challenges, and legal and sociopsychological obstacles. Thus, businesswomen without accounting and financial expertise could find it difficult to turn a profit (Nguyen 2006).

Most Vietnamese women fail to start enterprises due to a lack of registration knowledge. Another aspect is their choice not to expand their business, which family members commonly begin without profit in mind. Women seldom work in male-dominated fields like construction, politics, science, and technology but in textiles, footwear, health, and education. Male businesses outnumber female ones (Nguyen 2006).

Rural women believe they must put their husbands first to be excellent wives and mothers. Hence, women are seldom encouraged to network socially and professionally (Nguyen 2020). Traditional civilizations forbid women from walking alone or approaching strangers, making it hard for them to build business-starting ties and communication. As women are perceived as submissive, weak, and irrational, society struggles to realise their potential (The Vietnam Women Entrepreneurs Council 2007a, 2007b). Their spouse or another male family member must also provide the go-ahead (Nguyen 2006). Likewise, the International Labour Office (ILO), in 2008, found that lack of maternity protection, discriminatory property, marriage, inheritance laws, and other gender-based hurdles made it tough for Vietnamese women entrepreneurs to begin and enlarge their businesses.

According to Nguyen et al. (2016), women were unable to take advantage of business training owing to a lack of time flexibility as a result of prioritising domestic responsibilities (The Vietnam Women Entrepreneurs Council (VWEC) 2007a, 2007b). Providers do not care since women's involvement is poor. Some women delay business development due to low productivity and infrastructure, lose mobility, and face social limitations. Muslim women lacked mobility the most (Nguyen 2006).

In addition, regardless of the profit made, businesswomen engaged in foreign commerce must pay additional fees on top of the yearly increase in income tax. Although the items have not yet been exchanged, the VAT must be paid beforehand, and such causes an impact on small businesses' financial flow. Moreover, inconsistent enforcement of tax laws by authorities allows for unofficial payments and encourages bribery as a common business method (Benzing et al. 2005; Zhu et al. 2015).

Employing business advisers or consultants that may boost their company's performance and provide them with crucial ways is an option for women entrepreneurs—unfortunately, owing to the hefty price demanded knowledge, they are forced to pass up this chance (Nguyen 2006).

13.1.4 Thailand SMEs

SMEs make up 93.8% of all businesses in Thailand. Small businesses comprise 76% of all SMEs, while medium-sized businesses comprise 17.8% of all manufacturing facilities. According to the Thailand National Statistical Office (2007), SMEs comprise 76.1% of the manufacturing sector, with the biggest concentration occurring in the food and beverage, textiles, apparel, and wood products industries (Chuthamas et al. 2011). In 2012, there were 2.7 million SMEs or 98.5% of all businesses in Thailand. In the same year, SMEs comprised 80.4% of the workforce and 37% of the country's GDP (Yoshino et al. 2015). The percentage of Thai women who started their businesses in 2013 was 17.3%. (Avin and Kinney 2014). Nonetheless, female entrepreneurship is more prevalent amongst women in Thailand (19.3%), where it is extremely prevalent, compared to China (9.3%), Korea (12.2%), and India (8.7%), where it is considerably less prevalent (Franzke et al. 2022).

13.1.4.1 Motivation of Women Entrepreneurs in Thailand SMEs

Many studies have examined the motivation of Thai women entrepreneurs in SMEs. One implies that if husbands fail to provide, the wife must. Thai women seek freedom, need control, and do not want to work for others (Thakur and Walsh 2013). Chuthamas et al. (2011) assert that Thai businesswomen's participation is driven by opportunity rather than necessity. Contrary to Warangkana (2015), who discovered that Thai women entrepreneurs are driven by favourable encouragement from family and friends and a desire to help their families. Johnsson and Kongsinsuwan (2008) concur with Chuthamas et al. (2011) and Thakur and Walsh (2013) that Thai female entrepreneurs desire independence, autonomy, becoming their boss, and self-actualization. Hatcher et al. (2007) found that 62% of Thai female entrepreneurs were driven by opportunity, not necessity.

13.1.4.2 Challenges of Women Entrepreneurs in Thailand SMEs

There are three main categories of impediments to female entrepreneurship in Thailand. First are discriminatory sociocultural norms and practices such as household duties, unfavourable preconceptions, inequality, and/or employment discrimination (Thakur and Walsh 2013). Secondly, when businesswomen lack status, network, and trustworthiness, access to information and help is a significant obstacle (Thakur and Walsh 2013). Thirdly, women struggle to acquire funds due to lacking financial management skills (Thakur and Walsh 2013). The difficulty in acquiring loans is becoming greater for female business owners in service industries like the food industry since formal credit institutions assess collateral based on net fixed assets, even though their companies are often quite small. Additionally, Thai businesswomen find it difficult to leave their businesses since they often cannot enrol in training programmes to improve their business abilities. This is because tiny and micro businesses often only have one or two employees, and taking time off for training courses necessitates shutting down the company at that time (Thailand Gender & Development Institute 1998).

Pakistan SMEs

SMEs in Pakistan employ up to 250 people have a paid-up capital of up to Rs. 250 million and generate up to $250 million in annual sales (SMEDA 2003). It is acknowledged that SMEs account for the majority of jobs. Over 90% of all private industrial businesses employ less than 100 people, while 78% of Pakistan's labour population outside of agriculture works for SMEs. Moreover, approximately 30% of Pakistan's GDP, 25% of manufacturing export revenue, and 35%

of manufacturing value added are contributed by SMEs. Pakistani SMEs generate low-value items that depend on conventional technology, although producing a fifth of manufacturing exports (Mahmood 2011; Mahmood et al. 2012).

Early closures are a problem for Pakistani Businesses. Three barriers make it difficult for SMEs to establish themselves and survive in Pakistan: access to funding, access to inputs, and access to markets for the goods (SME Development in Pakistan 2013). Micro, small, and even medium-sized enterprises preferred sole proprietorships due to their inexpensive initial costs and ease of formation. Sole proprietorships and unregistered partnerships do not need government approval but must follow labour, tax, and other regulations (SME Development in Pakistan 2013).

Women in business are estimated to be around 50% in Pakistan. However, compared to males (49.14% in 2006–2007 and 49.34% in 2007–2008), women's economic engagement is still quite low (13.53% in 2006–2007 and 14.01% in 2007–2008). Statistics indicate that women's involvement in the labour market has been rising, although slower (from 11.4% in 1994–1995 to 16.0% in 2004–2005). At the same time, just 3% of the total 3.2 million firms are held by women. This is since 60% of working-age women are not reported and get pay that is either unpaid or below what males receive (Rehman and Roomi 2012).

Most Pakistani female business owners (47.7%) operate in the services sector and marketing, and most of their clientele are women. They venture into the education, beauty, and food industries, where men and women rarely interact (Roomi and Parrott 2008). Most Pakistani women entrepreneurs run "boutiques, bakeries, clothes, handicrafts, jewellery, and other similar micro and small companies" (Khan 2014).

Pakistani women's major reasons for not working were housework and parenting. Family assistance is a blessing, and women employ coping skills in a society that prioritises family above business. A typical Pakistani woman entrepreneur is in her 30 or 40 s has a university or high school degree (often in a field unrelated to business), focuses on one enterprise, and is unwilling to expand into other business areas (Roomi and Parrott 2008).

13.1.4.3 Motivation of Muslim Women Entrepreneurs in Pakistan SMEs

Pakistani women want economic freedom; thus, they become entrepreneurs (Khan 2014). Women entrepreneurs build their businesses to combine work and family life, and being self-employed lets them work from home and raise their kids. Women build businesses because of their spouses and families; others claim to have inherited their parents' successful businesses (Rehman and Roomi 2012).

Pakistani female entrepreneurs are inspired by personal ambition, the desire to help their families financially, and self-satisfaction (Mahmood et al. 2012). 48% of respondents wanted to maintain or increase their family's socioeconomic status (Roomi and Parrott 2008) and would usually establish businesses when their husbands died or retired.

Women create businesses for independence, stability, and fulfilment. Shabbir and Di Gregorio (1996) found that most women launch their businesses to have greater control over their jobs, hours, workplace, and coworkers. Male attitudes and job corruption have pushed women to establish businesses. Older women were mostly dissatisfied with the workplace due to gender discrimination and the rigidity of paid job hours which raised problems. Self-employment gives greater freedom in location (working from home or nearby), hours, and other factors than paid employment. To prove their socioeconomic value, they started a business where most were stay-at-home mums without employment experience.

External resources, such as money and location, and relational resources, such as family, friends, colleagues, suppliers, and consumers, may be separated into three groups. Due to the interaction between women's desires and systemic issues affecting startup success, a conceptual framework was constructed to explain why some women could start a firm despite seemingly unfavourable circumstances while others were not. 76% of female entrepreneurs mentioned financial reasons, ranging from "want to earn money" to "financial problems in the family" (Goheer 2003). Having a passion (53%), a good career (47%), recognition and financial independence (24%), and altruism (9%) were also important. 75% of women company owners cited family support as the most significant enabling element; 30% cited their abilities; 11% cited their dedication to cultivating a leisure activity; and 3% cited low-cost components of production (FOP).

13.1.4.4 Challenges of Muslim Women Entrepreneurs in Pakistan SMEs

The issue in Pakistan is that, compared to males, 55.8% of Pakistani women live below the poverty level (Mahmood 2011). Pakistani women had less money, worse health, and lower levels of education. Also, female-led SMEs have a greater closure rate than male businesses, resulting from personal issues, not professional ones. The higher rate of closure of female-owned firms should be read in the context of their primary incentive for starting a company, which is often self-employment. They also tend to dissolve their businesses quickly when it is in their best interests financially or personally (SME Development in Pakistan 2013).

The startup and development stages of women's entrepreneurship in Pakistan are hampered by several factors, including psychological, social, cultural, religious, economic, marketing, managerial, and educational (Khan 2014). The primary obstacles experienced by female entrepreneurs in launching their businesses are financial constraints, a lack of marketing expertise, networking difficulties, and work-family conflicts on a familial, social, and religious level (Khan 2014; Madiha et al. 2017).

Hence, amongst the most important hurdles for women to attain balance in a conservative Islamic culture in Pakistan are a lack of time, gender prejudice, and family duties (Anwar and Rashid 2011). With the stereotype that males are "breadwinners" and women are "housemakers," juggling work and home obligations is difficult for women in business (Anwar and Rashid 2011). When family time must be sacrificed because of a conflict between professional and commercial obligations, managing these responsibilities is somewhat more difficult (Rehman and Roomi 2012).

Additional hindrances include a lack of training, direction, capital shortages, and regional customs brought on by deeply ingrained discriminatory sociocultural beliefs and traditions, especially those embedded in the legislative and legal environment and institutional support systems (Roomi and Parrott 2008).

Similar to female entrepreneurs in other nations, Pakistani businesswomen struggle with a lack of information technology expertise, a lack of research and development capacity (Khalique et al. 2020), a lack of knowledge about financial resources, information, and not aware of the various available opportunities. Also, individuals are prejudiced, particularly when requesting loans from private sector institutions (Mahmood et al. 2012).

In a more thorough investigation, the difficulties are separated into two groups. Gender-related difficulties comprise the second group, whereas gender-neutral challenges are the first. There are two stages to gender-neutral problems. The first stage is when a business is just starting. Pakistani women entrepreneurs have to deal with issues including access to financing, a lack of business management skills, and governmental rules and regulations.

The second phase addresses growth-phase issues such as technical skills, marketability, and funding (Roomi 2006). Women's authority, networking, trust and credibility, and geographical mobility are gender-related startup challenges. Gender-related development challenges include accepting women's power and structural immobility complicating business startups. Women were also restricted from travelling and working due to social and cultural norms (Roomi and Parrott 2008; Mahmood 2011).

Another research divides spatial mobility into physical and ability restrictions and addresses transportation inadequacy-related physical limitations. Sociocultural, tribal, and religious factors restrict women's potential. Women may own real estate under Islamic Shariah; however, just 3% of plots in selected communities are held by women, limiting their access to collateral assets and credit. Female entrepreneurs rely on cooperatives, savings, and family support (Syed 2010).

In addition, female entrepreneurs encounter a significant gender-related issue where their trustworthiness is consistently underestimated. As a result, women managers must work harder to persuade clients, partners, and even staff members of their ability to operate a successful and efficient business. The respondents often acknowledged that one of the biggest impediments to the growth of their business management abilities was the lack of gender-specific training options (Roomi and Parrott 2008; Mahmood 2011).

Starting a business is difficult without work experience or training (Shabbir and Di Gregorio 1996). Male workers' views of a female boss were key, and they labelled their business a "hobby" and discredited the women's authority. Cultural constraints hindered supplier trust and regional mobility. Without employment experience/technical knowledge, and family support, women realised that establishing businesses was not the best way to achieve their objectives. Their husbands and families opposed them more than the other reasons. Without hindrance, women could establish businesses easier. However, determination helps most women overcome financial, employee, supplier, and customer challenges.

Goheer (2003) lists gender biases, marketing, family/social, funding, government, mobility, and other difficulties as startup barriers. In many countries, women struggle harder with the main constraints (interaction, poor bargaining position, knowledge gaps, and lack of mobility). 41% blamed the government for producing a toxic business environment, 31% blamed a lack of finance, and 20% blamed the lack of business development services. One-third of women business owners said poor funding or a lack of financial autonomy affects their business growth (Field, Jayachandran, and Pande 2010). Other issues include regulatory environment (23%), law and order (18%), lack of firm development services (10%), and economic deterioration (9%).

13.2 Methodology

This study uses quantitative and qualitative research methods (Amaratunga et al. 2002; Creswell and Creswell 2018). A questionnaire is created for the quantitative research using data from previous year's surveys. As a result, given the nature of the questionnaire, the researcher may or may not be present during distribution. Hence, the questionnaire is sent to the three chosen nations of Thailand, Vietnam, and Pakistan.

The semi-structured interview questions for the qualitative research are also designed based on previous year's studies. Unfortunately, owing to financial restrictions, only nearby nations like Thailand and Vietnam were used for the interviews. The sample study had eight interviewees from Thailand and twelve from Vietnam. Due to the absence of a women-led business name list directory and not many officially registered, a purposive sample approach was adopted. IBM SPSS software was used to analyse the survey questionnaire data, while NVivo was used to analyse the interview data descriptively (Braun and Clarke 2006; Jackson and Bazeley 2019).

13.3 Findings of the Quantitative Approach

13.3.1 Asian Pacific Muslim Women Entrepreneurs' Profile

The quantitative study's respondents are 90 Muslim SME businesswomen in the three selected countries. The findings indicate that the majority of the respondents in the selected Asian Pacific countries are between 31 to 40 years old (48.7%), from Pakistan (44.4%), and married (81.1%). Before they became CEOs (41.1%) for 1 to 5 years (78.7%), they worked in a small firm (48.9%) for 1 to 5 years (61.1%).

They mainly want to start their own business because they do not want to work for someone else (21.1%). They are hardworking businesswomen who work 9 to 11 h (36.7%). The majority of the business is the first startup (78.9%) of a sole proprietorship (55.6%) in the manufacturing industry (22.2%). However, they do not deal with innovation (72.2%), only distributing the products (40%). Most respondents have full-time employees between 1 and 5 people (48.9%), and so do part-time (65.6%). They obtained sales annually from RM60,001 to RM85,000 (46.7%), and the monthly profit is between RM5,001 and RM10,000 (48.9%). This could be why they do not have other branches (64.4%). The majority too obtained financial assistance from commercial banks (62.2%) for the amount of RM5,001 to RM10,000 (30%), which is payable in more than five years (40%), and they do not face any difficulty in paying the loans (38.9%). They also agreed that this financial assistance greatly helps their business (64.4%). Please refer to Table 13.1 for the details.

13.3.2 Motivation of Muslim Women SME Entrepreneurs in Selected Asian Pacific Countries

Most of the respondents in the selected Asian Pacific countries strongly agreed with the statements in the questionnaire regarding their motivation to embark on business.

13.3.2.1 Personal Motivation

Most respondents are motivated by personal factors like being able to use experience and training (54.4%), contributing to family income (50%), helping husband/family (47.8%), having interest in doing business (42.2%), wanting independence (41.1%), self-challenging (36.7%), not giving up easily (35.6%), and financial independence (25.6%).

13.3.2.2 Family Motivation

They are also motivated due to family factors such as the availability of financial assistance from family/friends (98.1%), family support (56.7%), and continuing the family tradition (27.8%).

13.3.2.3 Economics Motivation

Most respondents are only motivated by the economic factor to obtain additional income due to the rising cost of living (46.7%).

13.3.2.4 Social Motivation

They also want to be role models (37.8%), acquire recognition for working (37.8%), improve society (37.8%), generate jobs for the needy (37.8%), join women's associations (32.2%), and assist women to make money (31.1%).

Table 13.1 Asian Pacific Muslim women SMEs entrepreneurs' profile

Item	Responses	Frequency	Percentage
Age	31–40	44	48.7
Country origin	Pakistan	40	44.4
Marital status	Married	73	81.1
Previous experience	Employee in a small firm	44	48.9
Education	High school	26	28.9
Working experience	1–5 years	55	61.1
Current position	CEO	37	41.1
Duration	1–5 years	70	78.7
Reason to start your own business	Don't want to work for someone else	19	21.1
First startup	Yes	71	78.9
Length of working hours	9–11 h	33	36.7
Form of establishment	Sole proprietorship	50	55.6
Industry	Manufacturing	20	22.2
Type of business	Distributor	36	40
Innovation	No	65	72.2
Full-time employee	1–5	44	48.9
Part-time employee	1–5	59	65.6
Annual sales	RM60,001–RM85,000	42	46.7
Profit per month	RM5,001–RM10,000	44	48.9
Branches	No	58	64.4
Financial assistance	Commercial banks	56	62.2
Loans	RM5,001-RM10,000	27	30
Repayment period	More than five years	36	40
Difficulties in repayment	No	35	38.9
Financial assistance help	Yes	58	64.4

13.3.2.5 Market Motivation

Market factors like positive customer feedback (52.2%), strong market demand (51.1%), distribution channels (43.3%), technology facilities for quick/easy production of quality products (42.2%), good marketing channels (37.8%), cooperative global business network (33.3%), and more supply options (33.3%) motivate them.

13.3.2.6 Networking Motivation

For the networking factors, they are motivated due to a good relationship with the customer (65.6%), cooperative competitors for raw materials (28.9%), and cooperative competitors to tender for government jobs (25.6%).

13.3.2.7 Government Motivation

Furthermore, they are motivated due to government factors such as financial assistance (33.3%) and good relationships with local authorities (32.2%).

13.3.2.8 Legal Motivation

They are also motivated due to legal matters, especially in terms of Shariah, such as Muslim women's right to do business (46.7%), availability of permission from their husbands to do business according to Shariah Law (40%), and freedom to wear a headscarf for Muslim women entrepreneurs (40%). Please refer to Table 13.2 for the details.

13.4 Findings of Qualitative Study

According to interviewees, personal, familial, societal, government, and commercial reasons motivate Muslim SME businesswomen in Vietnam, Thailand, and Pakistan. Industrious, patient, eager, adaptive, sincere, brave, honest, optimistic, service-minded, quick thinkers, positive thinkers, act fast, are confident, daring to take chances, perfectionists, single, have a decent financial background, and public relations motivate women to be businesswomen. In comparison, family motivation consists of their inclination to support a family and continue their family heritage. They obtained customers' support and friends as a social factor, assistance from government agencies as a government factor, and customer demand as a market factor.

13.4.1 Challenges of Muslim Women SMEs Entrepreneurs in Asian Pacific Countries

From the survey, most of the respondents in the selected Asian Pacific countries strongly agree that the most challenging problem in being a Muslim businesswoman is increasing market competitiveness (35.6%), amongst other challenges of personal, family, economic, social, human capital, government, and legal.

Meeting customer preferences is the most frequently mentioned challenge by interviewees in other Asian Pacific countries, along with other personal, market, political, economic, and governmental challenges such as juggling time between family and business, responsibility to parents, competition with other entrepreneurs, staff retention, deception by customers, meeting customer preferences, high promotion costs, honest employee, high sale margin demand from distributors, limited suppliers, politics, high interest from the bank loan, currency fluctuations, high taxes and increment of labour wages and lack of government assistance.

Table 13.2 Motivational factors of Muslim women SMEs entrepreneurs in selected Asia Pacific countries

Motivational factors	Frequency	Percentage
Financial assistance from family/friends	89	98.1
Good relationship with customers	59	65.6
Family support	51	56.7
Able to use experience and training	49	54.4
Positive comments from customers	47	52.2
Strong market demand	46	51.1
Contribute to family income	45	50
Help husband/family	43	47.8
Muslim women's right to do business	42	46.7
Additional income due to the cost of living	42	46.7
Interest in doing business	38	42.2
Availability of technology facilities for quick/easy production of quality product	38	42.2
Desire for independence	37	41.1
Permission from husband to do business according to Shariah law	36	40
Freedom to wear the headscarf for Muslim women entrepreneurs	36	40
Model for others in the society	34	37.8
Availability of good marketing channels	34	37.8
Respect for working	34	37.8
Achieve a better position in society	34	37.8
Create employment for poor people	34	37.8
Self-challenging	33	36.7
Do not give up easily	32	35.6
Availability of distribution channels	30	33.3
Financial assistance from government	30	33.3
Cooperative global business network	30	33.3
Participate in women association	29	32.2
Good relationship with local authorities	29	32.2
Help other women to earn money	28	31.1
More supply options	28	31.1
Cooperative competitors for raw materials	26	28.9
Continue family tradition	25	27.8
Financial independence	23	25.6
Cooperative competitors for government jobs	23	25.6

13.5 Solutions

Henceforth, to survive in business, most Muslim businesswomen in the selected Asian Pacific countries strongly proposed three solutions through the survey. Firstly to give donations or charity (34.4%), to give tithe in order to cleanse up own property (38.9%) and to have a good relationship with customers or competitors or suppliers (38.9%).

Partner and family support, spiritual practise, expanding knowledge and skills, networking, obtaining high morale and experience as personal solutions, online marketing and advertising channel, expanding the business into the *halal* market and Shariah-compliant as marketing solutions, obtaining capital and loans from family members and banks as economical solutions, providing *zakat* and charity, possessing good customer relations are amongst other proposed solutions.

13.6 Conclusion

In the quantitative research, the motivating reasons of Muslim women to be entrepreneurs in Asian Pacific SMEs are classified into personal, family, economic, social, market, networking, government, and legal elements, with financial aid from family and friends topping the list. In contrast, the most challenging problem is the market factor. Thus, they propose solving their problems by having a good relationship with customers, competitors, or suppliers by giving tithes and donations or charity to society. However, from a qualitative study of 20 Muslim SME women entrepreneurs from Vietnam and Thailand, their motivation factors are personal, family, social, government, and market. They face personal, government, market, political and economic challenges. Consequently, they propose personal, market, government, economic and social solutions.

Acknowledgements We want to express our sincere appreciation to RP012C-13HNE for supporting this study.

References

Ahmad SZ (2011) Evidence of the characteristics of women entrepreneurs in the kingdom of Saudi Arabia. An empirical investigation. Int J Gend Entrep 3(2):123–143

Alina Z (2013) Micro-entrepreneurship and Islamic economic system. Bus J Entrep 1:91–101

Al-Saidi R, Belwal R, Al-Badi R (2011) Women entrepreneurship in the al-Batinah region of Oman: an identification of barriers. J Int Women's Stud 12(3):58–75

Amaratunga D, Baldry D, Sarshar M, Newton R (2002) Quantitative and qualitative research in the built environment. Work Study 51(1):17–31

Anwar MU, Rashid AG (2011) Female entrepreneurs: a review of the literature and proposed conceptual framework. In: Proceedings of 2nd international conference on business management. pp 1–30. (ISBN: 978–969–9368–06–6)

Avin RM, Kinney LP (2014) Trends in female entrepreneurship in Vietnam. In: Preliminary paper presented at the 23th annual conference on feminist economics. Sponsored by Iaffe, University of Ghana, Accra, Ghana, June 27–29

Benzing C, Chu HM, Callanan G (2005) A regional comparison of the motivation and problems of Vietnamese women. J Dev Entrep 10(1):3–27

Braun V, Clarke V (2006) Using thematic analysis in psychology. Qual Res Psychol 3(2):77–101

Bunck JM (1997) Women and post-cold war socialism: the cases of Cuba and Vietnam. Cuba in Transition. In: Annual proceedings of the association for the study of the Cuban economy (ASCE), vol 7 (30th November).

Chuthamas C, Aminul Islam M, Thiyada K, Dayang Hasliza MY (2011) Factors affecting business success of small and medium enterprises (SMEs) in Thailand. Asian J Soc Sci 7(5):180−190

Creswell JW, Creswell JD (2018) Research design: Qualitative, quantitative, and mixed methods approaches. SAGE Publications Inc., Thousand Oaks, California

David PF, Hung CM (2012) Entrepreneurs in China and Vietnam: motivations and problems. Int J Entrep 16(Special Issue):95–112

Field E, Jayachandran S, Pande R (2010) Do traditional institutions constrain female entrepreneurship? a field experiment on business training in India. Am Econ Rev 100(2):125–129

Franzke S, Wu J, Froese FJ, Chan ZX (2022) Female entrepreneurship in Asia: a critical review and future directions. Asian Bus Manag 21:343–372

Goheer NA (2003) Women entrepreneurs in Pakistan: how to improve their bargaining power. International Labour Organization (ILO), Geneva

Gooty S, Narekapudi N (2022) Enabling women's entrepreneurship in Asia and the Pacific through smart legislative and policy reforms. https://www.unescap.org/blog/enabling-womens-entrepreneurship-asia-and-pacific-through-smart-legislative-and-policy-reforms. Accessed 26 Feb 2023

Hampel-Milagrosa A, Pham VH, Nguyen AQ, Nguyen TT (2010) Gender related obstacles to Vietnamese women entrepreneurs. United Nations Industrial Development Organization, Vienna

Hatcher C, Terjesen S, Planck M (2007) Towards a new theory of entrepreneurship of culture and gender: a grounded study of Thailand's most successful female entrepreneurship. AGSE, Australia

Holmen M, Thaw TM, Saarelainen E (2011) Female entrepreneurship in Afghanistan. J Dev Entrep 16(3):307–331

Ilhaamie AGA (2017) Muslim women entrepreneurs' motivation in SMEs: a quantitative study in Asia Pacific countries. Asian Econ Financ Rev 7(1):27–42

Ilhaamie AGA, Rosmawani CH, Siti Arni B, Hasan Al-Banna M (2015) The motivation of Muslim women entrepreneurs in China SMEs: towards trading halal goods and service. In: Suhaili S, Mohd Abd. Wahab FMB, Nor 'Azzah K, Ahmad Sufyan CA (eds) Perspektif Industri Halal Perkembangan dan Isu-isu, pp 191–207. UM Press, KL

Ilhaamie AGA, Siti Arni B, Rosmawani CH, Hasan Al-Banna M (2014) Challenges of Malaysian muslim women in SMEs. IJMIT 5(6):428−433

International Finance Corporation (2011) Strengthening access to finance for women-owned SMEs in developing countries. Washington

International Labour Organization (ILO) (2011) Women's entrepreneurship development in Vietnam: learning from good practices. ILO Country Office for Vietnam, Vietnam

Jackson K, Bazeley P (2019) Qualitative data analysis with NVIVO, 3rd edn. SAGE Publications Ltd., London

Jawad S (2010) Women and small business entrepreneurship in Pakistan. In: Fielden S, Davidson M (eds) International research handbook on successful women entrepreneurs. Edward Elgar, Cheltenham & New York, pp 117–131

Johnsson A, Kongsinsuwan S (2008) Female entrepreneurs in Sweden and Thailand–differences and similarities in motivation. Master Thesis in Business Administration, Malardalens Hogskola, School of Sustainable Development of Society and Technology, Sweden

Kalyani B, Kumar D (2011) Motivational factors, entrepreneurship and education: a study with reference to women in SMEs. Far East J Psychol Bus 3(3):14–35

Khalique M, Ramayah T, Hinaa K Nassir Shaari JA (2020) Islamic entrepreneurship: a systematic review of future challenges and prospects of Pakistani SMEs. Int J Econ, Manag Account 28(2):277−295

Khan I (2014) Women chamber of commerce and industry. JWE 1(2):62–91
Le QV, Raven PV (2015) Woman entrepreneurship in rural Vietnam: success and motivational factors. J Dev Areas 49(2):57–76
Le DB, Nguyen DH, Pham NT, Do KY, Dao TT (2011) Creation of an enabling environment for women entrepreneur in Vietnam: mainstreaming gender issues in government policy on enterprise development. ILO, Hanoi
Lerner M, Brush C, Hisrich R (1997) Israeli women entrepreneurs: an examination of factors affecting performance. J Bus Ventur 12:315–339
Madiha SD, Shakoor A, Abdul R (2017) Small and medium-size enterprises in Pakistan: definition and critical issues. Pak Bus Rev 46–70
Mahmood S (2011) Microfinance and women entrepreneurs in Pakistan. Int J Gend Entrep 3(3):265–274
Mahmood B, Khalid S, Muhammad Sohail M, Babak I (2012) Exploring the motivation and barriers in way of Pakistani female entrepreneurs. Br J Educ, Soc & Behav Sci 2(4):353–368
Mat ABC, Mansor NB (2010) The significance of psychology and environment dimensions for Malaysian Muslim women entrepreneurship venturing. Int J Hum Sci 7(1):253–269
Meyer N, Mostert C (2016) Perceived barriers and success factors of female entrepreneurs enrolled in an entrepreneurial programme. Int J Soc Sci Humanit Stud 8(1):48–66
Mohsin AK, Lei H (2020) Factor analysis for the development of women entrepreneurship in SMEs: an empirical evidence from Bangladesh. Chin Bus Rev 19(1):22–35
Nguyen T (2006) Voices of Vietnamese women entrepreneurs. International Finance Corporation, Washington
Nguyen HA, Hoang TG, Du VHT, Hoang BT (2016) A review of women's entrepreneurship in Vietnam taking into account sociocultural norms and the institutional ecosystem. Working paper No.11/2016/August 2016. World Trade Institute
Nguyen LTT (2020) Sustainability model of Vietnamese women entrepreneurship. Doctorate Thesis of Business Administration, School of Business and Tourism, Southern Cross University
Osman I, Zainal Ariffin A, Ahmad NH, Husin A, Abu Bakar S, Tanwir ND (2009) Following the trail of women homestay entrepreneurs in Malaysia: understanding their motivation and empowerment. TEAM J Hosp & Tour 6(1):24–35
Perri DF, Chu HM (2012) Entrepreneurs in China and Vietnam: motivations and problems. Int J Entrep 16(1):93–112
Poon JPH, Thai DT, Naybor D (2012) Social capital and female entrepreneurship in rural regions: evidence from Vietnam. Appl Geogr 308–315
Quan LV, Peter RV (2015) Woman entrepreneurship in rural Vietnam: success and motivational factors. J Dev Areas 49(2):57–77
Rehman S, Roomi MA (2012) Gender and work life balance: a phenomenological study of women entrepreneurs in Pakistan. J Small Bus Enterp Dev 19(2):209–228
Roomi MA, Parrott G (2008) Barriers to development and progression of women entrepreneurs in Pakistan. J Entrep 17(1):59–72
Roomi MA (2006) Women entrepreneurs in Pakistan, profile, challenges and practical recommendations. Doctoral Researcher Paper, School of Management Royal Holloway, University of London, London
Sadi MA, Al-Ghazali BM (2010) Doing business with impudence: a focus on women entrepreneurship in Saudi Arabia. Afr J Bus Manage 4(1):001–011
Salfiya Ummah MAC, Choy CS, Sulaiha Beevi A (2021) Factors influencing business performance of small and medium enterprises: a case study in Sri Lanka. J Asian Financ, Econ Bus 8(4):0201–0212
Scheela W, Van Hoa TT (2004) Women entrepreneurs in a transition economy: the case of Vietnam. Int J Manag Decis Mak 5(1):1–20
Shabbir A, Di Gregorio S (1996) An examination of the relationship between women's personal goals and structural factors influencing their decision to start a business: the case of Pakistan. Amama J Bus Ventur 11:507–552

Shaikh Ali SH, Mansor N, Ahmad N (2004) Factors that motivate women to be the entrepreneur. In: Proceeding of the 3rd international conference on SMEs in a global economy, UiTM and University of Wollongong, Malaysia
Shmailan A (2014) Female entrepreneurs in Saudi Arabia: a comparison of barriers and motivations: moving from disenfranchisement to empowerment. Elit Res J Educ Rev 2(2):6–21
SME Development in Pakistan (2013) Issues and remedies, pp 1–40. https://www.yumpu.com/en/document/read/21465643/sme-development-in-pakistan-issues-and-remedies-government. Accessed 25 Feb 2023
SMEDA (2003) *What is SME definition?* http://www.smeda.org/index.php?option=com_fsf&view=faq&catid=3&faqid=48. Accessed 26 Feb 2023
Syed J (2010) Chapter 11: women and small business entrepreneurship in Pakistan. In: Fielden SL, Davidson MJ (eds) International research handbook on successful women entrepreneurs. Edward Elgar Publishing, Inc., USA
Thailand Gender & Development Institute (1998) Country briefing paper on women in Thailand
Thakur R, Walsh J (2013) Characteristics of Thai women entrepreneurs: a case study of SMEs operating in Lampang municipality area. J Soc Dev Sci 4(4):174–181
The Vietnam Women Entrepreneurs Council (VWEC) (2007a) Women's entrepreneurship development in Vietnam. International Labour Organization (ILO), Vietnam
The Vietnam Women Entrepreneurs Council (VWEC) (2007b) Women's entrepreneurship development in Vietnam. International Labour Organization in Vietnam, Vietnam
Ullah MM, Mohamed T, Yousuf F (2013) Women entrepreneurship: an Islamic perspective. Eur J Bus Manag (EJBM) 5(11): 253–269. Special Issue on Islamic Management and Business
Warangkana T (2015) Determinants of business performance among women entrepreneurs in Southern Thailand. Doctorate Thesis of Business Administration, Universiti Utara Malaysia, Sintok
Women Empowerment through Business Member Organizations: Vietnam's Landscape. Dutsch Employer Cooperation Programme. International Training Center.
Yoshino N, Farhad T-H, Phadet C, Baburam N (2015) The importance of SMEs in Thai economy. Retrieved from Asia Pathways, http://www.asiapathways-adbi.org/2015/07/importance-of-smes-in-the-thai-economy/. Accessed 25 Feb 2023
Yusof R (2011) How religious values affect entrepreneurial behaviours among Muslim businesswomen in the district of Pendang, Kedah. NIEW J 3:25–46
Zhu L, Orhan K, Chu HM, Chu A (2015) Female entrepreneurship: evidence from Vietnam. J Bus Entrep 26(3):103–128

Rethinking Economic Crises in Islamic Perspective: Innovating Approach in Systematic Literature Review

14

Winata Wira, Umar Burhan, Asfi Manzilati, and Multifiah

Abstract

Business cycles or crises occur when the economy falls after prosperity (Schumpeter 1939) or when genuine insecurity increases volatility, uncertainty, and contradictions in several domains (Haralambie 2011). According to (Colander et al. 2009), crises have made mainstream economics a systemic failure. We use Islamic viewpoints to investigate the crises' account. In this regard, many Muslim authors attribute the economic crisis to mainstream economic discipline's failure and suggest Islamic economics as a solution or alternative. We use a systematic literature review to evaluate 36 studies based on peer-reviewed journals and sources between 1990 and 2022. Banking and Finance, Economic Thinking, Spending and Consumption, and Human Resources and Governance were identified from the selected literature. It largely focuses on banking and finance, underlining the inherent flaws of the sector in the capitalist system and lessons for strengthening Islamic banking and finance to handle future cyclical shocks. The literature of other clusters in the review also shows that the lessons from the crisis have influenced Muslims' consumption and spending decisions, including credit decisions, enrichment of Islamic economic thinking, and best practises to survive in crisis times for Muslims or Islamic institutions.

W. Wira (✉) · U. Burhan · A. Manzilati · Multifiah
Economics Doctorate Department, Faculty of Economics and Business, Universitas Brawijaya, Malang, Indonesia
e-mail: winatawira@umrah.ac.id

W. Wira
Faculty of Maritime Economics and Business, Universitas Maritim Raja Ali Haji, Tanjungpinang, Indonesia

A. Rafiki et al. (eds.), *Open Innovation in Small Business*, Contributions to Environmental Sciences & Innovative Business Technology,
https://doi.org/10.1007/978-981-99-5142-0_14

Keywords

Economic crisis • Financial crisis • Islamic perspective • Systematic literature review • Islamic economics

14.1 Introduction

Cyclical oscillations are plainly visible in the history of national and global economies. Economic growth experiences ups and downs referred to as the business cycle. In Business Cycles (1939), Joseph Schumpeter argues that crises arise when the economy falls after experiencing prosperity. He also contended that business cycles are not simply random fluctuations in economic activity but the natural and unavoidable consequences of capitalism's dynamism and innovation.

According to Haralambie (2011), a *crisis* can be defined as a situation characterized by real insecurity, resulting in increased volatility and uncertainty and accompanied by numerous contradictions in various fields such as economic, social, political, ideological, military, and so on. What happens in the economic domain is a relative surplus of goods to people's purchasing power, or what economist John Maynard Keynes (1883–1946) referred to as a decline in aggregate demand. Due to rising costs, industries are forced to cut production, including laying off employees, resulting in soaring unemployment. On the other hand, external factors such as natural disasters, disease outbreaks, political turmoil, war, and others may also lead to economic crises. Using 2020 as an example, the economic crisis caused by the COVID-19 pandemic reduced global GDP growth to minus 3.4% that year, resulting in increased poverty, unemployment, and inequality in nearly all countries (See Global Economic Prospects. The World Bank, 2020 titled "*Pandemic, Recession: The Global Economy in Crisis*").

In the 1970 and 1980s, Hyman Minsky's (1919–1996) theory of financial instability hypothesis explained the economic crisis caused by financial market instability, namely that increasing leverage levels describe the economy, and the risk of financial collapse begins when the economy is stable. When a hedge borrower becomes a speculative borrower and then a Ponzi borrower, credit lending activity increases the level of risk. Many economists believe that the 2007–2009 financial crisis mirrored Minsky's theory of financial instability almost four decades before the event (See Bhattacharya et al. 2015; Boyer, 2013; Michail, 2021; Thomas, 2009; Whalen, 2011).

The recurrence of crises has undermined the resilience of economics as a discipline. Mainstream economic methods have failed to predict future economic upheavals (Hoover, 2016). Imperia (2011) claims that economic crises prove mainstream economic theory is ineffective. The financial crisis disgraced economic theory, according to Acemoglu (2009), since academics failed to verify three of its key assumptions, including that severe business cycles cannot increase. Second, free markets reduce self-interest. Third, established firms may be expected to police their actions to maintain reputation. The opposite is true. Colander et al.

(2009) say the economic crisis is a systemic failure of economics. Economists contributed to the policies that caused the crisis and failed to forecast it.

Islamic economic thought did not emerge as something entirely novel, to quote Islahi (2005), because the search for solutions to economic problems is a preoccupation for all societies at all times. Islamic economics is merely a framework derived from the tenets of the Holy Book Qur'an and its religious tradition in practices through the deeds of the Prophet Muhammad (p.b.u.h) (Bhuiyan et al. 2020). In contrast to the neoclassical anthropological construct of human economic *Homo Economicus* (Gattoo, 2017), many Muslim writers have increasingly discussed the possibility of a better human model called *Homo Islamicus*, that is, human beings whose economic behavior as both individuals and members of society is based on Islamic values, principles, and methods (Mahomedy, 2013). Many Muslim authors have attributed the economic crisis to the failure of mainstream economic disciplines and, concurrently, consider Islamic economics as a solution or alternative (See Askari & Mirakhor, 2015; Ejaz & Khan, 2014; Khan, 2013; Mahomedy, 2016; Malkawi, 2020; Mirakhor & Askari, 2017) to shed light on economic developments that are increasingly unable to avoid crisis (Chowdhury & Żuk, 2018).

Since the global financial crisis of 2007–2009, an increasing number of Muslim authors have raised critical voices from an Islamic perspective (Siddiqi et al. 2009). Numerous Muslim economists characterize Islamic economics as a distinct discipline with a divine source of knowledge (Visser, 2019), in which analysis and problem-solving entail not only logical reasoning and empirical observation, but also a doctrinal interpretation of God's revelation (Furqani & Haneef, 2015). Islamic economics seeks to expose the objective, logical, and revelation-based realism of every social phenomenon and economic fact (Bakar, 1984). It consists of a body of knowledge dominated by legal doctrine and dogmatic religious teaching, hence it is also frequently referred to as shari'ah economics (Arfah et al. 2020). Consequently, as Khan (2013) noted that Muslim economic scientists do not delineate which parts of Islamic economics are theological or social science. In point of fact, theology cannot be linked with social science since it is not arranged in a form or structure that can be tested; more specifically, there are currently no Islamic societies that meet the criteria for verification. In the meantime, as Rehman & Askari (2010) reveal, Islamic countries do not adopt Islamic practices as expected. This evidence suggests that the influence of Islam on the preponderance of economic and social performance indicators is relatively poor (Pryor, 2007).

Our research highlights the extent to which Islamic economists have authored their views and analyses on economic hardships such as the current economic crises. This initial importance of effort look forward to indicating on how Islamic economists might have touched a complex topic requiring multi-angle factors of crises, i.e., the causes, symptoms and recurrence, and repercussion as well. Even though Islamic economics has characteristics that contrast with those of the mainstream, only a few attempts have been made to establish the extent of its influence on theory development and policy contributions. In this regard, we seek to employ a more suitable approach: a systematic literature review.

Research based on systematic literature reviews has become a widely recognized way of developing knowledge today. This research approach relies on empirical and rigorous primary pieces of evidence so that its significant contributions work to serve the academic realm and policy practice. The long experience and competence of systematic reviews in the medical field (See Cook et al. 1997a, 1997b; Wolf et al. 2001) have fuelled the growth of similar practices in the social sciences (Petticrew & Roberts, 2008).

14.2 Methods

Systematic Literature Review provides an overview of the research on the issue and contextualizes it based on important topics. It also identifies research gaps and directions. It draws judgements on what is and is not known using valid and rigorous empirical evidence from existing research, contribution selection and evaluation, data analysis, and synthesizing outcomes (Denyer & Tranfield, 2009). Scientific, repeatable steps are used (Cook et al. 1997a, b). It uses a chosen implementation plan or methodology ((Josette Bettany-Saltikov & Cronin, 2013) to minimize bias from non-systematic reviews and single studies (Sweet & Moynihan, 2007).

Our study aims to conduct a systematic literature review on Islamic perspectives on economic crises, including financial crises, by focusing on the relevant literature. Perspective is a means of understanding a problem or life in general[1]; therefore, the Islamic perspective serves as a way of examining issues following Islam's methods, principles, and teachings. In conventional economics, analyzing and describing a problem from positive (what is) and normative (what should be) perspectives can result in distinct analyses and descriptions. Likewise, the normative Islamic perspective approach pertains to its divine sources of knowledge, namely, the Quran and the Prophetic Tradition (Sunnah). On the other hand, the population of Islam adherents, specifically the Muslims themselves, who have a diversified nature and diverse intrinsic backgrounds, is a positive image that may contrast.

As in most systematic review studies, the first step is to apply the research criteria for the scope of this review, which include papers generated and published in the Scopus database and trustworthy journal publishers listed in the Google Scholar database. Only articles published in peer-reviewed journals, so monographs, research notes, reviews, conference proceedings, and working papers are excluded from the study or rejected. The next step is conducting the review, which starts with selecting articles, in which case specific search strategies are applied.

[1] Look in Dictionary of https://www.britannica.com.

The timeframe covered by the systematic review of the relevant literature is from 1990 to 2022, during which there have been a number of economic crises. Typically, authors have utilized current data to examine past occurrences (empirical); yet, thoughtful conceptual investigations that are likely to have been conducted are not necessarily based on a single catastrophe. This is due to the fact that many studies in Islamic economic thinking are based on doctrinal interpretations of revelation.

The article search was conducted using a combination of terms in the keyword database, such as the following, to discover articles: (1) Economic Crisis OR Financial Crisis OR Economic Recession OR Economic Depression And Islam OR Muslim OR Moslem, (2) Islamic Economics OR Islamic Economic And Crisis OR Recession OR Depression, (3) Islamic Economics Crisis OR Islamic Financial Crisis OR Islamic Economic Crisis OR Islamic Economic Recession. More than 5,000 items were retrieved utilizing the application program *Publish or Perish* following a search (Fig. 14.1).

Through scanning, we produced articles that met the requirements. First, scrutinizing titles and keywords led to the deletion of 3,264 titles since they were:

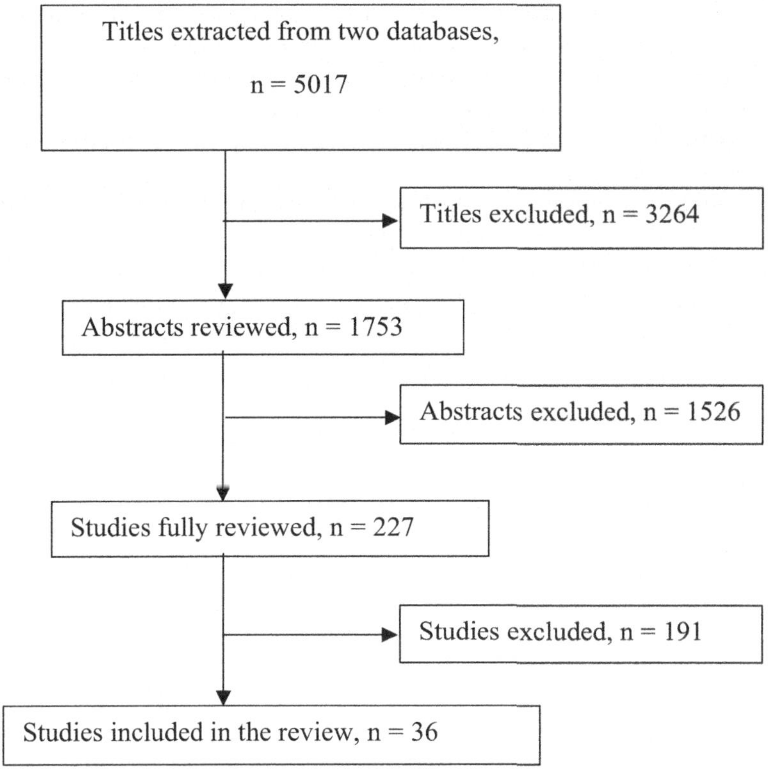

Fig. 14.1 Studies' selection stage

(i) duplication; and (ii) not relevant to the review focus. The subsequent phase entailed a rigorous abstract, introduction, and conclusion analysis that ruled out 1,526 abstracts. This stage involved thoroughly reading 227 studies and excluding 191 studies for the following reasons: review article, essay, report, bibliometric study, and lack of relevance to the study's objectives. In the final, we evaluated a total of 36 articles which consists of thirty-five research articles and one book chapter. As for the scope of review, we consider for the review decision only fully accessible papers written in English.

14.3 Findings and Discussion

14.3.1 Features of Published Works

With a search period spanning 1990 to 2022, our study discovered a shifting trend in the annual published works. Nonetheless, 2010 was the year in which the number of publications covering search categories with the keywords economic crisis and Islam or Muslim grew at the fastest rate. This period reflects the beginning of the global economic recovery following the financial crisis of 2007–2009. Many studies have shown the attraction of Islamic financing compared to the conventional global financial industry, which was supposed to hold responsible for the crisis. The representations of Islamic economists' critiques of the crisis grew concurrently with the academic output, which emphasized the rise of Islamic finance addressing the flaws of the global financial system (Hassan et al. 2021; Kayed & Hassan, 2011) (Fig. 14.2).

Most of the 36 publications studied were published in Western countries, mainly Europe (69%) and the United States of America (14%), with the remainder coming from Asian publishers. Publishers published the most articles in the United Kingdom, which was 15, followed by six articles from the Netherlands, five from the U.S., two from Saudi Arabia, and one each from Turkey, Malaysia, Pakistan,

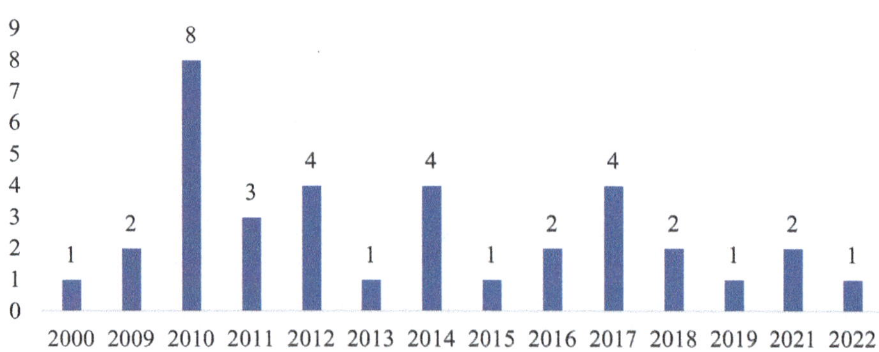

Fig. 14.2 Yearly number of published works in the review. *Source* Author's calculated numbers from google scholar

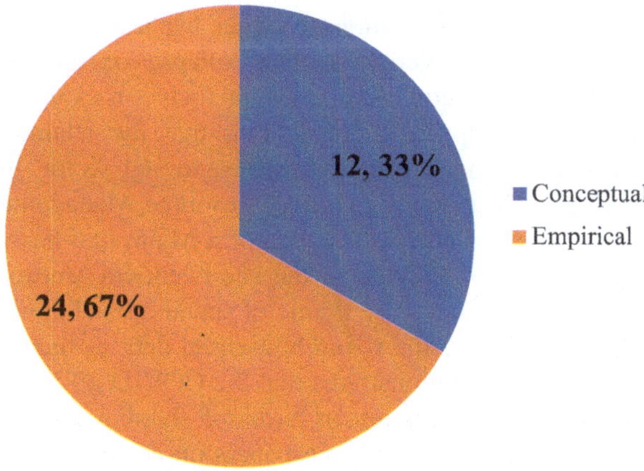

Fig. 14.3 Types of works reviewed. *Source* Authors' calculated numbers from google scholar

Singapore, Switzerland, Ukraine, Germany, and India. As a result, publishers in Islamic countries produced only five (14%) pieces (Appendix 1). Nonetheless, when looking at the first author's affiliation, most correspond to academic or public institutions based in Muslim nations. In addition, a comparison in terms of the types of studies conducted reveals that the proportion of empirically based studies is more dominant than the conceptual types (Fig. 14.3).

We found several topics from the papers reviewed, i.e., subjects related to cluster of Banking and Finance, Economic Thinking, Human Resources Management and Governance, and Spending and Consumption. Most papers, 30 in total, are on Banking and Finance, followed by three papers on economic thinking, two on spending and consumption and one human resource management and governance paper (Fig. 14.4).

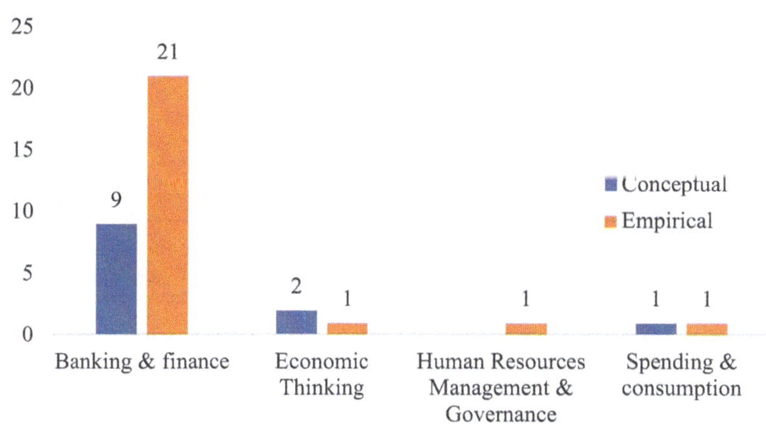

Fig. 14.4 Papers' related topics. Source: Author's calculated numbers from google scholar

Almost all publications connect banking and finance to the 2007–2009 economic crisis triggered by the collapse of the global financial system. Both conceptual and empirical research focus on examining historical data to demonstrate the comparative advantages that the Islamic banking and finance sector enjoyed in comparison to the conventional sector during the financial crisis. Yet, one research addresses Human Resource Management and Governance, explaining the relative outperformance of a Malaysian-based Islamic finance corporation when the financial crisis of the late twentieth century struck the country and other East Asian nations. The theme of spending and consumption provides light on the Islamic notion of the rising household debt as the crux of the last financial crisis. Another work demonstrates that the COVID-19 pandemic-led economic crisis had influenced the economy in terms of shaping impulsive behavior to consumption such as panic buying. Three papers on economic thought shed light on the applicability and transformative power of moral and ethics-based Islamic economic thinking to cope with current economic crises.

Most of the papers assessed generated references to the advancement of research undertaken by other authors. When compared to conceptual papers, the empiricals had the largest citation contribution. Bourkhis & Nabi (2013), Farooq & Zaheer (2015), Parashar & Venkatesh (2010), Olson & Zoubi (2017), Kenourgios et al. (2016), Alqahtani & Mayes (2018), Alam & Rajjaque (2010), Alqahtani et al. (2017), Rosman et al. (2014) and Ibrahim & Rizvi (2018) are the top ten. Meanwhile, citations contributed by conceptual papers are significantly sourced from Ahmed (2009), Smolo & Mirakhor (2010), Causse (2012) relative to Chapra (2011), Matthews & Tlemsani (2010), Zakaria et al. (2012), and Moisseron & Teulon (2014). While in terms of related topics, banking and finance, both empirical and conceptual, generated the most citations, followed by spending and consumption which have the fewest citations, and economic thought has only one (Fig. 14.5).

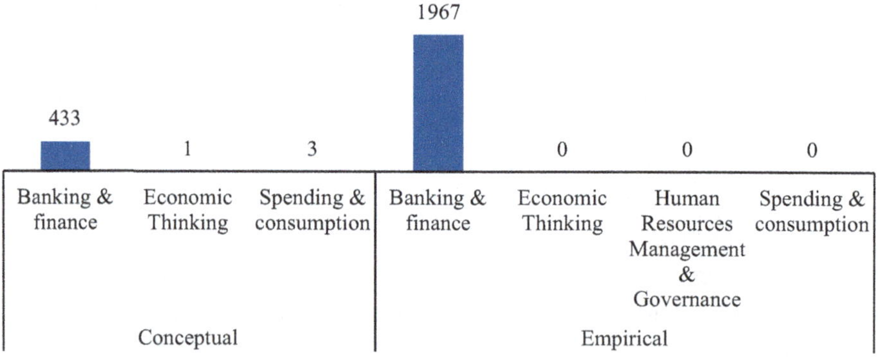

Fig. 14.5 Total papers citation in terms of related topics. *Source* Author's calculated numbers from google scholar

14.3.2 Islamic Banking and Finance: Learning from the Past Great Recession

Most Islamic studies on the economic crisis and its implications have focused on paving the way for a better financial system, which the authors call an alternative or, more prominently, a replacement, meaning a change from conventional banking and financial behaviors to an improvement based on Islamic values and religious, moral principles. Thus, the determination to learn from the global financial crisis is evident in the rise of most authorship of Islamic banking and finance.

Othman et al. (2012), the breakdown of the global financial system in 2007–2009 demonstrates that no economic power can thrive on schemes and instruments that defy divine laws. It is Islamic banking and financial system with usury prohibition as the main feature which may secure financial transaction and to promote fair and legal exchange. Chapra (2011) attributed the US subprime mortgage crisis on financial market discipline. While in Islamic finance, discipline refers to the risk-sharing agreement prior to loan availability. Matthews & Tlemsani (2010) utilize the financial crisis biblical account analogy corresponding to The Financial Tower of Babel which in the end collapses because the symbolic basis of assets do not have the value assumed. And as the price increases stop, Hyman Mynski's financial instability theory predicts the tower's fall. Ejaz and Khan (2014) attribute the crisis to inadequate legal "binding" between nominal and real transactions in modern banking, implying the scale as nominal-real economy gap and risk distribution mechanisms the crisis exposes.

Good accounts are limited to the short period when the crisis occurred. It was through greater capital adequacy ratios that Islamic banking survived 2005–2008 (Chazi & Syed, 2010). While Trad's et al. (2017) support Islamic banks' profitability and stability amid the 2008 crisis. Hasan & Dridi (2011) found that Islamic banks' business models limited the negative impact on profitability in 2008, but risk management shortcomings reduced their profitability compared to conventional banks in 2009. From 25 Islamic and 114 conventional banks in 10 dual-banking nations, Ibrahim and Rizvi (2018) assessed Islamic banks continued financing during the crisis while conventional banks' credit growth fell. Farooq and Zaheer (2015) discovered that Islamic bank branches were less likely to withdraw deposits during financial panics in Pakistan. Islamic banks with conventional operations tend to lend and withdraw deposits during financial panics. Islamic banks outperformed conventional banks in the 2006–2009 period, with profitability during the crisis influenced by capital adequacy, credit risk, financial risk, operational efficiency, liquidity, bank size, GDP, money supply growth rate, banking sector development, and inflation rate (Hussien et al. 2019). Djennas (2016) found that Islamic finance and traditional economic systems performed similarly during the key crises from 1980 to 2013, but through some specific components of the financial stress index, countries that adopt Islamic finance can avoid various crisis situations and economic downturns.

As for the capital market, Islamic Shariah-compliant sustainable funds can stabilize global financial markets during a crisis, according to Hassan (2009). From

December 2002 to December 2007, 136 Islamic equities funds outperformed Dow Jones World Index, Almadar GCC Index, and Dow Jones World Asia Pacific Index benchmarks for one, three, and five years. Hassan (2009)'s findings are crucial to Zaimovic and Dedovic (2021), who found that the Dow Jones Islamic Market World Index (DJIM) Islamic finance index performed well during the COVID-19 pandemic crisis, with positive returns and significantly lower return volatility than during the 2007–2009 financial crisis, suggesting that historical lessons can help predict future economic cycles.

Kassim and Kamil (2012) found that Malaysian Islamic unit trusts from January 2000 to December 2009 can be a good hedging technique when the market weakens. Comparable performance implies that Islamic stocks were a prominent safe-haven asset for G7 stock markets throughout the pandemic crisis (Arif et al. 2022). (Alam & Rajjaque, 2010) revealed that Shariah-compliant equities outperformed the other two portfolios in all assessments during crises: market portfolio (S&P 350) and the market excluding financial businesses' portfolio (S&P-Fin). And according to Kenourgios et al. (2016), Islamic equities and bonds may also reduce risk and instability during times of instability.

Nonetheless, the evidence implies that favorable findings benefiting populist view of the Islamic finance and banking should not be embraced too readily. In 2007–2008, Islamic banks in the Middle East, Africa, and Southeast Asia performed better, but when the real sector got more impacted the following year, they fell much more in profitability (Olson & Zoubi, 2017). Alqahtani and Mayes (2018) observed no significant differences between Islamic and conventional banking throughout the global financial crisis in 76 banks from the GCC (Gulf Cooperation Council) region from 2000 to 2013. Alqahtani & Mayes (2017) revealed GCC Islamic banks were more cost-efficient throughout the financial crisis but lost more afterward. Bourkhis and Nabi (2013) reported both the bank's health had remained relatively similar after the crisis. Rosman et al. (2014) examined 79 Middle East and Asian Islamic banks during the 2007–2010 financial crisis and found that most were inefficient or loss-making. Contagion factors cause Islamic bank returns to decline (Fakhfekh & Hachicha, 2014). Parashar and Venkatesh (2010) revealed that Islamic banks suffered more during the global financial crisis due to weaker capital ratios, leverage, and return on equity. Kassim and Shabri (2010) revealed that Islamic and conventional banking systems were equally vulnerable to financial shocks.

Given the diversity of view, very few seek a middle ground in learning from the conventional finance sector's debacle. Following the financial crisis, the West's perspective may change, allowing Islamic finance to develop globally (Ahmed, 2010) and to become a long-term solution to financial instability that can shield markets from crises (Matthews & Tlemsani, 2010). The interest cancellation policy as a crisis response by the G7 and G20 leaders might show alignment with Islamic banking (Trabelsi, 2011).

As Smolo and Mirakhor (2010) observed, despite the financial crisis's very limited impact on the Islamic financial sector, the capitalist financial system's main flaws are relevant and valuable for Islamic finance's development. Islamic banking and finance growth is part of conventional banking and finance growth. The Islamic finance industry must overcome its vulnerability to future crises by learning from conventional banking and finance's risk mitigation failures (Ahmed, 2009; Causse, 2012) (Table 14.1).

14.3.3 Islamic Economic Thinking

Schools of economic thought emerged from the mainstream economic approach to economic crises. Economics was relevant throughout a period of serious crises. Each crisis marks the end of a school of thought or theory while encouraging other schools to emphasize the most relevant analytical components. New ideas replace old ones in a regular cycle. At some point, the recurring crises would lead to the crisis of economics itself (Acemoglu, 2009; Colander et al. 2009; Lawson, 2009).

Islamic economics reflects Islamic religious principles. Islamic economic knowledge comes primarily from the Holy Qur'an, a divine revelation, unlike mainstream economics, which is based on empirical reasonable arguments. Allah commands Muslims to establish Muhammad Shallallallhu 'Alayhi Wasallam (p.b.u.h) the world's role model in Surah Al-Ahzab verse 21. This command puts the Sunnah—the Prophet's words and deeds—second Islam's source of knowledge after the Qur'an. in Surah An-Najm verses 3–4, Allah informs us, "..Nor does he speak from (his own) inclination, it is not but a revealed revelation," which makes the Sunnah of the Prophet nothing but the Qur'an in the form of actions. A hadith coming from Umm 'Aishah can seal the conclusion without any doubt, stating that the character of Prophet Muhammad (p.b.u.h) is nothing but the character of the Qur'an.

Ascarya's (2017) uses Islamic doctrines of usury prohibition (riba), repudiation of speculation (gharar), moral hazard, debt exploitation, adverse selection, and moral hazard to explain the financial crisis. His research shows that the big recession cannot be separated from Islamic theology and commandment. It clarifies many difficulties, such as interest-based monetary systems and fiat money, poor governance, and susceptible fiscal systems, which have caused inflation and growth. The empirical results extend the literature and clarify the crisis causes that violate Islamic law by using a single gold-standardized global currency and a profit-sharing scheme to mitigate crises effects on economic growth and inflation.

Moisseron' (2014) also lays emphasis on the downfall of global capitalism as a function of debt exploitation, greed, systemic financial hegemony, and speculation. In his view, the applicability of Islam to diverse problems at all ages is based on the progression of Islamic thought from one phase to the next, which corresponds to the challenges and dynamics of each generation in social encounters. Unfortunately, Islamic economic theory has yet to become a transparent model. Islamic economics, in Moisseron's view, has overemphasized banking and finance issues.

Western economists believed their state-directed financial regulation would suffice to avert economic crises, but independence from morals and ethics has

Table 14.1 List of selected articles in review in Islamic banking and finance

No	Authors and year	Titles
1.	Ahmed (2010)	Global financial crisis: an Islamic finance perspective
2.	Ahmed (2009)	Financial crisis: risks and lessons for Islamic finance
3.	Alam and Rajjaque (2010)	Shariah-compliant equities: empirical evaluation of performance in the European market during credit crunch
4.	Alqahtani and Mayes (2017)	The global financial crisis and Islamic banking: the direct exposure to the crisis
5.	Alqahtani and Mayes (2018)	Financial stability of Islamic banking and the global financial crisis: evidence from the gulf cooperation council
6.	Arif et al. (2022)	Pandemic crisis versus global financial crisis: are Islamic stocks a safe-haven for G7 markets?
7.	Bourkhis and Nabi (2013)	Islamic and conventional banks' soundness during the 2007–2008 financial crisis
8.	Causse (2012)	Islamic finance: an alternative finance or an antidote to the crisis of capitalism?
9.	Chapra (2011)	The global financial crisis: some suggestions for reform of the global financial architecture in the light of Islamic finance
10.	Chazi and Syed (2010)	Risk exposure during the global financial crisis: the case of Islamic banks
11.	Djennas (2016)	Business cycle volatility, growth and financial openness: does Islamic finance make any difference?
12.	Ejaz and Khan (2014)	The underlying cause of the global financial crisis: an Islamic perspective
13.	Fakhfekh and Hachicha (2014)	Return volatilities and contagion transmission between Islamic and conventional banks throughout the subprime crisis
14.	Farooq and Zaheer (2015)	Are Islamic banks more resilient during financial panics?
15.	Hasan and Dridi (2011)	The effects of the global crisis on Islamic and conventional banks: a comparative study
16.	Hassan (2009)	After the credit crunch: the future of Shari'ah compliant sustainable investing
17.	Hussien et al (2019)	The performance of Islamic banks during the 2008 global financial crisis Evidence from the GCC countries

(continued)

Table 14.1 (continued)

No	Authors and year	Titles
18.	Ibrahim and Rizvi (2018)	Bank lending, deposits and risk-taking in times of crisis: A panel analysis of Islamic and conventional banks
19.	Kassima and Kamil (2012)	Performance of Islamic unit trusts during the 2007 global financial crisis: evidence from Malaysia
20.	Kassim and Shabri (2010)	Impact of financial shocks on Islamic banks: Malaysian evidence during 1997 and 2007 financial crises
21.	Kenourgios et al. (2016)	Islamic financial markets and global crises: contagion or decoupling?
22.	Matthews and Tlemsani (2010)	The financial tower of babel: roots of crisis
23.	Olson and Zoubi (2017)	Convergence in bank performance for commercial and Islamic banks during and after the global financial crisis
24.	Othman et al. (2012)	Islamic banking: the firewall against the global financial crisis
25.	Parashar andVenkatesh (2010)	How did Islamic banks do during global financial crisis? banks and bank systems
26.	Rosman and Zainol (2014)	Efficiency of Islamic banks during the financial crisis: an analysis of middle Eastern and Asian countries
27.	Smolo and Mirakhor (2010)	The global financial crisis and its implications for the Islamic financial industry
28.	Trabelsi (2011)	The impact of the financial crisis on the global economy: can the Islamic financial system help?
29.	Trad, Rachdi et al. (2017)	Banking stability in the MENA region during the global financial crisis and the European sovereign debt debacle
30.	Zaimovic and Dedovic (2021)	World Economy and Islamic finance: comparison of government policies during the global financial crisis and the COVID-19 crisis

Source Author's compilation

generated recurrent turmoils. Despite the common belief of Islam's compatibility with capitalism, Islam emphasizes morality. Putting into context, Karim (2010) envisions the transformative process toward the new economic order due to Islamic finance expansion and welfare capitalism legacy in Muslim nations (Table 14.2).

14.3.4 Spending and Consumption

The rise in household debt makes one of the most important lessons from the global financial crisis and earlier crises, such as the 1997–1998 East and Southeast

Table 14.2 List of selected articles in review in Islamic economic thinking

No	Authors and year	Titles
1.	Ascarya and Iskandar (2017)	The root causes of financial crisis in Islamic economic perspective
2.	Moisseron and Teulon (2014)	Is morality-based Islamic economics an answer to the global financial crisis?
3.	Karim (2010)	The economic crisis, capitalism and Islam: the making of a new economic order?

Source Authors' compilation

Asian monetary crisis and the financial disaster in 2007–2009 (Zakaria et al. 2012). Islamic economics has a definite sharia approach to household debt. The household debt-induced economic catastrophe was structurally flawed. It stressed breaking Islamic debt norms, resulting in unmanageable family and macro debt.

Zakaria (2012) uses Islamic economics to explain the expanding household debt supply–demand relationship. In addition, government policy backing by lowering interest rates and encouraging lending to financial institutions without checks and risk management. Due to household debt buildup, most unqualified borrowers defaulted. Islam considers debt a final resort for urgent requirements. In contrast, the traditional view perceives debt as enhancing welfare. Banks operate on conventional debt, even though Islam forbids interest and profit. Thus, Zakaria et al. (2012) found that Islamic debt regulation or financing based on household consumption and responsible banking is important.

However, a strong Muslim family needs to be careful about what they buy. Family spending that is wise and simple is a key part of Islam, but it is becoming more and more difficult to follow in today's world of consumerism and ease of shopping. A study by Mukit, et. al. (2021) explains the purchasing awareness of consumers in eight Muslim countries by excluding needless consumption from their consumption list and successfully minimizing food waste at home in the days preceding the COVID-19 pandemic-induced economic catastrophe. Its empirical findings have helped fulfill the gap in crisis-related Islamic consuming literature. Allah says in Surah Ali Imran verse 186, "You will surely be tested in your possessions and in yourselves. And you will surely hear from those who were given the Scripture before you and from those who associate others with Allah much abuse. But if you are patient and fear Allah-indeed, that is of the matters [worthy] of determination" (Table 14.3).

14.3.5 Human Resources Management and Governance

Budhwar and Fadzil (2000) learned how one Southeast Asian Islamic financial organization thrives in times of crisis during the last decade of twentieth century. Their analysis found that the effective blend of domestic work culture and Islamic principles resulted in the unique competitiveness of the object under study,

Table 14.3 List of Selected Articles in Review in Islamic spending and consumption

No	Authors and year	Titles
1.	Mukit et al. (2021)	The economic influence on consumers buying behavior in Islamic countries: evidence from the COVID-19 economic crisis
2.	Zakaria et al. (2012)	Burgeoning household debt: an Islamic economic perspective

Source Authors' compilation

Tabung Haji Malaysia (THM). The theory of corporate resources in attaining sustained competitive advantage (Barney 1991) explains THM's competitive position, including its positive value, uniqueness and scarcity in the industry, imperfect imitability, and inability to be substituted by competitors.

Their work outlined THM's service profile, which was highly valued by its customers, efficient, and follows the principles of equality and fairness. Management practices are unique, emphasizing employees' valued treatment, such as providing opportunities for Hajj, post-retirement care, or in the event of termination of employment. Islamic principles of riba, halal, ummah, musyawarah, muhasabah, and other Shariah concepts function seamlessly. Direct interviews with multi-level management and multi-functional staff increased their research contribution. In addition, involving expert consultation enhanced the analyses. Their work documents Islamic institutions' crisis response and offers hope for future cyclical disturbances.

14.4 Innovating Approach to Our Systematic Literature Review

We recognize the dual challenges posed by the subject of our study. First, it departs from an interdisciplinary mindset in which the economics perspective confronts the study of Islam, which emphasizes the high content of divine morality and ethics in economic practice. Therefore, the usefulness of a systematic literature review with topic-specific question constructs from a single field may not appear as envisaged as in the most case. Secondly, as a discipline in the making, Islamic economics, as argued by Khan (2013), is almost devoid of critical theories as its conventional counterpart, including, of course, the theory of crises or business cycles. It is evident in most of the studies we evaluated, inseparable from doctrinal and conceptual interpretations. As expected, empirical research encompasses most banking and finance-related research. As one might assume, empirical studies span practically all banking and finance research, exhibiting the deductive application of measures and criteria borrowed from established Western financial theories with Islamic adjustments. The I slamization of knowledge has contributed to Islamic banking and finance, a project Islamic scientists launched to incorporate western social science into Islam (Dass, 2016). Islamic economics is often called "interest-free banking" (Vali et al. 2014).

The systematic review exercise adopted in our study focuses on uncovering economic crisis authorship that includes elements from the Islamic population. The Islamic perspective, in this case, seeks to reflect what the authors have produced from their work investigating the economic crisis and how the interpretation is made by incorporating Islamic principles and values into the economic practices of Islamic societies and institutions today. We relied on published works, although not necessarily only literature from published works could be reviewed. Applying this strategy of inclusion only to published literature helps to show how far the picture of the review topic has evolved in a critical and inclusive academic conversation.

The systematic review serves as a means to synthesize the primary research findings, from which we have evaluated enormous volumes of information and, thus, as Gough et al. (2017) argue, argue that they present a comprehensive picture of what is known and what is not, or what works and what does not, regarding the reviewed issue. Advances in policy practice and knowledge development stem from the progress that literature has made so far. Indeed, obstacles are associated with applying for systematic reviews in social fields. According to Tranfield et al. (2003), several methods of knowledge construction exist in social science, allowing for diverse methodological approaches. Petticrew and Roberts (2008) argue the rationale behind conducting reviews with various research methodologies, such as those from the social sciences, which include various qualitative designs. These authors also suggest overcoming methodological issues in the broader-social area-motivated reviews. Incorporating qualitative designs into systematic evaluations remains a formidable obstacle (Dixon-Woods et al. 2006). Nonetheless, the concentration of systematic reviews' first applications in the medical area and their extension into the social sciences demonstrate that efforts for innovation can accommodate more diverse applications.

14.5 Conclusion

The authorship on Islamic banking and finance reflects the majority of works investigating the unfolding of economic crises in the normative view of Islam. Since the Great Recession of 2007–2009, criticism from the Islamic establishment has grown louder, highlighting the failure of the world economic system through financial modus operandi as the main culprit. The collapse of the world financial system at the same time saw the emergence of Islamic finance industry that was growing significantly in terms of growth. The nature of financial capitalism, which justifies the accumulation of profits in the name of maximum utility and individual rationality, confronts the principles of Islamic economics, which favor justice and equal opportunity to grow by mutual agreement in risk sharing. Claims of moral superiority derived from doctrinal interpretations of religion somehow justify the repetition of crises as a consequence of violating divine commands, accompanied by the *a pri ori* proposition that economic policies inspired by religious consciousness can produce true prosperity and a bulwark against potential disasters or possible crises. Crisis, in religious theology, is often interpreted as punishment for the sins of transgressing the divine commandments. It is vital to identify and live economic practices that do not contradict God's law or sharia to prevent recurring crises in the future. The thesis is that the Islamic finance and banking sector, as a function of Islamic economics, is a worthy substitute to protect the world economy from the turmoil of future economic cycles. However, the axiom of Islam as the one-fits-all solution (*rahmatan lil 'alamin*) in reality requires *aposteriori* verification from a scientific base.

The increase in empirical and conceptual studies concentrating on the financial and banking sectors has been an essential feature of the literature on economic crises and critical reflection on Islam or Muslims on the one hand. Most of them explicitly attempt to prove the extent to which the Islamic sector can provide different evidence when the conventional sector suffers in times of crisis. While some studies find the expected results, especially in times of ongoing crisis, testing of later data shows the opposite pattern (Alqahtani & Mayes, 2017, 2018; Bourkhis and Nabi, 2013; Hasan & Dridi, 2011; Olson & Zoubi, 2017; Rosman et al. 2014). In light of these data, the populist notion of Islamic finance and banking competence over their non-Islamic (conventional) counterparts during the current financial crisis should not be adopted too quickly. At the same time, the Islamic finance and banking sector can optimize the necessary learning from past crises to become more resilient to potential uncertainties, especially from the major weaknesses of the conventional financial sector (Smolo & Mirakhor, 2010) as well as from the sector's vulnerabilities and failure to mitigate risks (Ahmed, 2009).

We also explore additional findings from the literature pertinent to the scope of the investigated subject, namely Islamic economic thought. Ascarya (2017)'s study is notable because his empirical investigation demonstrates how an economic operating model based on Islamic principles can prevent signs of a crisis by suppressing the weight of the reverse behavior on the two measures of inducing inflation and restraining or harming economic growth. This empirical endeavor is also consistent with Moisseron and Teulon (2014)'s conclusion that global capitalism has failed because it maintains an economic sustainability that encourages debt exploitation, greed, financial hegemony, and speculation. The studies of Zakaria et al. (2012) and Mukti et al. (2021) provide important insights into how the economic crisis not only fosters a prudent spending attitude among individual Muslims, but also an awareness to avoid extravagant behavior (*isyraf*) in response to the crisis. At times of crisis, the competitive advantage of Islamic institutions demonstrates the combination of Islamic work culture with domestic cultural values as a potent synergistic chemistry (Budhwar & Fadzil, 2000).

Acknowledgements This research was made possible by the support of the Research and Service to Community Institute (LP3M) of the Raja Ali Haji Maritime University, based in Tanjungpinang of the Riau Islands, Indonesia.

Appendix 1 Journal publishers and country

Publishers	Number of studies selected	Country base
Emerald group publishing limited	11	UK
Elsevier	6	Netherlands
Routledge Taylor and Francis group	3	UK
John Wiley & Sons	2	USA
King Abdulaziz university scientific publishing center	2	Saudi Arabia
Wiley online library	2	USA
Borsa Istanbul Anonim Sirketi	1	Turkiye
Bussiness perspectives	1	Ukraine
https://www.degruyter.com	1	Germany
Inderscience publishers	1	Switzerland
International digital organization for scientific information	1	Pakistan
Macmillan publishers Ltd	1	UK
Penerbit Universiti sains Malaysia	1	Malaysia
Serials publication	1	India
The clute institute	1	USA
World scientific publishing company	1	Singapore

References

Acemoglu D (2009) The crisis of 2008: lessons for and from economics. Crit Rev 21(2–3):185–194

Ahmed A (2010) Global financial crisis: an Islamic finance perspective. Int J Islam Middle East Financ Manag 3(4):306–320

Ahmed H (2009) Financial crisis: risks and lessons for Islamic finance. ISRA Int J Islam Financ. Emerald Group Publishing Limited, UK

Alam N, Rajjaque MS (2010) Shariah-compliant equities: empirical evaluation of performance in the European market during credit crunch. J Financ Serv Mark 15(3):228–240

Alqahtani F, Mayes DG (2018) Financial stability of Islamic banking and the global financial crisis: evidence from the gulf cooperation council. Econ Syst 42(2):346–360

Alqahtani F, Mayes DG, Brown K (2017) Islamic bank efficiency compared to conventional banks during the global crisis in the GCC region. J Int Finan Markets Inst Money 51:58–74

Alqahtani F, Mayes DG (2017) The global financial crisis and Islamic banking: the direct exposure to the crisis. Banks Bank Syst

Arfah A, Olilingo FZ, Syaifuddin S, Dahliah D, Nurmiati N, Putra AHPK (2020) Economics during global recession: Sharia-economics as a post COVID-19 agenda. J Asian Financ, Econ Bus. Korea Distribution Science Association (KODISA), South Korea

Arif M, Naeem MA, Hasan M, Alawi SM, Taghizadeh-Hesary F (2022) Pandemic crisis versus global financial crisis: are Islamic stocks a safe-haven for G7 markets? Econ Res-Ekon Istraz. Routledge Taylor and Francis Group, UK

Ascarya A, Iskandar D (2017) The root causes of financial crisis in Islamic economic perspective. Ekonomi Islam 1930(02):1–11

Ascarya (2017) The real determinants of financial crisis and how to resolve it in Islamic economics perspective of Eco. Int J Econ Res. Serials Publication, India
Askari H, Mirakhor A (2015) The next financial crisis and how to save capitalism.
Bakar O (1984) The question of methodology in Islamic science. Muslim Educ Q 2(1):16–30
Barney J (1991) Firm resources and sustained competitive advantage. J Manag 17(1):99–120
Bettany-Saltikov J, Cronin P (2013) How to do a systematic literature review in nursing. A step-by-step guide. Nurse education in practice, vol 13. McGraw-Hill Open University Press
Bhattacharya S, Goodhart CAE, Tsomocos DP, Vardoulakis AP (2015) A Reconsideration of Minsky's financial instability hypothesis. J Money, Credit, Bank 47(5):931–973
Bhuiyan AB, Ismail AG, Noor AHM, Solaiman M, Ali MJ (2020) Islamic economics methodology patterns of reasoning and the structure of theories. Int J Shari'ah Corp Governance Res 3(2):10–24
Bourkhis K, Nabi MS (2013) Islamic and conventional banks' soundness during the 2007–2008 financial crisis. Rev Financ Econ 22(2):68–77
Boyer R (2013) The global financial crisis in historical perspective: an economic analysis combining Minsky, Hayek, Fisher, Keynes and the regulation approach. 3(3):93–139
Budhwar P, Fadzil K (2000) Globalization, economic crisis and employment practices: lessons from a large Malaysian Islamic institution. Asia Pac Bus Rev 7(1):195–198
Causse G (2012) Islamic finance: an alternative finance or an antidote to the crisis of capitalism? In: International symposia in economic theory and econometrics. Emerald Group Publishing Limited, UK
Chapra MU (2011) The global financial crisis: some suggestions for reform of the global financial architecture in the light of Islamic finance. In: Thunderbird international business review. John Wiley & Sons, US
Chazi A, Syed LAM (2010) Risk exposure during the global financial crisis: the case of Islamic banks. Int J Islam Middle East Financ Manag. Emerald Group Publishing Limited, UK
Chowdhury A, Żuk P (2018) From crisis to crisis: capitalism, chaos and constant unpredictability. Econ Labour Relat Rev 29(4):375–393
Colander D, Goldberg M, Haas A, Juselius K, Kirman A, Lux T, Sloth B (2009) The financial crisis and the systemic failure of the economics profession. Crit Rev 21(2–3):249–267
Cook DJ, Greengold NL, Ellrodt AG, Weingarten SR (1997a) The relation between systematic reviews and practice guidelines. Ann Intern Med 127(3):210–216
Cook DJ, Mulrow CD, Haynes RB (1997b) Systematic reviews: synthesis of best evidence for clinical decisions. Ann Intern Med 126(5):376–380
Dass MA (2016) Islamization of social sciences: a critique of contemporary scholars. Doctoral dissertation, Aligarh Muslim University
Denyer D, Tranfield D (2009) Producing a systematic review. In: Buchanan DA, Bryman A (eds) The SAGE handbook of organizational research methods. The sage handbook of organizational research methods
Dixon-Woods M, Bonas S, Booth A, Jones DR, Miller T, Sutton AJ, Young B (2006) How can systematic reviews incorporate qualitative research? a critical perspective. Qual Res 6(1):27–44
Djennas M (2016) Business cycle volatility, growth and financial openness: does Islamic finance make any difference? Borsa Istanbul Rev 16(3):121–145
Ejaz N, Khan H (2014) The underlying cause of the global financial crisis: an Islamic perspective. Econ Pap 33(1):45–54
Fakhfekh M, Hachicha N (2014) Return volatilities and contagion transmission between Islamic and conventional banks throughout the subprime crisis: evidence from the DCC-MGARCH model. Int J Manag Financ Account 6(2):133–145
Farooq M, Zaheer S (2015) Are Islamic banks more resilient during financial panics? Pac Econ Rev 20(1):101–124
Furqani H, Haneef MA (2015) Methodology of Islamic economics: typology of current practices, evaluation and way forward. In: Islamic economics: theory, policy and social justice, vol 23
Gough D, Oliver S, Thomas J (2017) An introduction to systematic reviews. Sage
Haralambie G (2011) The global crisis and cyclical theory. Theoret Appl Econ 18(11)

Hasan M, Dridi J (2011) The effects of the global crisis on Islamic and conventional banks: a comparative study. J Int Econ Commer Policy 2(2):163–200

Hassan A (2009) After the credit crunch: the future of Shari'ah compliant sustainable investing. Humanomics 25(4):285–296

Hassan MK, Khan A, Paltrinieri A (2021) Islamic finance: a literature review. In: Islamic finance and sustainable development: a sustainable economic framework for Muslim and Non-Muslim countries, pp 77–106

Hoover KD (2016) The crisis in economic theory: a review essay. J Econ Lit 54(4):1350–1361

Hussain Gattoo M, Gattoo H (2017) Modern economics and the Islamic alternative: disciplinary evolution and current crisis. Int J Econ, Manag Account 25(2):173–203

Hussien ME, Alam MM, Murad MW, Wahid A (2019) The performance of Islamic banks during the 2008 global financial crisis evidence from the Gulf cooperation council countries. J Islam Account Bus Res. Emerald Group Publishing Limited, UK

Ibrahim MH, Rizvi SAR (2018) Bank lending, deposits and risk-taking in times of crisis: a panel analysis of Islamic and conventional banks. Emerg Mark Rev. Elsevier, Netherlands

Imperia A, Maffeo V (2011) As if nothing were going to happen: a search in vain for warnings about the current crisis in economic journals with the highest impact factors impact factors: Andrea imperia and Vincenzo Maffeo. In: The global economic crisis: new perspectives on the critique of economic theory and policy, p 19. Routledge, London

Islahi AA, Jeddah SA (2005) Contribution of Muslim scholars to economic thought and analysis. Scientfc Publisher Center, King Abdul Aziz University, Jeddah

Karim WJ (2010) The economic crisis, capitalism and Islam: the making of a new economic order? Globalizations. Routledge Taylor and Francis Group, UK

Kassim SH, Kamil S (2012) Performance of Islamic unit trusts during the 2007 global financial crisis: Evidence from malaysia. Asian Acad Manag J. Penerbit Universiti Sains Malaysia, Malaysia

Kassim SH, Shabri M (2010) Impact of financial shocks on Islamic banks: Malaysian evidence during 1997 and 2007 financial crises. Int J Islam Middle East Financ Manag. Emerald Group Publishing Limited, UK

Kayed RN, Hassan MK (2011) The global financial crisis and Islamic finance. Thunderbird Int Bus Rev 53:551–564. Springer

Kenourgios D, Naifar N, Dimitriou D (2016) Islamic financial markets and global crises: contagion or decoupling? Econ Model. Elsevier, Netherlands

Khan M (2013) What is wrong with Islamic economics? what is wrong with Islamic economics?

Lawson T (2009) The current economic crisis: its nature and the course of academic economics. Camb J Econ 33(4 SPEC. ISS.):759–777

Mahomedy AC (2013) Islamic economics: still in search of an identity. Int J Soc Econ 40(6):556–578

Mahomedy AC (2016) The ontological foundations of modern economics and Islam: how do they compare. https://saef.ukzn.ac.za/

Malkawi M (2020) Fall of capitalism and rise of Islam. https://books.google.com/

Matthews R, Tlemsani I (2010) The financial tower of babel: roots of crisis. Int J Islam Middle East Financ Manag. Emerald Group Publishing Limited, UK

Michail N (2021) Financial instability BT-money, credit, and crises: understanding the modern banking dystem. In: Michail N (ed), pp 137–154. Springer International Publishing, Cham

Mirakhor A, Askari H (2017) Ideal Islamic economy: an introduction. Springer

Moisseron JY, Teulon F (2014) Is morality-based Islamic economics an answer to the global financial crisis? J King Abdulaziz Univ, Islam Econ. King Abdulaziz University Scientific Publishing Center, Saudi Arabia

Mukit MMH, Nabila NJ, Abdel-Razzaq AI, Shaznin KF (2021) The economic influence on consumers buying behavior in Islamic countries: evidence from the COVID-19 economic crisis. Economics. Germany. https://www.degruyter.com/.

Olson D, Zoubi T (2017) Convergence in bank performance for commercial and Islamic banks during and after the global financial crisis. Q Rev Econ Financ. Netherlands, Elsevier

Othman R, Aris NA, Azli RM, Arshad R (2012) Islamic banking: the firewall against the global financial crisis. J Appl Bus Res. The Clute Institute, US
Parashar SP, Venkatesh J (2010) How did Islamic banks do during global financial crisis? Banks Bank Syst 5(4):54–62
Petticrew M, Roberts H (2008) Systematic reviews in the social sciences: a practical guide. Syst Rev Soc Sci: Pract Guid 4(4):1–336
Pryor FL (2007) The economic impact of Islam on developing countries. World Dev 35(11):1815–1835. https://doi.org/10.1016/j.worlddev.2006.12.004
Rehman SS, Askari H (2010) How Islamic are Islamic countries? Glob Econ J 10(2)
Rosman R, Wahab NA, Zainol Z (2014) Efficiency of Islamic banks during the financial crisis: an analysis of middle eastern and Asian countries. Pac Basin Financ J. Elsevier, Netherlands
Schumpeter JA (1939) Business cycles, vol 1. Mcgraw-hill, New York, pp 161–174
Siddiqi MN, Chapra MU, Khan MF, Seidu AM, Masri RYA, Metwally MM (2009) Issues in the international financial crisis from an Islamic perspective. Islamic economic research center. King Abdulazis University, Jeddah
Smolo E, Mirakhor A (2010) The global financial crisis and its implications for the Islamic financial industry. Int J Islam Middle East Finance
Sweet M, Moynihan R (2007) Improving population health; the uses of systematic reviews
Thomas I (2009) Hypothesis as an explanation of the crisis working paper, vol 11.
Trabelsi M (2011) The impact of the financial crisis on the global economy: can the Islamic financial system help? J Risk Financ 12(1):15–25
Trad N, Rachdi H, Hakimi A, Guesmi K (2017) Banking stability in the MENA region during the global financial crisis and the European sovereign debt debacle. J Risk Financ. Emerald Group Publishing Limited, UK
Tranfield D, Denyer D, Smart P (2003) Towards a methodology for developing evidence-informed management knowledge by means of systematic review*. Br J Manag 14(3):207–222
Vali S, Nasr R, Vali S, Nasr R (2014) Islamization of knowledge: a critical overview. Islam Stud 30(3):387–400
Visser H (2019) The Islamic economy: its origin, its world view and its claims. Cent Eur Rev Econ Manag 3(4):53–89
Whalen CJ (2011) Financial instability and economic security after the great recession. Edward Elgar Publishing.
Wolf FM, Shea JA, Albanese MA (2001) Toward setting a research agenda for systematic reviews of evidence of the effects of medical education. Teach Learn Med 13(1):53–60
Zaimovic A, Dedovic L (2021) World economy and Islamic finance: comparison of government policies during the global financial crisis and the COVID-19 crisis. J King Abdulaziz Univ, Islam Econ 34(1):79–92
Zakaria RH, Kader RA, Jaafar NIM, Marican S (2012) Burgeoning household debt: an Islamic economic perspective. Middle East J Sci Res. International Digital Organization for Scientific Information, Pakistan

Sustainability Indicator Performance Measurement Framework for the Malaysian Defence Industry Offset Program

15

Mohammad Noor Abdullah, Norlaila Mazura Hj. Mohaiyadin, Hafizah Mat Nawi, Mohd Nor Yahaya, and Ariffin Ismail

Abstract

The Framework of Sustainability Assessment (FSA) in businesses is a topic of considerable pressing interest to the nation, academics, and business practitioners. Unfortunately, the field remains unstructured mainly, with no indicators framework for evaluating Malaysian defence industry offset performance. This study aims to construct a systematic indicator selection approach for sustainability, identify indicators suitable for measuring sustainability in defence industry offset programs and propose a new dimensional or categorisation of indicators to address sustainability in a holistic approach. This study executes systematic reviews by identifying common and consistently used sustainability indicators that are suitable for measuring defence industry offset programs. The selected indicators are organized and classified into different dimensions and sub-categories. The study has taken a holistic approach to defence industrial sustainability based on the bespoke approach, revising the three most common categories of Environmental, Financial and Social into three new dimensions—Community, Economy, and Organization (CEO). This study should provide an overview of the possibilities available to government agencies and businesses who wish to embed sustainability indicators to establish goals, monitor success, and assist managers in identifying what types of anticipation need to be emphasised in the future as part of a continuous improvement and iteration process.

M. N. Abdullah (✉) · N. M. Hj. Mohaiyadin · H. M. Nawi · M. N. Yahaya · A. Ismail
Faculty of Management and Defence Studies, National Defence University of Malaysia, Kuala Lumpur, Malaysia
e-mail: mohammadnoor@upnm.edu.my

A. Rafiki et al. (eds.), *Open Innovation in Small Business*, Contributions to Environmental Sciences & Innovative Business Technology,
https://doi.org/10.1007/978-981-99-5142-0_15

Keywords

Sustainability • Performance measurement • Defence industry • Offset program • Malaysia

15.1 Introduction

An offset is a countertrade agreement where a company compensates another country for a hard currency purchase of an unspecified goods (Kenton 2021). As reported by Technology Depository Agency (2018), Malaysia has implemented the inaugural countertrade program leveraging government procurement starting in 1987, while Ariffin and Arof (2019) argued that countertrade can be used as a mechanism to facilitate reciprocal trade between two nations, which involves in importing products from abroad and receiving payment in the form of goods equal to the imported items. In commerce, offsets can be referred to as countertrade, industrial collaboration, industrial participation, business value development and governmental procurement (Malm 2016).

Malaysia has been practising the offset program as among the national economic development tools to significantly increase Malaysia's global marketability by expanding its own industry, technology, and economy. As stated by Balakrishnan and Matthews (2009), the rise of the Malaysian defence offset program is principally associated with the acquisition of defence equipment and has been used as a dominant mechanism in promoting Malaysian defence industrialisation. The Malaysian offset program was instituted as Countertrade Program in 1987. Since then, it has evolved into Countertrade and Offset Policy in 2005 and 2014, and the Ministry of Finance (MoF) changed the national offset policy to Industry Collaboration Program (ICP). The new offset policy was issued and published as Policy and Guideline on Industry Collaboration Program (ICP) in government procurement (2nd edition). The most current offset policy circulated by the Malaysian Treasury in 2022 is acknowledged as Policy and Industrial Collaboration Program Guidelines (ICP) in Government Procurement. The Malaysian ICP program comprises of three groups; the Economic Enhancement Program (EEP), Offset and Countertrade. EEP is applied to local companies while Offset and Countertrade programs are applied to foreign companies. Technology Depository Agency (TDA) is the agency to operationalise a national ICP management framework incorporated on 11 March 2015 under the prerogative of MoF. The Malaysia ICP authority, TDA monitor, analyse, and manage all matters related to the ICP implementation in Malaysia (Technology Depository Agency, n.d.).

Offset is a type of economic regulation imposed by the purchasing nation to satisfy a contractual need for the sale of security products set forth by the government of the purchasing nation (Yedvav et al. 2022). In other words, as a foreign company successfully secured the contract to supply defence goods, the company must reimburse part of the value via means other than monetary. As for

Matthews and Anicetti (2021), a procurement country's requirement is that the foreign arm supplies must offer additional offsets to make up for the astronomically high procurement costs results in an offset. Thus, the compensation procedures that are necessary as a prerequisite for buying equipment from an Original Equipment Manufacturer (OEM) or Local Prime Company can be referred to as offset arrangements. Referring to MoF, the ICP program was derived from the government procurement contract, a procurement contract by a Government Linked Company (GLC) or any other procurement that utilises funds provided or issued by the Malaysian Government. Not all government contracts will induce the ICP program; it shall only be triggered if the contract is reaching the predetermined threshold value. Wessner and Wolff (1997) mentioned that offset is not free and comes with a price tag in which the cost of implementation of an offset program is borne by the OEM or the seller. The potential of securing a sale is the primary motivation for suppliers to engage in offsets. Rheinmetall (n.d.) defined the offset program as a reciprocal trade agreement between a nation buying industrial goods and a corporation exporting those commodities along with referral programs given in return for certain purchases without cost or complimentary.

Additionally, it is claimed that technology innovation can be enhanced through the offset agreement as it can offer significant opportunities in promoting innovation. Generally, nations are increasingly anticipated to rely on offsets for the technology transfer of advanced manufacturing skills in hoping to gain economic competitive advantages through the processes of the business world. Although the objectives of sellers and procurers in the defence offset business appear to have conflict interests, a focus on a "win–win" is an innovative strategy that might uncover the shared value. If offsets are used innovatively, it can assist the nation to develop progress towards expertise in advanced manufacturing and related services, which over time could result in a change of its economic capacity. SME Aerospace (SMEA) is an excellent case of successful Malaysian aerospace company that was formed through an innovative offsets program venture with BAE Systems to produce Hawk pylons, a structure on the wing of an aircraft used for supporting an engine or carrying a weapon, fuel tank, or other load. Since operating in 1990, SMEA has built up its capacity and is presently producing aircraft parts, components, and sub-assemblies, integrated into BAE Systems' global military and civil supply chain, as announced by SMEA in her website. This shows that SMEA is able to develop diversified industrial manufacturing capabilities which then results in expanding their company operations throughout various nations.

Moreover, innovation is vital in securing continuous success for any organization. According to Pichlak and Szromek (2021), innovation plays a key role in promoting sustainability. Innovation is generally defined as the adoption of new or noticeably improved product attributes, modified delivery methods, marketing strategies, or organisational methods in company operations, workplace structures, or external relations. Besides, the increased focus on sustainability has a big impact on performance management. In general, regardless of the performance assessment system used, a firm must translate its overall objectives into sustainability-related activities in each important performance area if it wants to improve its sustainability performance. Following that, the organisation must create precise measurement indicators so that it can assess how well it is accomplishing its objectives in each important area.

There are several reasonable justifications for pursuing this research. Firstly, according to the Federal Auditor General's Report, no evaluation has been carried out to assess the impact of Ministry of Defence (MinDef) procurement on the achievement and sustainability of the offset program beyond its completion period. From literature analysis, researchers have identified and produced a mass of sustainability indicators for industry manufacturing enterprises but none for primarily addressing the indicators used to measure the Malaysian defence industry offset program. The absence of this proceeding has motivated this study to examine the Malaysian defence industry offset program. Secondly, despite the huge spending involved in public funds in MinDef transactions for defence procurement, the outcome regarding their short- and long-term impacts on the local defence industry is still undetermined. With the emergence of this huge offset program credit value and variation range of industrial and commercial compensation activities, it is imperative to develop a set of sustainability indicators as a facilitating mechanism for the Malaysian bureaucratic and defence industry to evaluate the sustainability of their respective offset program. Finally, formulating a sustainability indicator measurement framework for the Malaysian Defence Offset Program is a topic of considerable pressing interest, both to the nation, academics and business practitioners. Unfortunately, the field remains largely unstructured with no indicators framework for evaluating defence industry offset performance. This research was commenced to create new measurement of a novelty theoretical framework to identify the elements of indicators suitable for the defence industry used in measuring the effectiveness of their respective offset program.

Therefore, the aim of this study is to develop a set of feasible sustainability indicators for measuring the Malaysian defence industry offset program. The researcher will formulate to accommodate an uncomplicated yet practical, straightforward, and user-friendly indicator framework that is harmonious between comprehensiveness and manageability. Through this innovation, this paper will deliver importance functions, firstly it allows the identification of an issue that is a mandatory civil policy interference through the preference of an applicable policy mechanism. Secondly, it provides a new metrics of performance indicators identification that is based on an innovative theoretical approach to assess the progress and sustainability of an offset program in meeting organisational goals. The study objectives are (1) to construct a systematic indicator selection approach for sustainability, (2) to identify indicators suitable for measuring sustainability in defence industry offset programs and (3) to propose a new dimensional or categorisation of indicators to address sustainability in a holistic approach. Furthermore, this paper intends to address three research questions. (1) Research question one (RQ1): How to construct a systematic indicator selection approach for sustainability? Research question two (RQ2): What are the indicators suitable for measuring sustainability in defence industry offset programs? Research question three (RQ3): How to establish a new dimensional or categorisation of indicators to address sustainability in a holistic approach?

15.2 Literature Review

The literature review was utilised as a foundation and platform to overview the issues for evaluating the outcome and long-term business sustainability. This was followed with content analysis to identify common and consistently used sustainability indicators suitable for measuring defence industry offset programs. The selected indicators were then organised which later classified into various dimensions as well as sub-categories.

15.2.1 Industry Sustainability

In 1987, Dr. Gro Harlem Brundtland, the former Prime Minister of Norway, introduced the concept of sustainable development in her Brundtland Report, where she explained that sustainable development is defined as development that meets current needs without jeopardising future generations' ability to meet their own needs. Jarvie (2016) went on to discuss how social fairness, economic progress, and the environment are all intertwined with sustainable development in the Brundtland Report. According to Manulak (2022), Brundtland Report published by World Commission on Environment and Development has popularised the concept of sustainable development, which was also known as the "Brundtland Commission." Since then, numerous governments, corporate professionals, and academics have made efforts to interpret the conceptual goal of sustainable development into use in the real-world situations.

Alonzi (n.d.) in her article "Proposal for NGOs" defined sustainability as an organization's capacity to carry out its objective or programme far into the future. Even if every endeavour must come to an end ultimately, its effects should endure. Alonzi (n.d.) explained that the donor also wants to gain insight into how the initiative and its effects will last beyond their direct engagement in it. A project or organisation can be sustainable in three primary categories: organisational, financial, and community sustainability. As an illustration, in business sustainability, the completed project should be able to continue operating and generating profits for the ensuing few years.

According to The Straits Times (2015), the CEO of Bursa Malaysia was implementing new guidelines to assist listed issuers in improving their sustainability-related disclosures to meet the evolving information requirements of various stakeholders. According to the Chief Regulator Officer of Bursa Malaysia, there is no single definition of sustainability. However, within the framework of Bursa Malaysia, sustainability is defined as conducting business responsibly by considering the effects on the economy, environment, and social context, all the while ensuring long-term viability. It was subsequently stated that Bursa Malaysia Securities Bhd. had improved sustainability reporting criteria in the Main Market Listing Standards and the ACE Market Listing Requirements with the intention of elevating the sustainability practises and disclosures of listed issuers (Chung 2022).

As reported in Bursa Malaysia (2015), as companies pay more attention to how their operations affect the economy, environment, and society, sustainability is becoming a bigger focus in business management. A holistic approach to business management is advocated to ensure corporate continuity and long-term gains by taking into account financial, economic, environmental, and social (EES) risks and possibilities. The statements issued by the bureaucracy have shown sustainability as a business imperative in Malaysia. In terms of the defence industry offset program, sustainability is how the offset program beneficiaries (local defence industrialist) carry out the project activities even after the completion of the offset contract. The Contract shall be considered complete when all terms and conditions have been complied with by the OEMs and the OEMs have discharged all reciprocal offset obligations successfully.

15.2.2 Measurement Effectiveness of Sustainability

Papulová et al. (2021) explained that since the end of the twentieth century, the managers' preferred areas of interest have been connected to the success of their organisations and enterprises remained competitive and progressive in the long run; thus, implementation of a suitable performance measurement system (PMS) is essential. Companies' business activities need to be systematically and consistently evaluated and measured. According to Balakrishnan (2007), effectiveness is achieving the desired or intended results. The achievement of objectives and the progression towards reaching them should serve as the basis for evaluating effectiveness. Effectiveness also may assist in assessing a nation's offsets objective(s).

The Federal Auditor General's Report on the MinDef ICP program released on 22 March 2022 concluded that there were no criteria for evaluating the outcome and impact to measure the achievement of the long-term objectives of the MinDef ICP program. Hence, the sustainability of the defence ICP program could not be assessed. The MoF, which is the nation's ICP custodian, agrees with the audit view and concedes to amend the ICP Policy and Guidelines to address the issues for assessing and evaluating the outcome and long-term sustainability of MinDef ICP programs. In this audit review, Auditor General has identified 29 MinDef main procurement contracts worth RM 31.378 billion that have exceeded the ICP threshold value and are thus subjected to the ICP program. Table 15.1 contains a summary of ICP obligated from 2010 to 2020 valued at the time of the contract.

The Auditor General has concluded that the Ministry of Defence (MinDef) has not comprehensively evaluated the Industrial Collaboration Programme (ICP's) results and its long-term effects. The report also recommended measures to address

Table 15.1 Main procurement contracts that reached the ICP threshold value

No	Main procurement contract	Value of main procurement contract (RM million)	ICP implementation year
1	Pilatus PC MK II Training Aircraft	138.46	2015–2017
2	Eurocopter EC725	1,640.00	2011–2019
3	In-Service Support dan Extended In-Service Support Submarine	694.35 & 2,000.00	2010–2017
4	Combat Net Radio	175.53	2014–2017
5	AV-4 Vehicle Lipan Bara Armored AV-4	140.90	2016–2020
6	Helicopter Rental	155.43	2017–2019
7	V-Shorad Starstreak	897.00	2015–2019
8	Seventh Amendment of Carrier Aircraft	245.57	2019–2022
9	AV-8 Armored Vehicle	7,550.00	2012–2022
10	Littoral Combat Ship	9,000.00	2014–2023
11	Submarine Refit	1,160.00	2015–2022
12	Phase 1 & 2 Earth Satellite System	673.60	2016–2021
13	In-Service Support EC725	215.00	2017–2020
14	Maintenance, Repair and Overhaul (MRO) C130 Hercules	200.00	2017–2020
15	Lease of Secure Satellite Communication Channel	519.29	2013–2029
16	Littoral Mission Ship	1,200.00	2017–2023
17	Naval Strike Missile Surface to Surface Missile	610.84	2018–2024
18	105MM Light Gun LG1	263.15	2018–2021
19	Supplying Petroleum Materials to MAFs	875.00	2017–2020
20	Aircraft Combined Maintenance VIP/VVIP Royal Malaysian Airforce	139.60	2019–2022
21	Six-Armed Reconnaissance Multipurpose Helicopters	321.90	2015–2017
22	Maintenance, Repair and Overhaul (MRO) Sukhoi SU-30MKM	371.70 (Additional 1) 322.50 (Additional 2) 495.00 (Additional 4)	2016–2019 (Additional 1 & 2) 2019–2021 (Additional 4)

(continued)

Table 15.1 (continued)

No	Main procurement contract	Value of main procurement contract (RM million)	ICP implementation year
23	Cyber Defence Operation Centre Project for the Malaysia Armed Forces	337.08	2015–2018
24	Construction of Nuri PUTD Aircraft Hangar	123.75	2018–2021
25	Military Quarter (RKAT) Camp Bera, Pahang	188.18	2018–2021
26	RKAT Kem Segamat, Johor	142.83	2018–2020
27	Road construction from Ba'kelalan to Bario Phase 2 (B), Sarawak	116.20	2016–2018
28	Construction of a new road from Nanga Merit to Sungai Baleh Bahagian Kapit, Sarawak	148.24	2016–2019
	Upgrading Log Roads and Building New Road from Belaga to Nanga Merit Kapit Division, Sarawak	316.41	2016–2020
Total		31,377.51	

Source Malaysian National Audit Department Portal (2022), The Federal

the issues for the criteria for assessing the ICP's outcomes and impacts. Moving on, the need to evaluate the sustainability of the ICP program is aligned and in conformity with the latest ICP Policy and Guidelines 2022 as well as the previous ICP policies, which perceived the necessary crieria for OEM as the ICP provider to persuade the government that the proposed ICP project has long-term sustainability from the perspective of operating the acquired assets, national development, and contribute to the nation's economic growth, both during and beyond the designated ICP implementation period.

Despite the huge value of outlay involved in MinDef transactions in defence procurement, the outcome has been dismal and still mired by continuous challenges (Balakrishnan and Johar 2021). Furthermore, Roslan et al. (2018) mentioned that no empirical study has been carried out to assess the impact of this defence offset procurement on the social welfare benefit in Malaysia. It is detrimental to the defence industrialist consensus that there is no check and balance to the implementation and success of these defence offset procurement practices.

According to Prtak (2019), initiatives to measure productivity development across the entire economy are essential and must take into account the public sector, governmental institutions, and services. It is also the only sustainable approach to boost long-term living standards. Since the goods and services are frequently provided through non-market administrative channels or implemented directly without payment from users or beneficiaries, measuring productivity in the public sector is challenging. Additionally, the given items and services are not sold at market rates, which limits the analysts of the output valuation that an end product offers.

The above scenario protracted by Prtak (2019) and Alonzi (n.d.) is best suited for Malaysian local defence companies in which the implementation cost of the offset program is granted without charges or given for free. The MinDef of the United Kingdom's initiative to develop its productivity assessment methodology is identified as the Public Sector Efficiency Group (PSEG) conceptual framework. This model was selected as it defines productivity clearly and, in a way, is compatible with military capability data. It also explicitly recognises the important difference between outputs and outcomes.

15.2.3 Sustainability Indicators

Krajnc and Glavič (2003) stated that today's business achievement is no longer measured by economy or finances. A more holistic measurement will be through sustainability indicators. Indicators also have various functions. Companies used indicators to establish objectives and monitor consequent success and assist the managerial visualise the forms of anticipation that needed to be emphasised in future. Tan et al. (2015) added that indicators are one category of sustainability assessment tools useful to summarise and screen complicated data into useful information and trail performance progressively. Regulating a set of common indicators can contribute to providing a coherent assessment framework and allow better local sustainability benchmarking.

Balakrishnan (2007) stated that those indicators determine how effective offsets have been as a tool for technological and industrial advancement in the targeted regions. Waas et al. (2014) classified the following applications of indicators:

a. Assessing trends and conditions related to objectives and goals.
b. Reflecting the condition of an organisation (system).
c. Giving advanced early warning information.
d. Anticipating future conditions and trends.
f. Comparing across places and situations.
g. Focusing on what is happening in a large system.

However, Azapagic and Perdan (2000) protracted indicators of sustainable production would enable the identification of more sustainable options through the following:

a. Comparison of company identical products made by different companies.
b. Comparison of different processes producing the same product.
c. Benchmarking of divisions within the enterprise.

d. Rating of a company's business performance versus other companies in the same sector.
e. Assessing growth towards sustainable development of a sector.

15.2.4 Conceptualisations of Sustainability Indicator Performance Measurement Framework for Defence Industry Offset Program

Tan et al. (2015) revealed the existence of ambiguous definitions and the absence of a clear concept of sustainability, which has been attributed to the disparate conceptualisations of sustainability and no appropriate methods for its practical measurement. It is perceived that there are vast indicator frameworks existing for measuring sustainability. However, the setbacks are that they are either too high-level to manage or too sophisticated to be adopted by SMEs. The researchers also proposed establishing a systematic indicator selection method to identify a common set of sustainability indicators from the existing indicator frameworks to recommend an improved categorisation of indicators to address different aspects of sustainability holistically. The selected indicators are then connected and designated into different dimensions and sub-categories.

The Malaysian government as the offset program owner would like to see the project able to continue over a while even after the OEMs leave. Since the offset program is complimentary and comparable to projects sponsored by Non-Governmental Organisation (NGOs), which are costless, this research on sustainability will pursue the conception modelled by Alonzi (n.d.) which has constructed project sustainability into three main categories namely organisational, financial and community. To sufficiently address defence industrial sustainability, this paper adopted a holistic approach based on the Alonzi (n.d.) model but applied a bespoke approach and revised the three categories of organisation, financial and community into a newly designed dimension, which includes Organisation, Economy and Social (OES).

First is economic sustainability, which will be subdivided into financial sustainability and business sustainability. Financial sustainability refers to how the project or organisation will continue to get the funding needed after the grant has finished. For business sustainability, the concern is whether the project generates sufficient profit or financial gains and continues to function on earned income. Second is the organisational sustainability or offset activities performance. The recognised offset activities stipulated in the Policy and Industrial Collaboration Program Guidelines (ICP) in Government Procurement 2022 are not limited to human capital development, transfer of technology/knowledge/skills, content development local, research, development and commercialisation, access to local markets and direct investment. The entire cost and programs will be provided and borne by the OEM. Finally, is social sustainability; one example of social sustainability is the turnover of employees and how they are still attached and devoted to carrying out the project activities even after the OEMs leave.

15.3 Methodology

15.3.1 Systematic Sustainability Indicator Selection Approach

Taking into consideration of above literature research, this study will develop a procedure for a set of common indicators identification and selection for the defence industry. The procedure is presented in Fig. 15.1. The classification approach and final set of selected indicators will be presented in the Results and Discussion section.

This study initiates the development of a productivity assessment methodology identify as Defence Offset Program Efficiency Conceptual Framework, as shown

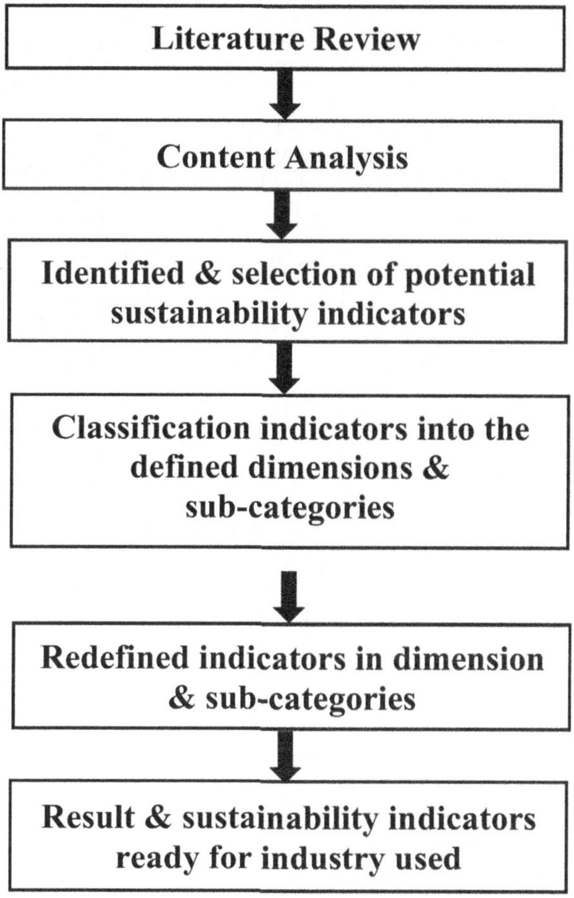

Fig. 15.1 Systematic identification and selection method. *Source* Author's proposed concept

in Fig. 15.2. This model was selected since it defines productivity literally and also easily recognises the important difference between outputs and outcomes.

In the multi-level models, processed data and indicators are created from primary data at the base of the pyramid by aggregation or selection out of particular aspects that have been assessed as particularly relevant. The weighted and aggregated indices are at the apex of the data pyramid.

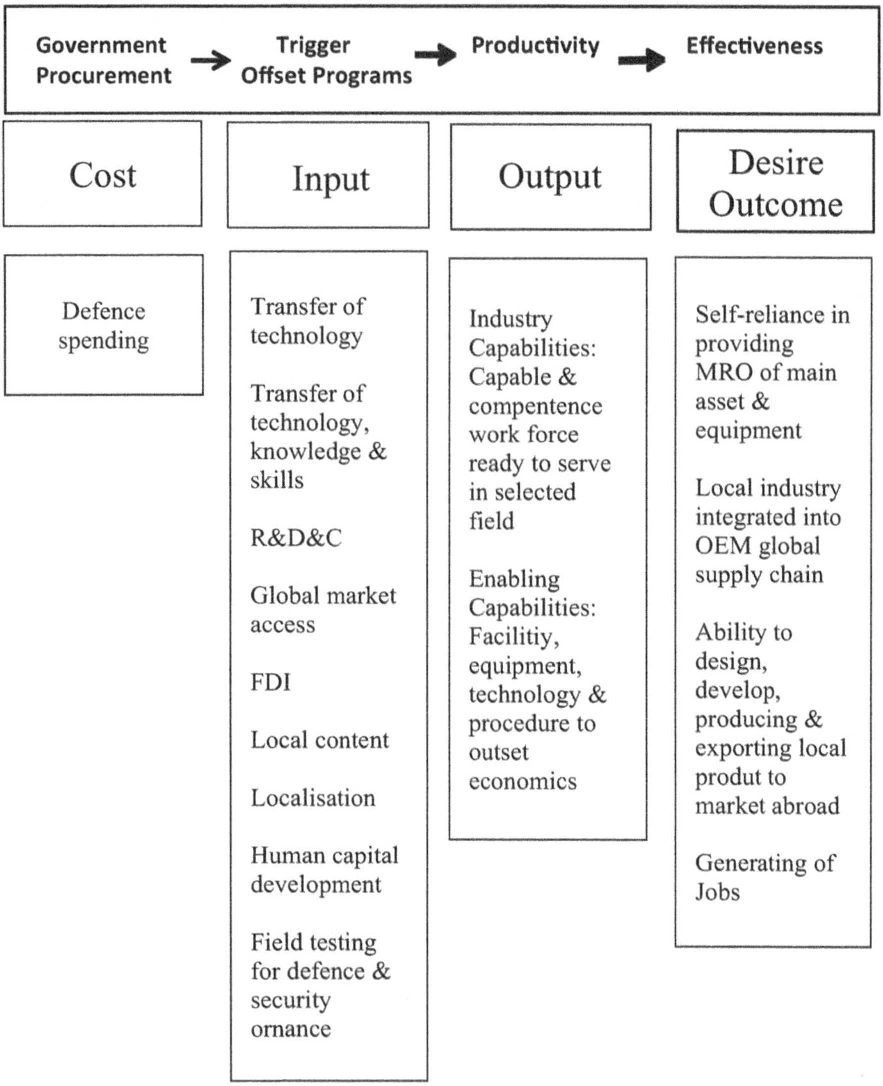

Fig. 15.2 Defence offset program efficiency conceptual framework. *Source* Author's proposed concept

The information age has contributed to the rapid growth of data that often overwhelms our ability to process information and this selected data set is often called an indicator. Ott (1978) defines an indicator is a method created to reduce a large amount of data down to its most basic form. In summary, an index is intended to simplify, and naturally, some information is lost during the process of simplifying. If the index is appropriately constructed, it is hoped that the missing data won't significantly affect the response. This study used the indicator as an instrument or tool to measure and estimate the forthcoming trends of performances and for the acquisition of information about possible difficulties occurring. So, indicators are linked to prediction, while Indices measure the performance of a group of activities.

Krajnc and Glavič (2003) also recommended that an enterprise should be uncomplicated and easy to implement based on available data and commonly measures aspects of production before moving toward more complex indicators. This study conducted a systematic review to explore potential indicators within to the Malaysian Treasury directive on Policy and Industrial Collaboration Program (ICP) Guidelines and an article published by Technology Depository Agency (TDA), which provides offset management services including the implementation and identification technologies required by the nation. This study should provide the government and companies with new metrics on indicators used to illustrate the progress of an offset program in meeting a range of economic, social and organisational goals.

15.3.2 Elements of Indicators

According to Veleva and Ellenbecker (2001), the primary elements that assist in distinguishing indicators from primary data, parameters, goals, or issues are the indicator dimensions. A measurement unit, measurement period, and boundaries are necessary to construct an indicator. To allow the indicators to be accepted by practitioners and promote better understanding, the four elements from Veleva and Ellenbecker (2001) were adapted as follows:

(a) Unit of measurement—the metric used to represent an indicator (such as kilograms, kilowatts, dollars, per cent and days).
(b) Period of measurement—the period for tracking and calculating an indicator (such as yearly, bi-yearly and monthly).
(c) Boundaries—determines how far a company wishes to go in measuring the indicators (such as product line, facility, suppliers, distributors and the entire life cycle of a material or product).
(d) Quantification method—the formula used to calculate an indicator, whether to use the total amount or per unit of product or any other factors to normalise the performance.

15.4 Result and Discussion

15.4.1 Sustainability Indicator Framework for Defence Industry Offset Program

Considering a good deal of effort has been dedicated by researchers to develop dimensions of sustainability indicators for economic and social, this study will be confined to the less participation dimension, which is the organisation. A systematic indicator selection method using six steps path for identifying sustainability indicators was made based on The Data Pyramid Model theoretical framework. As a result, 30 newly identified indicators with references to the nationally recognised indicator framework were formulated as shown in Table 15.2. The overview of the organisational chart for indicators categorisation by dimension is presented in Fig. 15.3. This study will benefit the Malaysian offset authority and defence industry in measuring their respective offset performance beyond their completion period. This conceptual sustainability indicator framework also may provide a standard guideline for the government and companies with new metrics to assess and illustrate the progress of their offset program in meeting organisational goals.

15.5 Conclusion

This paper has focused primarily on addressing the indicators used for measuring the Malaysian defence industry offset program where the PSEG conceptual framework was selected owing to its ability to clearly define productivity and, in a way, compatible with the requirements of military, public sector and government agencies. Furthermore, the difficulty of productivity measurement for the public sector is that the goods and services are often implemented directly without payment by users or beneficiaries. The above scenario is best suited for Malaysian local defence companies in which the implementation cost of the offset program is granted without charges or given for free. This study is purely confined to the dimension of the organisation, considering a good deal of effort has been dedicated by researchers to develop dimensions on sustainability indicators for economic and social. This study offers a range of 30 newly identified indicators with references to nationally recognised indicator frameworks. A systematic indicator selection method using six steps path for identifying sustainability indicators was based on the Data Pyramid Model theoretical framework. This study will benefit the Malaysian offset authority and defence industry for measuring their respective offset performance beyond their completion period in a holistic manner. This conceptual sustainability indicator framework may also provide a standard guideline for governments and companies with new metrics to assess and illustrate the progress of an offset program in meeting a range of economic, social, and organisational goals.

Table 15.2 Summary of sustainability indicators dimension for organisation (offset activity performance).

Sub-dimension (yearly)	Indicators		Unit	Suggested quantification method
1. Research, development and commercialization (R&D&C), transfer of company technology (ToT)/transfer of knowledge (ToK)	1.1 IPR transfer and commercialization through JV/ partnership with local 1.2 Training and skills development course 1.3 Tools, equipment, laboratory and workshop set-up 1.4 IPR development and sharing 1.5 Technology commercialization and roll out/ start-up company 1.6 Drawings, manuals and training and documentations	} (1.1–1.3)	RM	RM Total amount of R&D&C, ToT and ToK invested for sustainability effort
2. Global market access	2.1 Captive Market Access 2.2 Market access assistance and OEM Supply Chain Acceptance	}	$\frac{RM}{RM}$	$\frac{\text{Total company sales}}{\text{Total industry sales}}$

(continued)

Table 15.2 (continued)

Sub-dimension (yearly)	Indicators		Unit	Suggested quantification method
3. Human capital development	3.1 Technical transfer, skills and competency development for professional Services 3.2 On-the-job training 3.3 Knowledge transfer and skills development for professional services 3.4 Non-technical transfer and skills development 3.5 Training and Skills Development Courses-general 3.6 Higher learning placement programme at local/international institution (master, PhD)	}	%	$\frac{\text{Employee trained in sustainability}}{\text{Total number of employee}}$
4. Investment	4.1 Cash 4.1.1 Equity/ working capital investment to local entity 4.1.2 Project financing (Principal only) 4.1.3 Principal guarantee for standby letter of credit (SLOC) 4.2 Non-Cash return 4.2.1 Equipment, special tools, infrastructure, acquisition of IP, etc.	}	RM	Payback period/net present value/internal rate of return/accounting rate

(continued)

Table 15.2 (continued)

Sub-dimension (yearly)	Indicators		Unit	Suggested quantification method
5. Local content	5.1 Local Services/ Sourcing 5.1.1 Design and systems integration work 5.1.2 High-skill works 5.1.3 Installation, assembly, testing, commissioning and project and management 5.1.4 General works 5.2 Local Product 5.2.1 Parts and components, main equipment, test equipment	}	RM	Total amount of local content invested for sustainability effort
5.3 Plant facility	5.3.1 Plant equipment and machinery 5.3.2 Tools, jigs and fixtures 5.4 Logistics 5.4.1 Integrated logistic support (ILS) 5.4.2 Forwarding, haulage and transportation, storage and warehouse 5.4.3 Local insurance 5.5 Partnership/ Consortium 5.5.1 Work scope awarded to local partner 5.6 Professional and high 5.6.1 The minimum skill required is certificate	}	RM	Total amount of local content invested for sustainability effort

Source Author's proposed framework

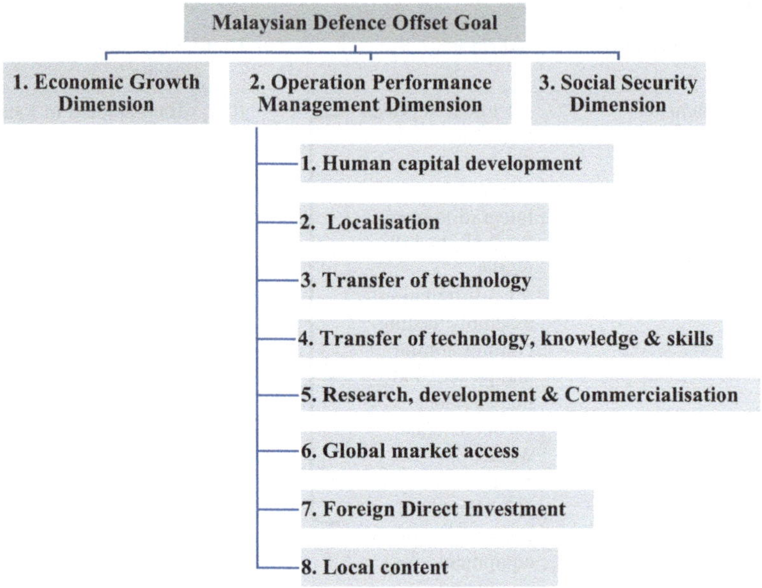

Fig. 15.3 Organisational chart of indicators categorisation by dimension. *Source* Author's proposed concept

References

Alonzi A (n.d.) What is project sustainability? Proposals for NGOs. https://proposalsforngos.com/what-is-project-sustainability/. Accessed on 15 March 2023

Ariffin Z, Arof AM (2019) Challenges in developing indigenous combat management systems (CMS) for Malaysian maritime defence and enforcement services. Mar Front 10(1):1–7

Azapagic A, Perdan S (2000) Indicators of sustainable development for industry: a general framework. Process Saf Environ Prot 78(4):243–261

Balakrishnan K (2007) Evaluating the effectiveness of offsets as a mechanism for promoting Malaysian defence industrial and technological development. https://dspace.lib.cranfield.ac.uk/handle/1826/2504?show=full. Accessed on 15 March 2023

Balakrishnan K, Johar TN (2021) Malaysian defence industry: context, challenges and the way forward. J Def Secur 14(1):1–18

Balakrishnan K, Matthews R (2009) The role of offsets in Malaysian defence industrialisation. Def Peace Econ 20(4):341–358

Bursa Malaysia (2015) Sustainability reporting guide. https://www.bursamalaysia.com/sites/5bb54be15f36ca0af339077a/content_entry5ce3b5005b711a1764454c1a/5ce3c83239fba2627b286508/files/bursa_malaysia_sustainability_reporting_guide-final.pdf?1570701456. Accessed on 15 March 2023

Chung H (2022) Bursa Malaysia adds climate change reporting to sustainability reporting framework for companies on Main, ACE markets. The Edge Markets. https://www.theedgemarkets.com/article/bursa-malaysia-adds-climate-change-reporting-sustainability-reporting-framework-companies. Accessed on 15 March 2023

Jarvie ME (2016) Brundtland report. Encyclopedia Britannica. https://www.britannica.com/topic/Brundtland-Report. Accessed on 15 March 2023

Kenton W (2021) Countertrade: definition, types and examples. Retrieved from Investopedia. https://www.investopedia.com/terms/c/countertrade.asp. Accessed on 15 March 2023

Krajnc D, Glavič P (2003) Indicators of sustainable production. Clean Technol Environ Policy 5:279–288

Malaysian National Audit Department Portal (2022) The federal auditor general's report 2020 1st series industrial collaboration program management. https://lkan.audit.gov.my/laporan/manage/788. Accessed on 15 March 2023

Malm A (2016) Technology transfer within related offset business: From an aircraft production perspective. PhD dissertation. Linköping University Electronic Press

Manulak MW (2022) The Brundtland commission and the seeds of change. In: Change in global environmental politics: temporal focal points and the reform of international institutions. Cambridge University Press, Cambridge, UK, pp 139–175

Matthew R, Anicetti J (2021) Offset in a post-Brexit world. RUSI J 166(5):50–62

New Straits Times (2015) Bursa Malaysia coming up with guidelines to help companies. https://www.nst.com.my/news/2015/11/111329/bursa-malaysia-coming-guidelines-help-companies. Accessed on 15 March 2023

Ott WR (1978) Environmental indices: theory and practice. Ann Arbor Science Publishers Inc., Ann Arbor

Papulová Z, Gažová A, Šlenker M, Papula J (2021) Performance measurement system: implementation process in SMEs. Sustainability 13(9):4794

Pichlak M, Szromek AR (2021) Eco-innovation, sustainability and business model innovation by open innovation dynamics. J Open Innov Technol Mark Complex 7(2):149

Prtak MD (2019) Measuring defence productivity: a first step. Retrieved from Gov. UK. https://quarterly.blog.gov.uk/2019/10/30/measuring-defence-productivity-a-first-step/. Accessed on 15 March 2023

Rheinmetall (n.d.) Offset Agreements. https://www.rheinmetall.com/en/rheinmetall_ag/group/compliance/kompensationsgeschaefte_1/index.php. Accessed on 15 March 2023

Roslan NA, Uli J, Yahaya MN, Nair MKS (2018) Malaysian military armoured vehicle procurement and its impact on employees' societal welfare (quality of life). Int J Acad Res Bus Soc Sci 8(6):1160–1169

Tan HX, Yeo Z, Ng R, Tjandra TB, Song B (2015) A sustainability indicator framework for Singapore small and medium-sized manufacturing enterprises. Procedia CIRP 29:132–137

Technology Depository Agency (n.d.) About ICP. https://www.tda.my/about-icp/. Accessed on 15 March 2023

Veleva V, Ellenbecker M (2001) Indicators of sustainable production: framework and methodology. J Clean Prod 9(6):519–549

Waas T, Hugé J, Block T, Wright T, Benitez-Capistros F, Verbruggen A (2014) Sustainability assessment and indicators: tools in a decision-making strategy for sustainable development. Sustainability 6(9):5512–5534

Wessner C, Wolff A (1997) National research council. Policy issues in aerospace offsets: report of a workshop. The National Academies Press, Washington, DC

Yedvav H, Kordova S, Fridkin S (2022) Offset obligation in defense projects: schedule, budget, and performance implications. Sustainability 14(18):11186

A Comprehensive Review on Innovation in Small Business

16

Ahmad Rafiki, Nindya Yunita, and Muthya Rahmi Darmansyah

Abstract

This chapter aims to elaborate the history and root of research on innovation in small business. The study adopts content analysis technique where various sources are referred including reports, official documents and research articles. There are issues related to the innovation and SMEs which can be learned and analyzed. Opportunities brought by innovation to the SMEs are need to be revealed and used as valuable information for strategic planning and initiatives.

Keywords

Innovation • Small business • Islam

16.1 Introduction

Small and medium-sized businesses (SMEs) and other entrepreneurs today face a number of significant obstacles, such as intense competition, consumers' constant preference for higher-quality but less expensive goods, and the requirement to remain adaptable to sudden changes in the industrial landscape (Lesakova et al. 2017). Since enhancing innovation performance is the most important thing for a business and policymakers, constant innovation is essential to addressing these issues and enhancing business quality (Sharmelly 2017; Ortiz-Villajos and Sotoca 2018; Yigitcanlar et al. 2019; Capozza and Divella 2019). Innovation makes a market more competitive (Pudjiarti and Suharnomo 2018). Continuous innovation is essential for business existence, but also for individual firms. Moreover, it may result in social and economic changes. A rise in GDP (gross domestic product),

A. Rafiki (✉) · N. Yunita · M. R. Darmansyah
Faculty of Economics and Business, Universitas Medan Area, Medan, Indonesia
e-mail: hahmadrafiki@gmail.com

A. Rafiki et al. (eds.), *Open Innovation in Small Business*, Contributions to Environmental Sciences & Innovative Business Technology,
https://doi.org/10.1007/978-981-99-5142-0_16

which also aids in lowering unemployment in each nation, is extremely likely if SMEs are managed innovatively (Ejdys 2016; Di Cintio et al. 2017; Tekola and Gidey 2019).

The concept of "firm innovation" has undergone numerous definitions and operationalizations (Story et al. 2015). Firm innovation has been defined as, on the one hand, an organizational culture that reflects the firm's openness to new concepts and desire to seek unconventional methods of doing things (Kyrgidou and Spyropoulou 2013). This behavioral perspective emphasizes further that businesses with higher levels of innovation are more likely to alter and adopt new practices (Calantone et al. 2002; Menguc and Auh 2006). Others, on the other hand, contend that a firm's capacity for innovation relates to its inclination to create novel goods or services (Garcia and Calantone 2002). However, different definitions emphasize originality, innovation, newness, experimenting in new product development activities, and/or the quantity of new goods and services (Story et al. 2015). With this, innovation and government support are considered important in the success of micro, small and medium enterprises or MSMEs in recovering from the impact of the Covid-19 pandemic.

16.2 Literature Review

16.2.1 Definition of Innovation Culture

The culture of innovation is the result of acculturation of corporate cultural values with innovations made by the company. Corporate culture is related to the values and principles that are held during the business process. Culture is also a habit and beliefs and practices that are repeated (Añón Higón and Driffield 2011). Meanwhile, innovation is a form of developing ideas and implementing something new to increase the company's competitive value and remain relevant to the times.

With this, the culture of innovation can be interpreted as a step for a business to incorporate an innovative attitude into the corporate culture to increase innovation in business operations. The culture of innovation is a measure of business success today because a business that has a culture of innovation will always have a competitive advantage compared to its competitors (Colombo et al. 2012; Tan et al. 2009).

16.2.2 Innovation and SMEs

There are four categories of innovation that SME managers or owners need to focus on (Taneja et al. 2016), namely: product innovation, process innovation, organizational innovation, and marketing innovation. Focusing creative resources and assets in the four innovation categories properly is one of the keys to increasing the innovative level of a business.

Digitalization can impact innovation for several reasons, including the use of e-commerce to launch new products, the use of broadband internet to capture and process knowledge developed elsewhere and to manage the flow of knowledge within and between companies (Airaksinen et al. 2008). Digital transformation opens up possibilities for organizations to take advantage of the interrelationships between people, data, information and knowledge by leveraging digital connections. Organizations that have undergone digital transformation are able to introduce new practices and innovative initiatives into their business operations (Díaz-Chao et al. 2015; Sahut and Peris-Ortiz 2014). Therefore, digital transformation leads to the emergence of new ideas and communication between business partners in the value chain.

Another opinion was proposed by Ardito et al. (2021) who explained that digital technology could support innovation performance in MSMEs. They can take advantage of digitalization to develop and manage the flow of information through digital technology. Research by Karina and Astuti (2022), Yang et al. (2012) showed that digital orientation is a determining factor for the success of digital innovation in small and medium enterprises.

Major changes are brought about by digital transformation, which is fueled by information technology innovation. Digital transformation is characterized by sophisticated innovations across all social systems, shifting market fundamentals, reconfiguring boundaries, and fundamentally altering business models. The change in digital strategy can deliver benefits for MSMEs managers in increasing their profits (Tang et al. 2021). Previous research on digital transformation with the results of a positive effect on MSMEs' performance through competitive advantage is proposed by Sainio et al. (2012).

In the meantime, competitive advantage partially mediates the link between SMEs' performance and digital transformation. Kwon and Lim (2012) mentioned that there is a correlation between the success of SMEs and competitive advantage, which is mediated by product innovation. In creating competitive advantage, owner-managers of SMEs must consider innovation aspect.

Innovational capability is considered as a valuable asset for companies to provide and maintain competitive advantage (Lawson and Samson 2001). Innovation capability facilitates companies to quickly introduce new products and adopt new systems; it is important to give meaning to ongoing competition (Rajapathirana and Hui 2018).

Companies and business actors must strive for innovation in order to win the competition in an increasingly competitive globalization era. Innovation is a creative process that requires a new invention that is developed into a product or service with economic value from an item or service (Fritsch 2017; Schumpeter 1934). Akpan et al. (2021) affirmed that innovation has an important role in influencing company performance because it is related to basic activities or activities related to products/processes. Therefore, companies are required to be able to translate and incorporate emerging ideas to produce decisions related to efforts to innovate within companies to deliver benefits for the improvement of company performance. Numerous researches have been conducted on the effect of digital transformation on the performance of MSMEs and revealed that digitalization can positively affect company performance, including those suggested by Brynjolfsson and Yang (1997), Mithas et al. (2005), Duarte Alonso and Bressan (2016), Link and Scott (2012), Menguc and Auh (2006).

According to Kessler and Chakrabarti (1996), the speed of innovation can be influenced positively or negatively by strategic orientation and organizational capability factors. While Stefan and Bengtsson (2017) exposed the factors related to the types of innovation that contribute to increasing the competitiveness of MSMEs. In addition, it is confirmed that there is limited number of research on how innovation affects the relationship of digital transformation to the competitive advantage of MSMEs and simultaneously improves their performance. However, Kessler and Chakrabarti (1996) affirmed that there is growing recognition that the speed of innovation is important to create and maintain corporate competitiveness in the midst of rapidly changing business environment.

Figure 16.1 explained the root of SMEs innovation technology which starts with motivation, socialization, and identification of needs. Then, there are 5 aspects (entrepreneurship workshop, entrepreneurship internship, business practices, business consultation, and entrepreneurial incubator) that need to be emphasized in the process before focusing on two main important dimensions; improvement of new entrepreneurs and increasing technology innovation competency. This flow is practically adopted by SMEs to achieve growth and sustainability in business. Nowadays, associating every aspect of business to digital/technology or even artificial intelligence is a must.

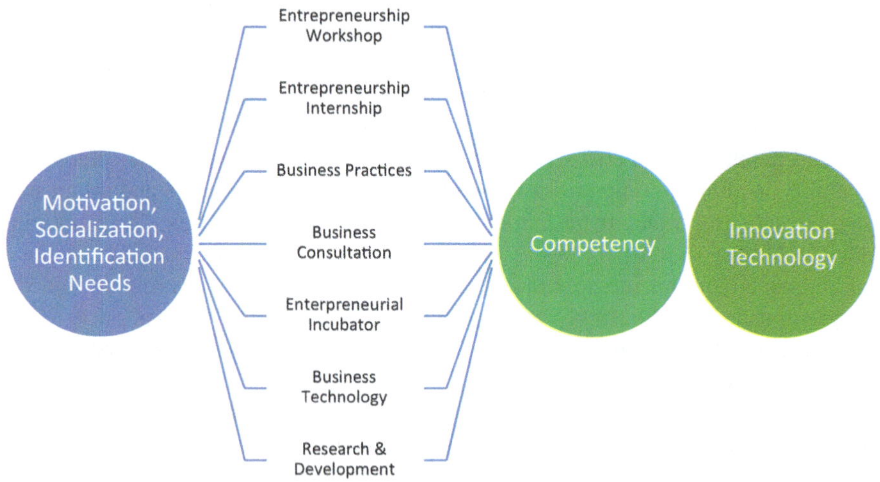

Fig. 16.1 SMEs technology innovation development model. *Source* Authors' adapted model from Indrawati et al. (2020)

16.3 Methods

This study adopts a content analysis with additional few items on bibliometric analysis. Bibliometric analysis manifests interconnections among the articles whereby the data was retrieved from Scopus, a reputable and comprehensive database. The primary reason for choosing Scopus is its availability, nearly 60 percent larger than the Web of Science (WoS) database (Zhao and Strotmann 2015).

16.4 Discussion

Based on some findings from techniques in bibliometric analysis, the research on innovation and small business has been done by many researchers in various countries. There are 7,933 documents sourced from reputable database called Scopus by searching through keywords; “innovation”, “small business”. As depicted in Table 16.1, the highest hit on the number of documents is in the year 2022 with a total number of 765 documents. It is also shown that the number of documents increase incrementally since 2002, which consider the interest of studies on this topic is perceived positively.

As depicted in above Table, there are 10 affiliations that produced articles on innovation and small business. In fact, the University of North Carolina leads the list of affiliations, followed by Delft University of Technology and University of Manchester. Most of the affiliations are located in the Western countries as well as developed countries. It is also indicating a concern for the developing countries to explore more on innovation context particularly in small business.

Moreover, it shows that Link A. N. became the most active authors who engage in research on innovation and small business with a total of 56 documents so far, while nine other authors have more less produced 10–23 documents. This table also indicated that the research on the topic of innovation and small business has been a focus of researchers from various countries (developing and developed countries).

Furthermore, United States leads the list of dominant country with 1454 documents that studied on innovation and small business, then followed by countries such as United Kingdom, China, Italy and others. The interests of researchers on innovation and small business in United States and United Kingdom are obviously superior (see the number of documents) compared to other countries.

Meanwhile, it is shown that the research on innovation and small business have been written in article form which is published in journals (61.9%), compared to the conference papers (20.5%), book chapters (8.5%) and others. The research on innovation and small business mostly through publication of articles is known as a tight process (review) and thus more important and valuable information could be understood and adopted by business practitioners.

Lastly, the table shows that the research on innovation and small business is related more on business, management and accounting (27%), followed by

Table 16.1 A summary of top ten research output, affiliation and document types

No	Year/ number document	Author/ number of documents	Country/ number of documents	Affiliation/ number of documents	Document type/number of documents	Subject area/ number of documents
1	2023 (253)	Link, A. N. (56)	United States (1,454)	The University of North Carolina (59)	Article (4,909)	Business, Management and Accounting (4,020)
2	2022 (773)	Audretsch, D. B. (23)	United Kingdom (840)	Delft University of Technology (40)	Conference Paper (1,623)	Social Sciences (1,874)
3	2021 (679)	Scott, J. T. (21)	China (395)	The University of Manchester (37)	Book Chapter (673)	Engineering (1,816)
4	2020 (627)	Hashemian, H. M.(16)	Italy (369)	Universitat de Valencia (31)	Review (316)	Economics, Econometrics and Finance (1,704)
5	2019 (573)	Rahman, H. (16)	Spain (319)	Universidade do Minho (31)	Book (167)	Computer Science (1,348)
6	2018 (466)	Ramos, I. (14)	Russian Federation (314)	University of Cambridge (31)	Conference Review (103)	Decision Sciences (797)
7	2017 (436)	Doroshenko, Y. A.(12)	Australia (298)	LUT University (29)	Note (62)	Environmental Science (680)
8	2016 (475)	Colvin, D. P. (11)	Germany (296)	Universidade de Sao Paulo (28)	Short Survey (43)	Enegery (457)
9	2015 (358)	Bouwman, H. (10)	Indonesia (282)	Indiana University Bloomington (28)	Editorial (21)	Earth and Planetary Sciences (247)
10	2014 (344)	McAdam, R. (10)	India (240)	HSE University (27)	Erratum (12)	Mathematics (234)

social sciences (12.6%), engineering (12.2%), economics and finance (11.5%) and others. Uniquely, the research on engineering also engages with innovation, or in other words, innovation covers all aspects/sections including engineering, computer science, decision science etc.

16.4.1 Islam and Innovation

Islam does not only explain about worshiping Allah but also explains various ways to make its people happy in this world and in the hereafter. In life, humans will not be separated from thinking activities. Every human being must use his reasoning power to think about everything he goes through in life. Islam also does not prohibit the reason from being used to carry out creativity or innovation in work and fulfilling one's life. Islam in fact orders us to manage the existing natural resources as well as those possible which can be beneficial to anyone such as humans, animals or other living things.

Innovation is thinking and doing something new to add or create useful values, both social and economic benefits. To generate innovative behavior one must view innovation fundamentally as a manageable process. Islam strongly supports innovation in any way except in the realm of Aqidah which is the absolute foundation and requires following the instructions of the Prophet PBUH. Innovation is very urgent in the business world, because a business is not said to be successful when it just stays put, without producing a change. Innovation is highly organized, has processes, principles, types, sources, goals and cycles to achieve better things.

Today the development of the digital economy is increasing rapidly and has become a hot topic of discussion among academics and practitioners. The digital economy is defined as the adoption of technology (ICT) into various sectors of economic activity which contributes to the development of new activities in the private sector and the public sector. Technology and online enter the business sector which makes the production process more efficient and reduces various costs.

Technological developments have proven capable of increasing economic capacity at the aggregate level and increasing market capacity for industries at the micro level. In the business sector itself, digital startups are able to provide unexpectedly large effects or impacts in the context of income and welfare. The phenomenon of successful startups such as GO-Jek, which has developed various lines of services, has been able to dispel the image of motorcycle taxi drivers as merely transporting people, but also turning them into delivery services for various services. Fresh and brilliant ideas and those that lead to empowerment are what entrepreneurs need to grow, including Islamic entrepreneurs.

Seeing the potential above, Islamic business actors (in the context of implementing Islamic law) need to aggressively develop business ideas and innovations that use technology in their application. In the context of startups—digital startups—the first step taken is to build and come up with startup ideas that are interesting and empower people with the principles of Islamic law. Furthermore, invite young people who are interested in digital startups that are based on Islamic business law to take advantage of government programs in the development of digital startups in Indonesia. In addition, the government needs to facilitate the development of digital startup communities as a place to tell stories, share, innovate and collaborate in developing their startups.

Inspiration for innovation is also present in the story of Prophet Noah who was ordered by him to build a boat to be used to save himself when flooded, while

Table 16.2 Number of documents by country/territory

No	Year/number document	Country/number of documents
1	2022 (23)	United States (61)
2	2021 (28)	United Kingdom (37)
3	2020 (29)	Malaysia (34)
4	2019 (20)	Indonesia (21)
5	2018 (20)	Russian Federation (12)

having never existed and knew the method and form before. Islamic highlighted the innovation in the Qur'an as follows:

> And say: "Work (righteousness): Soon will Allah observe your work, and His Messenger, and the Believers: Soon will ye be brought back to the knower of what is hidden and what is open: then will He show you the truth of all that ye did." (Al-Qur'an, At-Taubah: 105).

There are 309 documents sourced from a reputable database called Scopus by searching through keywords; "innovation", "Islam". As shown in Table 16.2, the highest hit on the number of documents is in the year 2020 with a total number of 29 documents. In fact, the interest of research on innovation and Islam started in 2004. However, the increase of number of articles could be considered potentials to be explored.

As shown in Table 16.2, it is shown that the research on innovation and Islam dominated by researchers in two non-Muslim countries such as United States and United Kingdom with 61 and 37 documents respectively, is then followed by two Muslim populous countries (Malaysia and Indonesia). This indicates that the interest of research on innovation and Islam does not limit to certain countries but recognized as inseparable terms or in other words, it is proven that Islam considers innovation as a crucial element to be explored and adopted.

16.5 Conclusion

SMEs must build an innovative culture in the work environment so that can explore business values that will become a competitive advantage in every aspect of the business. There are various reasons for the need to immediately implement a business culture in various aspects. *Firstly*, human resources aspect. A culture of innovation will shape teams to think creatively and innovatively. By attracting better team talent, the organization could run effectively. Every person's skill set and inventive mindset are needed in business. These people will take action, test things out, and are always looking for a problem to solve. These are the skills and attitudes our businesses need to survive.

These creative individuals naturally desire a creative environment as well since they want the freedom to create, experiment, express them, and bring about significant improvements. This will encourage an inventive mindset, and new actions will also emerge in tandem with the growing innovation culture. Moreover, it also is required to retain the best talent. Employees need an environment to be able to develop skills and discover new talents. Retaining talent depends on how often the team can get used to innovating. The more they can test new ideas and learn on the job, the more satisfaction they will get from their work, and the more engaged they will be.

Secondly, marketing aspect. A culture of innovation will create creativity in creating marketing campaigns that are unique and different from those of competitors. There are many marketing strategies and tactics that can be used to attract the attention of consumers. However, it takes creativity and innovation to make marketing campaigns more prominent, unique and attractive.

Moreover, innovation has been shown to give businesses a competitive edge. In other words, an innovative culture must be established in order for the company to stand out among its many rivals. Furthermore, it needs to make the businesses think beyond the own market. This explains the importance of exploring product potential and seeking differentiation from the current business so that the business remains relevant. This process can only be achieved by consistently innovating to create new added value in creating products.

Thirdly, operational aspect. A culture of innovation will bring various benefits in operational aspects. For example, creating efficiency. Innovation forces the company to make the appropriate adjustments, which frequently increases operational efficiency. Employees start to establish an innovative culture that supports finding tasks that are done more effectively, more quickly, or both. The competition in business is won by those who can streamline operations. Moreover, the business practitioners must be accustomed of using technology. Most SMEs are used to carrying out operations as routine jobs that require a lot of preparation and effort even though this routine work can be streamlined with the use of technology. There are many more examples of innovation from the use of technology in business that can make business operations more efficient. Therefore, it is important for culture innovation in operational aspects.

Fourthly, financial aspect. Finance is an important aspect to ensure the business can run well. Innovation can be applied to financial aspects to make financial conditions healthier. One of them is through making business more transparent. The use of business finance processing application technology is part of innovation. The use of this application will make income and expense reports reportable transparently and can be presented in real-time. This will give business owners full control over financial supervision. The use of this technology will be difficult if they do not cultivate innovation as early as possible.

Fifthly, sales aspect. Sales are at the heart of sustainable business growth. Without sales, business operations will be disrupted due to a lack of cash flow. The companies need to find new niche markets. With the embedded innovation culture, it will perform a number of consumer studies and looking for new, more lucrative niche markets, then will create even more potential to grow the market.

The innovation culture will provide various benefits in various aspects of business. Every owner-manager of SMEs must be more advanced by prioritizing the principles of innovation and making it a new culture in business. A sustainable business must begin with innovation initiatives and thus SMEs can grow to support the country's economy.

References

Akpan IJ, Soopramanien D, Kwak DH (2021) Cutting-edge technologies for small business and innovation in the era of COVID-19 global health pandemic. J Small Bus Entrep 33(6):607–617

Añón Higón D, Driffield N (2011) Exporting and innovation performance: analysis of the annual small business survey in the UK. Int Small Bus J 29(1):4–24

Airaksinen A, Panizza A de, Vrije EB, Hagsten E, Leeuwen G van, Franklin M, Maliranta M, Kotnik P, Stam P, Rouvinen P, Farooqui S, Quantin S, Svanberg S, Clayton T, Barbesol Y (2008) Information society: ICT impact assessment by linking data from different sources. Micro, August. https://doi.org/10.13140/RG.2.1.3704.2966

Ardito L, Coccia M, Messeni Petruzzelli A (2021) Technological exaptation and crisis management: evidence from COVID-19 outbreaks. R&D Manag 51(4):381–392

Brynjolfsson E, Yang S (1997) The intangible benefits and costs of investments: evidence from financial markets, pp 1–21

Calantone RJ, Cavusgil ST, Zhao Y (2002) Learning orientation, firm innovation capability, and firm performance. Ind Mark Manag 31(6):515–524

Capozza C, Divella M (2019) Human capital and firms' innovation: evidence from emerging economies. Econ Innov New Technol 28(7):741–757

Colombo MG, Laursen K, Magnusson M, Rossi-Lamastra C (2012) Introduction: small business and networked innovation: organizational and managerial challenges. J Small Bus Manag 50(2):181–190

Di Cintio M, Ghosh S, Grassi E (2017) Firm growth, R&D expenditures and exports: an empirical analysis of Italian SMEs. Res Policy 46(4):836–852

Díaz-Chao Á, Sainz-González J, Torrent-Sellens J (2015) ICT, innovation, and firm productivity: new evidence from small local firms. J Bus Res 68(7):1439–1444

Duarte Alonso A, Bressan A (2016) Micro and small business innovation in a traditional industry. Int J Innov Sci 8(4):311–330. https://doi.org/10.1108/IJIS-06-2016-0013

Ejdys J (2016) Entrepreneurial orientation vs. innovativeness of small and medium size enterprises. J Eng Proj Prod Manag 6(1):13–24

Fritsch M (2017) The theory of economic development–an inquiry into profits, capital, credit, interest, and the business cycle. Reg Stud 51(4):654–655

Garcia R, Calantone R (2002) A critical look at technological innovation typology and innovativeness terminology: a literature review. J Product Innov Manage Int Publ Prod Dev Manage Assoc 19(2):110–132

Indrawati H, Caska, Suarman S (2020) Barriers to technological innovations of SMEs: how to solve them? Int J Innov Sci 12(5):545–564. https://doi.org/10.1108/IJIS-04-2020-0049

Karina AP, Astuti RD (2022) The role of digital orientation, digital capability and digital innovation on the relationship of environmental dynamism towards FMCG SME's performance. In:

Proceeding of the international conference on family business and entrepreneurship, vol 3, no 1
Kessler EH, Chakrabarti AK (1996) Innovation speed: a conceptual model of context, antecedents, and outcomes. Acad Manag Rev 21(4):1143–1191
Kyrgidou LP, Spyropoulou S (2013) Drivers and performance outcomes of innovativeness: an empirical study. Br J Manag 24(3):281–298
Kwon SH, Lim YW (2012) A study for rejection and acceptance for information technology innovative products: based on smart phone usage intention of general mobile phone users. J Korea Soc Comput Inf 17(1):219–226
Lawson B, Samson D (2001) Developing innovation capability in organisations: a dynamic capabilities approach. Int J Innov Manag 5(03):377–400
Link AN, Scott JT (2012) Employment growth from the small business innovation research program. Small Bus Econ 39:265–287
Lesakova L, Gundova P, Kral P, Ondrušova A (2017) Innovation leaders, modest innovators and non-innovative SMEs in Slovakia: key factors and barriers of innovation activity. Organizacija 50(4):325–338
Menguc B, Auh S (2006) Creating a firm-level dynamic capability through capitalizing on market orientation and innovativeness. J Acad Mark Sci 34(1):63–73
Mithas S, Ramasubbu N, Krishnan MS, Sambamurthy V (2005) Information technology infrastructure capability and firm performance: an empirical analysis. Ann Arbor 1001:48109–51234
Ortiz-Villajos JM, Sotoca S (2018) Innovation and business survival: a long-term approach. Res Policy 47(8):1418–1436
Pudjiarti ES, Suharnomo S (2018) Does institutional intervention play a role in small business clustering? An empirical evidence from semarang, Indonesia. Calitatea 19(163):52–59
Rajapathirana RJ, Hui Y (2018) Relationship between innovation capability, innovation type, and firm performance. J Innov Knowl 3(1):44–55
Sahut JM, Peris-Ortiz M (2014) Small business, innovation, and entrepreneurship. Small Bus Econ 42:663–668
Sainio LM, Ritala P, Hurmelinna-Laukkanen P (2012) Constituents of radical innovation—exploring the role of strategic orientations and market uncertainty. Technovation 32(11):591–599
Schumpeter JA (1934) The theory of economic development. Harvard University Press, Cambridge, Massachusetts
Sharmelly R (2017) Crafting a winning innovation strategy. Strateg Dir 33(3):8–11
Stefan I, Bengtsson L (2017) Unravelling appropriability mechanisms and openness depth effects on firm performance across stages in the innovation process. Technol Forecast Soc Chang 120:252–260
Story VM, Boso N, Cadogan JW (2015) The form of relationship between firm-level product innovativeness and new product performance in developed and emerging markets. J Prod Innov Manag 32(1):45–64
Tan J, Fischer E, Mitchell R, Phan P (2009) At the center of the action: innovation and technology strategy research in the small business setting. J Small Bus Manag 47(3):233–262
Tang C, Xu Y, Hao Y, Wu H, Xue Y (2021) What is the role of telecommunications infrastructure construction in green technology innovation? A firm-level analysis for China. Energy Econ 103:105576
Taneja S, Pryor MG, Hayek M (2016) Leaping innovation barriers to small business longevity. J Bus Strateg 37(3):44–51. https://doi.org/10.1108/JBS-12-2014-0145
Tekola H, Gidey Y (2019) Contributions of micro, small and medium enterprises (MSMEs) to income generation, employment and GDP: case study Ethiopia. J Sustain Dev 12(3):46–81
Yang J, Liu H, Gao S, Li Y (2012) Technological innovation of firms in China: past, present, and future. Asia Pacific J Manag 29:819–840

Yigitcanlar T, Sabatini-Marques J, da-Costa EM, Kamruzzaman M, Ioppolo G (2019) Stimulating technological innovation through incentives: perceptions of Australian and Brazilian firms. Technol Forecast Soc Change 146:403–412

Zhao D, Strotmann A (2015) Analysis and visualization of citation networks. In: Synthesis lectures on information concepts, retrieval, and services, vol 7, no 1, pp 1–207

Family Business and Innovation on Internationalisation Process

17

Abdullah N. Almana and Timothy Connerton

Abstract

This study aims to elaborate the importance of family business and innovation based on the internationalisation process. Some important elements are discussed such as the nature of family business, characteristics of family business and critical success factors of family business. Then it is followed by the relationship of family business and internationalisation and finally on the family firm innovation. The content analysis is adopted to elaborate the subject of family business, innovation and internationalisation and refer it precisely to the previous literature.

Keywords

Family business • Innovation • Internationalisation

17.1 Introduction

The globalisation of local organisations has a positive influence on the economic and national image of a country (Jabar et al. 2016). Mohamad et al. (2008) and George et al. (2005) indicate that not many organisations dare to engage in internationalisation; organisations face various difficulties, such as difficulty obtaining financial help, capital shortage, a limited number of trained personnel, and lack of knowledge. Because they started with little money, investors may be concerned about the organisations' long-term viability, especially given their lack of expertise, experience, marketing network, scalability, and overseas market possibility. In addition, there are concerns about the responsibilities of supporting agencies

A. N. Almana (✉) · T. Connerton
Business School Lausanne, Chavannes-Près-Renens, Switzerland
e-mail: almana.abdullah@googlemail.com

A. Rafiki et al. (eds.), *Open Innovation in Small Business*, Contributions to Environmental Sciences & Innovative Business Technology,
https://doi.org/10.1007/978-981-99-5142-0_17

in assisting organisations through channels, programmes, and various activities connected to the internationalisation process that might create competencies and propose change processes.

Internationalisation is concerned with the firm's interactions with its worldwide environment. It is exhibited in the creation and usage of diverse international activities, notably inward, outward, and cooperative operations. It derives from the development and utilisation of personnel's cognitive and attitudinal preparedness.

As a consequence of population shifts, globalisation and the information technology revolution, which has made it simpler for consumers to buy alternatives from nations hundreds of kilometers away, family companies are increasingly challenged with numerous difficulties throughout the globe. In order to survive and thrive, family companies must learn to think beyond their immediate surroundings. Families with the agility and flexibility to adapt to new technologies and environments will emerge victoriously.

Meanwhile, many family businesses seem to avoid or underestimate internationalisation due to the risk of losing control via changes in capital structure, such as paid-up capital. Furthermore, many researchers have delved into the question of family business internationalisation or cross-border investments in places other than the Middle East (Kontinen and Ojala 2012). Despite the fact that all family businesses, regardless of size, sector, or market, have similar characteristics, family firms in the Middle East have a particular profile, which offers some unique and intriguing difficulties.

Based on earlier research, Olejnik and Swoboda (2012) classified internationalisation into three primary types. First, the Uppsala Model suggests that the internationalisation process occurs gradually, with any commercial entity or organisation progressively embracing the internationalisation process. Second, because the corporation is born worldwide, internationalisation may happen quickly, which shows how penetration into the foreign market can be expanded when the present business climate is considered, and quick measures are implemented. Finally, internationalisation can occur when a company is born worldwide, known as radical.

According to Wright and Ricks (1994), internationalisation is a firm-level activity that crosses national boundaries. Internationalisation happens when a corporation tries to adapt its business processes to a global environment, such as resource management and strategy (Calof and Beamish 1995). Local enterprises begin to extend their operations to other countries at this point to reach new markets, maintain viability, and seek out better opportunities. Researchers from many countries have looked at these antecedents to identify the factors that drive companies to join the global market. The massive number of organisations conducting business online has prompted Sinkovics and Sinkovics (2013) to investigate the role of the Internet as an alternative to internationalisation.

Internationalisation provides opportunities for growth, obtaining new skills, access to new resources, and long-term profitability growth (George et al. 2005; Hazarbassanova 2016; Hsu et al. 2013; Yiu et al. 2007; Zhang et al. 2016; Adomako et al. 2017). However, due to a lack of resources, management skills, and expertise, most businesses are unable to successfully reach global markets (Onkelinx et al. 2016). Lu and Beamish (2001) emphasised that the many challenges of internationalisation are associated with the liability of foreignness and newness.

17.2 Literature Review

17.2.1 The Concept of Internationalisation

Buckley et al. (1992) defined internationalisation as a constant process in which relationships are continually established, developed, maintained, and dissolved to achieve the firm's objectives, or as a cumulative process in which relationships are continually established, developed, maintained, and dissolved to achieve the firm's objectives (Johanson and Mattsson 2015).

Financial, managerial, market-oriented (covering both national and international marketplaces), and industry and company characteristics are among the major categories studied by researchers concentrating on obstacles to internationalisation by exporters and/or non-exporters (Morgan et al. 2019). Barriers to internationalisation exist at any level of the process; they may vary in severity depending on the firm's degree of internationalisation (Katsikeas and Morgan 1994).

The existing literature on internationalisation key motivations divides them into four categories: decision-maker characteristics, company-specific variables, environmental factors, and firm features (Katsikeas and Piercy 1993). For instance, Cavusgil (1984) suggested that internal and external impulses should be addressed in the choice to internationalise an organisation, agreeing that enterprises are likely to be driven by various stimuli depending on the stage of internationalisation.

Corporate internationalisation has become a prominent topic in international entrepreneurship, particularly in emerging nations. The influence of a firm's internal characteristics on its degree of internationalisation is the subject of international entrepreneurship (Zhang et al. 2016; Yiu et al. 2007). Emerging-market enterprises strive to expand into new markets outside of their home country to obtain a competitive edge and avoid adverse institutional conditions at home (Ramon-Llorens et al. 2017; Hazarbassanova 2016).

According to Karagozoglu and Lindell (1998), the key barriers include a lack of management knowledge and competence, human elements, such as foreign business abilities, and views of the business environment. High internationalization costs, including conducting market research abroad, procuring legal consulting services, translating documents, adapting products to foreign markets, travel expenses, and increased business and financial risk.

A lack of entrepreneurial, managerial, and marketing skills, bureaucracy and red tape, as well as difficulty in getting information and expertise, are among the top five restraints and obstacles to internationalisation, according to Szabo's study (technology equipment and know-how). According to Leonidou (1995), non-exporting businesses' most important problems are locating and analysing foreign markets, finding and gaining professional representation abroad, and completing export documentation/procedures.

According to Singh et al. (2010), other export impediments include competitive pressures in foreign markets and difficulties obtaining qualified employees to execute the export role. Leonidou stated that among the exporters' challenges are caused by excessive red tape, tardy payment by overseas purchasers, and weak economic conditions in foreign markets, all of which appear to be operational concerns and mostly tied to external conditions (Leonidou 1995). Meanwhile, Bilkey (1978) suggested that finances, foreign government limitations, limited understanding of foreign sales techniques, inadequate distribution, and a lack of international market contacts were typical export issues. Furthermore, exporting is further hampered by inadequate resources, managerial skills, linguistic incapacity, cultural differences, and mental distance (Fletcher 2004; Miesenbock 1988; O'Farrell et al. 1998).

According to Fletcher (2004), scholarly literature on the primary elements influencing internationalisation is thorough. It separates managerial traits, organisational features, external barriers, and external incentives to participate in business abroad. Meanwhile, studies on organisational factors highlighted the readiness to produce goods for international markets, technical edge, and willingness to investigate international markets (Bilkey 1985; Cavusgil 1984). Marketing operations by rivals in markets abroad, perceptions of increased risk in overseas markets, understanding the industry and how it functions, cost concerns, lack of export training, and government aid are all examples of external barriers (Bilkey 1985). Finally, the most significant external motivations are government-provided export incentives, government oversight of demand variables, a drop in local demand or surplus capacity, and lower production costs.

This study is also based on previous studies on internationalisation (Ratten and Tajeddini 2017; Braga et al. 2017; Meneses et al. 2014; Kampouri et al. 2017; Rafiki et al. 2018; Kontinen and Ojala 2012; Davidkov and Yordanova 2016).

17.2.2 Family Business

A family business is a company whose majority shareholders and management are controlled by family members; it is hoped that the family's descendants will follow in their footsteps later as managers. Meanwhile, a company is called a family business if it consists of two or more family members who supervise corporate finance. An organisation is classified as a family business if at least two generations are involved in the business, and they influence company policy.

Because there is no clear consensus among scholars and practitioners on what the term "family business" signifies, the literature has utilised a range of definitions. Despite the absence of a universally accepted definition of a family company, various studies have indicated that family businesses differ from an otherwise equivalent organisation on various levels depending on the critical role the owning family plays in the business operations. Ownership, family, and business are a system made up of three overlapping but separate subsystems that grow independently at different times but may influence one another.

17.2.3 Characteristics of Family Businesses

In that they are both set up to create profits and riches, family companies and non-family enterprises have a lot in common. Growth is a crucial strategic choice that family companies and non-family enterprises make. Furthermore, family companies significantly influence economic growth via the creation of jobs, productivity, and innovation. Apart from the features listed above, family firms are distinguished by several other characteristics that position them strategically ahead of non-family enterprises as follows:

a. **Longer-term thinking and a broader perspective**

Many family companies were built to serve as a last option for any family member who failed to make it outside of the firm. They have a long-term strategy to pass the business down through the generations, ensuring that no one in the family goes hungry. They are typically not under the same demands as public corporations to deliver returns to shareholders and investors.

b. **An entrepreneurial mind-set**

It is claimed that family companies are more enterprising than other sectors of the economy. According to the PWC family business study, 63% of respondents said family firms were more entrepreneurial than other sectors of the economy. They went on to say that with each new generation, family companies are able to reinvent themselves (Graves and Thomas 2008).

c. **Quicker and more flexible decision-making**

Family companies are frequently connected with the family that runs them, and the loyalty that comes with being the family's leader is brought into the workplace every day. This devotion extends to areas where significant company choices must be made; all employees turn to the steward for guidance, which is rarely questioned once granted.

d. **A greater commitment to jobs and the community**

One of the most appealing aspects of family companies is their commitment to the community by providing employment. They often despair when disciplining or reducing the salary of any of their employees because they sympathise with them. According to PWC's worldwide poll, 77% of respondents felt that family businesses have a greater sense of duty to promote employment than non-family businesses.

e. **A more personal approach to business based on trust**

The trust element is one of the key benefits of family companies over non-family firms. Because it is a family-owned business, the affection and desire to defend one another created at home is brought to the firm. As a result, family firms have higher trust among their employees and managers than non-family enterprises. According to recent research, 74% of family company owners in the Middle East feel that family firms have better culture and values than non-family enterprises. Because family companies sometimes identify their commercial failures with their family name, they work hard to ensure that their clients positively perceive them at all times (Akrout and Diallo 2017).

17.2.4 Family Business Critical Success Factors

Like every other organisation or endeavour, there are critical success factors that help family businesses to prosper if adhered to. Failure to adopt some or all of these critical success factors often leads to the failures or lack of continuity of the family business beyond its founder or its first-generation owners. Some of these factors include (Memili et al. 2010):

a. **Strong leadership**

One of the most important concepts in organisational behaviour is effective leadership; without it, goals are often elusive, no matter how well planned. A family business is not an exception to this rule. The need for strong leadership from the generation that controls the company is a key success factor.

b. **Alignment of successors and incumbents**

The steward must ensure that both the founding and successor generations agree on the business direction. This should include the incumbent's desire to stand down, the successor's willingness to take over, family members' agreement to keep the company in the family, individual roles accepted, and succession planning (Sharma et al. 1997).

c. **Education of the entrepreneur**

The entrepreneur's educational level significantly influences the family business's performance and sense of success. However, it has been suggested that the founder's appraisal of the firm as a success is the most essential component in continuity, since successful enterprises continue, are sold, or are given after the owner-manager departs. Whatever managers' perception, in this era, education is a critical part of achieving success in any organisational form, family or otherwise.

d. **Family cohesion**

At the beginning of every family business, the single idea is to preserve the business inside the family, keep it running in the original territory, and preserve family harmony. In order to achieve the idea, there is a need for cohesion among all the various stakeholders within the family. Cohesion is vital for collective action because it helps a group deal with risk and uncertainty, both common in global growth. The steward needs to maintain harmony through effective leadership and problem-solving.

e. **Trust issues**

During the early phases of a family business's growth, the trust inherent in most family connections helps the company significantly minimise transaction costs. The need for this monitoring results from an absence of trust among business stakeholders. Thus, trust may be an invaluable asset in global growth.

f. **Acceptance of technology**

Any family business that wishes to continue its existence would have to acknowledge the inevitability of technological disruption to its traditional processes. This is especially true in family firms engaged in manufacturing who have to grapple with competition from mass-produced goods, which often turn out to be cheaper than hand-made or traditionally prepared ones. One of the most pressing concerns for family companies is the loss of their socio-emotional worth; to prevent such losses, they are ready to take a larger risk to their performance, but they reject hazardous business actions that might aggravate existing problems.

g. **Strategic planning**

In order to survive and continue to provide a source of employment and wealth for the family, family business owners must plan strategically with regards to business expansions, diversification, and de-escalation of commitment. Making strategic decisions requires the capacity to discern external influences, appraise resources, and devise novel techniques to harness skills for long-term competitive advantage. It has also been discovered that strategic planning in small family businesses includes an 'entrepreneurial spirit' that directly impacts their engagement in entrepreneurial conduct. Through adequate strategy, family businesses have remained long after the demise of their founders.

h. **Willingness to change**

As they say, change is the only thing constant in life, hence every family business must anticipate change to the status quo sooner or later and make adequate preparations to cope with it. Some family businesses have taken advantage of these environmentally induced changes to innovate some very good strategies globally, such as the franchise system used by the Kentucky Fried Chicken or Tesco to create a presence in countries thousands of kilometres away. These franchises provide a steady income stream for the families even when local competition stifles profitability. It has also been said that family businesses are as distinctive and eccentric as the families that influence them and that the most successful ones defy standard management approaches.

i. **Limiting exposure**

Every business often tries as much as possible to protect itself through efficient risk identification and management and the continuous scanning of the business environment for emerging threats. But the founders of family businesses take it a notch in the way they protect their businesses because the lines between the business and family have been blurred by the filial relationship between the management and employees. The stewardship of family business owners and managers is based on their socio-emotional commitment to the firm, which may be quite strong since the company can provide stability, social contribution, belonging, and family status. Many founder-managers of family businesses see their business failure as synonymous with family failure. This has made family businesses avoid many risk-taking ventures, such as expansions, bank debts, diversifications, and internationalisations. Family-business owners tend to be averse to debt and the possibility of large losses in case of loan failures, preferring sources for additional funding from family members and relatives or through reinvestments of profits.

j. **Good succession management**

One of the most important issues that can make or break a family business is achieving a seamless succession process from one generation to another or from one steward to the other. While conflicts are unavoidable during these succession bids, it has been argued that a good succession is not the one that avoids conflicts but the one that successfully manages them.

Furthermore, family enterprises are frequently formed for the sole purpose of providing a source of income for the founder's immediate family and, to a significant degree, other close members of the immediate community. However, as mentioned previously, the combination of various factors has fuelled the growing urge to internationalise family enterprises (see Table 17.1). These factors (which

Table 17.1 A summary of the major factors influencing family business internationalisation

No	Factor	Author/year	Country
1	Commitment to internationalise	Graves and Thomas (2008)	Australia
2	Nature of target location	Fenhaber et al.	U.S.A
3	Background of manager	Carpenter and Fredrickson (2001)	U.S.A
4	Inter-firm relationship	Dana (2001)	New Zealand
5	Fear of loss of control	Romano (2000)	Australia, New Zealand
6	Degree of external ownership	George et al. (2005), Bell et al. (2004)	Sweden, UK
7	Management team international experience	Reuber and Fischer (1997), Athanassiou and Nigh (2002)	Canada, U.S.A
8	Network relationship	Zahra et al. (2000)	U.S.A
9	Company size	Bloodgood et al. (1996)	U.S.A
10	Information on the foreign market	Welch and Wiedersheinm (1980)	Australia
11	Human capital availability	Manolova et al. (2002), Oviatt and McDougall (1995)	U.S.A
12	Financial resource availability	Graves and Thomas (2008), Bloodgood et al. (1996)	Australia, U.S.A

Source By author himself

the author searched and extracted from articles of various journals) that drive internationalisation have been explored in other regions outside of Saudi Arabia.

Table 17.1 above lists the factors considered and selected to be the variables of this study, namely financial resources, human capital (education, experience, training), networking, company size, manager experience/background, and information, which are related to knowledge or education. These factors are justified based on the respective theories and previous studies related to the internationalisation process.

17.2.5 Family Business and Internationalisation

Familial has been used to describe the traits of a family company. The term refers to the causal linkages between a company's resources and talents and the owner's family. Familial is defined as "the unique bundle of resources a given business holds" due to the interaction system between the family, its members, and the corporation. Familial may lead to attributes difficult to duplicate, and it can help family companies survive and grow in tough economic times.

Family-owned businesses are more likely than non-family businesses to exhibit stewardship attitudes for the company's long-term success (including both staff

Table 17.2 Number of documents by country/territory

No	Year/number document	Country/number of documents
1	2023 (9)	Italy (66)
2	2022 (51)	Spain (65)
3	2021 (34)	United States (60)
4	2020 (38)	United Kingdom (37)
5	2019 (36)	Australia (27)
6	2018 (32)	India (25)
7	2017 (34)	France (23)
8	2016 (34)	Portugal (23)
9	2015 (19)	China (20)
10	2014 (23)	Germany (20)

and consumers). If the family business is to survive for future generations, stewardship attitudes must be adopted. Simultaneously, there is a risk of management entrenchment, resulting in the company's extinction (Miller et al. 2016).

Managers' commitment and devotion towards the company is the subject of stewardship theory. It is founded on the premise that, in the position of steward, the manager feels a great sense of responsibility for the company and sets a higher value on collective well-being than individual well-being to boost organisational performance (Miller et al. 2016).

Based on some findings from techniques in bibliometric analysis, the research on family business and internationalization has been done by many researchers in various countries. There are 408 documents sourced from reputable database called Scopus by searching through keywords; "family business", "internationalization". As depicted in Table 17.2, the highest hit on the number of documents is in the year 2022 with a total number of 51 documents. It is also shown that the number of documents increase incrementally since 2004, which consider that the interest of studies on this topic is perceived positively.

As depicted in Table, Italy leads the top ten list of dominant countries that published quality articles with a total number of 66 documents, followed by Spain and United States with a total number of 65 and 60 documents respectively. The interests of researchers on family business and internationalization in developed countries is obviously superior (see the number of documents) compared to developing countries.

17.3 Method

This study use a content analysis by reviewing past research studies, which then expected to come out with a proposed framework of family business and innovation on the internationalisation process.

17.4 Discussion

Many researchers agreed that family involvement in companies makes family businesses different from non-family firms (Miller et al. 2016). A single-family member controls a family business, particularly in making important business decisions. Some researchers interpret family involvement in ownership and management. The family's existence factor at the time of succession from within family members.

Family involvement in management has been seen as a red flag in the globalisation of family businesses (Kontinen and Ojala 2012). According to George et al. (2005), internal owners appear to be risk-averse, resulting in a reduction in the scale and scope of internationalisation; on the other hand, external owners—such as institutional owners and venture capitalists—significantly increase the scale and scope of internationalisation.

Although family businesses are frequently characterised as slow and risk-averse in internationalisation, they may do so swiftly under specific conditions, such as during a generational change (Graves and Thomas 2008). A correlation between poor managerial skills, a resistance to absorb outside expertise, and a lack of bridging network ties may also exist (Graves and Thomas 2008).

The process through which businesses participate in worldwide markets is referred to as internationalisation (Rexhepi et al. 2017). The owners of family companies are constantly seeking new models and tactics to help them successfully internationalise their enterprises. Many studies have demonstrated that businesses must employ different techniques and strategies to access global markets than the target home markets (Ratten et al. 2017). Internationally, family businesses operate under substantially different norms and regulations than local markets. Every difference that a family business owner could come upon must be properly investigated. They must consider factors, such as general strategy, human resource strategy, consumer behaviour, local laws and regulations, policies, location, local and global brand recognition, and so on. The entrance models and strategies they adopt will affect the success of their business (Rexhepi et al. 2017).

Across the globe, family companies are fighting for survival. Technology and globalisation have made it possible for rivals and international enterprises to compete. In addition, if loved ones fail to win the survival war, they may lose their sources of income, money, security, livelihood, and socio-emotional equilibrium. This is especially essential in the Middle East, where family firms account for about 80% of all enterprises.

Two factors that aid in the internationalisation of family enterprises are a long-term outlook and the ability to make quick decisions. Furthermore, it has been observed that family businesses with a drive to utilise information technology, creative skills, commitment towards internationalisation, and the capacity to share power and use existing resources, are more likely to succeed in international expansion (Kontinen and Ojala 2012). In general, the advent of new generations has been seen as good for internationalisation, but the generational transition has sometimes had little effect or even proven harmful to internationalisation (Graves and Thomas 2008).

In order to bring about change, innovators must stray from traditional business practices. The family engagement has a positive influence on the creation of new goods and initiatives. Due to its role in competitiveness and globalisation, innovation is critical for the success of family businesses. There are two types of innovation: gradual and radical. Smaller modifications that grow over time are referred to as incremental innovation. Radical innovation refers to groundbreaking and novel market developments. Due to their conservative character, family businesses are more engaged with incremental improvements. When it comes to implementing innovation into their internationalisation process, family businesses may adopt either exploration or exploitation strategies. It is a complex interplay when firm innovation and internationalisation relate to firm performance (Freixanet and Federo 2022).

17.5 Conclusion

Expansion into foreign markets gives a significant potential for development and value generation, but it also introduces distinct hurdles not seen in home markets (Lu and Beamish 2001). As a result, researchers working on internationalisation issues have attempted to outline the primary catalyst and obstacles to internationalisation.

References

Akrout H, Diallo MF (2017) Fundamental transformations of trust and its drivers: a multi-stage approach of business-to-business relationships. Ind Mark Manag 66:159–171

Adomako S, Opoku RA, Frimpong K (2017) The moderating influence of competitive intensity on the relationship between CEOs' regulatory foci and internationalisation. J Int Manag 23(3):268–278

Athanassiou N, Nigh D (2002) The impact of the top management team's international business experience on the firm's internationalization: social networks at work. MIR: Manag Int Rev 157–181

Bell J, Crick D, Young S (2004) Small firm internationalization and business strategy: an exploratory study of 'knowledge-intensive' and 'traditional' manufacturing firms in the UK. Int Small Bus J 22(1):23–56

Bilkey W (1978) An attempted integration of the literature on the export behaviour of firms. J Int Bus Stud 9(1):33–46

Bilkey WJ (1985) Development of export marketing guidelines. Int Mark Rev 2(1):31–40. https://doi.org/10.1108/eb008268

Bloodgood JM, Sapienza HJ, Almeida JG (1996) The internationalization of new high-potential US ventures: antecedents and outcomes. Entrep Theory Pract 20(4):61–76.

Braga V, Correia A, Braga A, Lemos S (2017) The innovation and internationalisation processes of family businesses. Rev Int Bus Strategy 27(2):231–247

Buckley PJ, Pass CL, Prescott K (1992) The internationalisation of service firms: a comparison with the manufacturing sector. Scand Int Bus Rev 1(1):39–56

Calof JL, Beamish PW (1995) Adapting to foreign markets: explaining internationalisation. Int Bus Rev 4(2):115–131
Carpenter MA, Fredrickson JW (2001) Top management teams, global strategic posture, and the moderating role of uncertainty. Acad Manage J 44(3):533–545
Cavusgil ST (1984) Differences among exporting firms based on their degree of internationalisation. J Bus Res 12(2):195–208. https://doi.org/10.1016/0148-2963(84)90006-7
Dana LP (2001) Networks, internationalization & policy. Small Bus Econ 16:57–62
Davidkov TP, Yordanova DI (2016) Exploring the Bulgarian family organisations' reluctance to internationalize. Int J Bus Glob 17(1):123–148
Freixanet J, Federo R (2022) The complex interplay of firm innovation, internationalization and learning capability in driving firm performance: a configurational analysis. J Strateg Manag 15(4):766–790. https://doi.org/10.1108/JSMA-11-2021-0220
Fletcher D (2004) International entrepreneurship and the small business. Entrep Reg Dev 16:289–305
George G, Wiklund J, Zahra AS (2005) Ownership and the internationalisation of small firms. J Manag 31(2):210–233. https://doi.org/10.1177/0149206304271760
Graves C, Thomas J (2008) Determinants of the internationalisation pathways of family firms: an examination of family influence. Fam Bus Rev 21(2):151–167
Hazarbassanova DB (2016) The value creation logic and the internationalisation of internet firms. Rev Int Bus Strategy 26(3):349–370
Hsu WT, Chen HL, Cheng CY (2013) Internationalisation and firm performance of organisations: the moderating effects of CEO attributes. J World Bus 48(1):1–12
Jabar FHA, Tajuddin N, Paino H (2016) Internationalisation of small and medium enterprises. In: Proceedings of the ASEAN entrepreneurship conference 2014. Springer, Singapore, pp 151–157
Johanson J, Mattsson LG (2015) Internationalisation in industrial systems—a network approach. In Knowledge, networks and power. Palgrave Macmillan, London, pp 111–132
Kampouri K, Plakoyiannaki E, Leppäaho T (2017) Family business internationalisation and networks: emerging pathways. J Bus Ind Market 32(3):357–370
Karagozoglu N, Lindell M (1998) Internationalisation of small and medium-sized technology-based firms: an exploratory study. J Small Bus Manag 36:44–59
Katsikeas CS, Morgan RE (1994) Differences in perception of exporting problems based on firm size and export market experience. Eur J Mark 28(5):17–39. https://doi.org/10.1108/03090569410062014
Katsikeas CS, Piercy NF (1993) Long-term export stimuli and firm characteristics in a Europaen LDC. J Int Mark 1(3):23–47
Kontinen T, Ojala A (2012) Internationalisation pathways among family-owned organisations. Int Mark Rev 29(5):496–518
Leonidou LC (1995) Export barriers: non-exporters' perceptions. Int Mark Rev 2(1):4–25
Lu JW, Beamish PW (2001) The internationalisation and performance of organisations. Strateg Manag J 22(6–7):565–586. https://doi.org/10.1002/smj.184
Manolova TS, Brush CG, Edelman LF, Greene PG (2002) Internationalization of small firms: personal factors revisited. Int Small Bus J 20(1):9–31
Memili E, Eddleston KA, Kellermanns FW, Zellweger TM, Barnett T (2010) The critical path to family firm success through entrepreneurial risk taking and image. J Fam Bus Strat 1(4):200–209
Meneses R, Coutinho R, Carlos Pinho J (2014) The impact of succession on family business internationalisation: the successors' perspective. J Fam Bus Manag 4(1):24–45
Miller D, Steier L, Le Breton-Miller I (2016) What can scholars of entrepreneurship learn from sound family businesses? Entrep Theory Pract 40(3):445–455
Miesenbock K (1988) Small businesses and exporting: a literature review. Int Bus Stud 6(2):43–61
Mohamad ANA, Rashed ZN, Abd Rahman R (2008) Potensi, cabaran dan hala tuju industry kecil & sederhana (IKS) dalam era pembangunan negara. Prosiding Perkem III 1:662–674

Morgan NA, Whitler KA, Feng H, Chari S (2019) Research in marketing strategy. J Acad Mark Sci 47(1):4–29
O'Farrell P, Wood P, Sheng J (1998) Internationalisation by business service organisations: an inter-industry analysis. Int Small Bus J 16(2):13–33
Olejnik E, Swoboda B (2012) Organisations' internationalisation patterns: descriptives, dynamics and determinants. Int Market Rev 29(5):466–495
Onkelinx J, Manolova TS, Edelman LF (2016) The human factor: investments in employee human capital, productivity, and organisations internationalisation. J Int Manag 22(4):351–364
Oviatt BM, McDougall PP (1995) Global start-ups: entrepreneurs on a worldwide stage. Acad Manage Perspect 9(2):30–43
Ramon-Llorens MC, García-Meca E, Duréndez A (2017) Influence of CEO characteristics in family firms internationalization. Int Bus Rev 26(4):786–799
Rafiki A, Hidayat SE, Al-Mana A (2018) Factors affecting small and medium family businesses' internationalisation in Saudi Arabia. Int J Islam Market Branding 3(1):45–64
Ratten V, Dana LP, Ramadani V (2017) Internationalisation of family business groups in transition economies. Int J Entrep Small Bus 30(4):509–525
Ratten V, Tajeddini K (2017) Innovativeness in family firms: an internationalisation approach. Rev Int Bus Strategy 27(2):217–230
Reuber AR, Fischer E (1997) The influence of the management team's international experience on the internationalization behaviors of SMEs. J Int Bus Stud 28:807–825
Rexhepi G, Ramadani V, Rahdari A, Anggadwita G (2017) Models and strategies of family businesses internationalisation: a conceptual framework and future research directions. Rev Int Bus Strategy 27(2):248–260
Romano P (2000) Sustainable development: a strategy that reflects the effects of globalization on the international power structure. Hous J Int'l L 23:91
Sharma P, Chrisman JJ, Chua JH (1997) Strategic management of the family business: past research and future challenges. Fam Bus Rev 10(1):1–35
Singh G, Pathak RD, Naz R (2010) Issues faced by oganisations in the internationalisation process: results from Fiji and Samoa. Int J Emerg Mark 5(2):153–182
Sinkovics N, Sinkovics RR (2013) The internet as an alternative path to internationalisation? Int Mark Rev 30(2):130–155
Welch LS, Wiedersheim-Paul F (1980) Initial Exports – A Marketing Failure? J Manage Stud 17(3):333–344
Wright RW, Ricks DA (1994) Trends in international business research: twenty five years later. J Int Bus Study 42(3):294–306
Yiu DW, Lau CM, Bruton GD (2007) International venturing by emerging economy firms: the effects of firm capabilities, home country networks, and corporate entrepreneurship. J Int Bus Stud 38(4):519–540
Zahra SA, Ireland RD, Hitt MA (2000) International expansion by new venture firms: international diversity, mode of makret entry, technological learning, and performance. Acad Manage J 43(5):925–950
Zhang X, Ma X, Wang Y, Li X, Huo D (2016) What drives the internationalisation of Chinese organisations? The joint effects of international entrepreneurship characteristics, network ties, and firm ownership. Int Bus Rev 25(2):522–534

Shariah Governance Framework for Islamic Donation-Based Crowdfunding Platforms in Malaysia

18

Nurul Aini Muhamed, Muhammad Iqmal Hisham Kamaruddin, Wan Nur Fazni Wan Mohamad Nazarie, Aimi Fadzirul Kamarubahrin, and Rafisah Mat Radzi

Abstract

This study attempts to examine current *Shariah* governance practices and propose a *Shariah* governance framework for Islamic donation-based crowdfunding (DCF) platforms in Malaysia. A series of interviews with six Islamic DCF platforms in Malaysia was conducted. For reporting purposes, this study employed a single-case study approach. The interview was conducted and completed at the end of August 2022 via an online approach through Google Meet. The findings highlight *Shariah* governance practices for Islamic DCF platforms covering *Shariah* governance structure, roles of the parties involved and types of *Shariah* contracts used in the operation. Based on these findings, a *Shariah* governance framework for Islamic DCF is proposed covering *Shariah* contracts, roles of parties, establishment of the *Shariah* committee as well as *Shariah* governance functions. This study is among the pioneer studies that explore *Shariah* governance framework for Islamic DCF platforms in Malaysia. The study findings and recommendations are useful for Islamic DCF platforms as well as related parties like regulators, funders, potential project managers and the public in general to understand *Shariah* governance framework for Islamic DCF platforms in Malaysia.

Keywords

Shariah governance • *Shariah* governance framework • Islamic donation-based crowdfunding (DCF) • Islamic finance

N. A. Muhamed · M. I. H. Kamaruddin (✉) · W. N. F. W. M. Nazarie · A. F. Kamarubahrin
Faculty of Economics and Muamalat, Universiti Sains Islam Malaysia, Nilai, Malaysia
e-mail: iqmalhisham@usim.edu.my

R. M. Radzi
School of Distance Studies, Universiti Sains Malaysia, Penang, Malaysia

A. Rafiki et al. (eds.), *Open Innovation in Small Business*, Contributions to Environmental Sciences & Innovative Business Technology,
https://doi.org/10.1007/978-981-99-5142-0_18

18.1 Introduction

Crowdfunding nowadays is accepted as one medium for funds collection. The development of Industrial Revolution 4.0 has encouraged the growth of technology-based services such as crowdfunding. This digital technology adoption enables the collections of funds from mass numbers of funders through the electronic platforms, and later to be distributed to beneficiaries (Hossain and Oparaocha 2017). Benefiting from digital technology, crowdfunding platforms can disclose information on funds collections, where stakeholders can view numbers of collection concurrently (Salido-Andres et al. 2021). This feature attracts the users who demand for disclosure and transparency of information.

Crowdfunding is established in different types of establishments to serve the different objectives of the stakeholders. In general, there are four main types of crowdfunding which are: (i) donation-based crowdfunding (DCF), (ii) reward-based crowdfunding (RCF), (iii) equity-based crowdfunding (ECF), and (iv) lending-based crowdfunding (LCF) (Hossain and Oparaocha 2017; Rijanto 2018). Under DCF, no financial returns given to funders, such as in the case of raising donations for a disaster relief. Meanwhile, under RCF, contributors contribute in exchange for a reward that comes in a form of product or service of the beneficiaries' company. While, under ECF, investors have ownerships rights on the shares of the business and become entitled for its financial returns in the form of dividends. Lastly, under LCF, contributors contribute to crowdfunding on the basis of providing loans and repayment is required (Dushnitsky et al. 2016).

Similar to other financial-related products or services, crowdfunding also can be divided into conventional and Islamic models. Unlike the conventional crowdfunding model, Islamic crowdfunding must adhere and comply with *Shariah* principles. This includes intentions of crowdfunding projects (*halal*) as well as being free from prohibited activities (*riba'*, *maysir*, *gharar* and other *haram* activities) (Achsien and Purnamasari 2016; Faudzi et al. 2021). Besides, Islamic DCF is normally involved with Islamic social finance funds like *zakat*, *waqf* and *sadaqah* in crowdfunding activities (Sulaeman 2020).

Since before crowdfunding become a phenomenon around the world as an alternative source of funding, the Muslim community has been practically conducting crowdfunding activities through Islamic social finance funds. In this case, before technological advancement, Islamic countries established *zakat* and *waqf* institutions as one the mechanisms for wealth creation and distribution specially to address socio-economic problems and well-being (Kamaruddin et al. 2018). These institutions were collecting *zakat*, *waqf* and other Islamic social finance funds from the public and distributing back to the Muslim society through various socio-economic activities and projects.

With technological advancement, the emergence of crowdfunding is not only for socio-economic purposes, but also used to raise capital to support innovations, entrepreneurial ideas, and ventures (Hossain and Oparaocha 2017). In this case, crowdfunding helps in providing seed capital, testing the unique ideas by the public as well as establish the engagement with consumers (Younkin and Kashkooli 2016).

Realising its promising growth, many countries, such as the US, UK, China, Singapore and Malaysia have taken steps to provide adjustments in their regulations to accommodate the operation of crowdfunding (Abdullah 2016). Malaysia is the first country in Southeast Asia that has put in place a regulated framework for crowdfunding (Wahjono et al. 2015). In 2015, the Securities Commission Malaysia issued the *Guideline on Recognized Markets* which became the main reference for the establishment of the registered ECF. In 2019, the SC revised its crowdfunding initiative by introducing the Malaysia Co-Investment Fund (MyCIF), which combines both ECF and LCF together. Currently, there are nine (9) ECF and ten (10) LCF platforms in Malaysia.

The situation is different from DCF. As has been criticised, there is no specific regulation that is developed to control and monitor donation-based crowdfunding in Malaysia. The only applicable law that can be applied for DCF is the Trustee Act 1949. This wide-ranging act also guide charity-based organisations and other non-profit organisations. Due to this limitation, this study is called to examine current *Shariah* governance practices and propose a *Shariah* governance framework for Islamic DCF especially in Malaysian context.

After this section, this study has focused on the background of DFC, Islamic DCF and its *Shariah* governance practices in Malaysia. Further, this study continues with the explanation on the methodology employed. Next, presentation of the research findings, discussion and proposal of *Shariah* governance framework for Islamic DCF in Malaysia are being covered. Finally, a summary, recommendations and implications of the findings are highlighted in the last section. This study provides novelty contribution by proposing *Shariah* governance framework for Islamic DCF.

18.2 Literature Review

18.2.1 Background of DCF

Crowdfunding uses online platforms as intermediaries to raise funds from funders and distribute them to people in need. This approach gains benefits from online information and connections, where people are connecting through the internet. Crowdfunding has advantage in tapping the collective efforts of a large pool of individuals—primarily online via social media and crowdfunding platforms—and leverages their networks for greater reach and exposure' (Hossain and Oparaocha 2017). Crowdfunding concept was emerged in 2006, starting from the business-based crowdfunding and later was expanded to DCF and ECF (Piliyanti 2019).

Differently from other types of crowdfunding, DCF is based on intrinsic motivation (Chen et al. 2021) and concepts of charity and almsgiving without asking for returns. In general, there are three parties involved in every type of crowdfunding which are: (i) platforms, (ii) donors/funders, and (iii) beneficiaries/project managers. Platforms as intermediaries will channel funds contributed by the funders to the beneficiaries/project managers. Return is not becoming push for funders' involvement, thus there is no obligation for the beneficiaries for paying back. DCF can be formed in two ways either as commercial platforms or established as non-profit organisations wing (Fig. 18.1).

DCF is significant as it can help those from underserved and unbankable groups of community to receive funding and use them for their needs, especially during misfortune events (Figueroa-Armijos and Berns 2021). These include natural disasters such as flood, earthquake or due to other needs such as getting support for education and medical purposes.

Meanwhile, beneficiaries of Islamic DCF is much more extensive in comparison with other restricted funds, such as *zakat* and *waqf* funds. While people under the poverty line can get *zakat* and government assistance from *zakat* institutions (in the case of *asnaf*), *waqf* and Department of Social Welfare (such as B40 and people with disabilities), those people who are not within this circle will possibly be left out from getting the financial support, especially during the pandemic. These include category of citizens at T20 who may be impacted by the misfortunes during the pandemic, such as losing their jobs or impacted by the death of family members. The T20 group may move to M20, or B40 during pandemic. As such, a more flexible way to get unrestricted financial assistance is required.

Crowdfunding thus not only benefits the beneficiaries, but also reduces the reliance on the government assistance. DCF in Malaysia faces with some challenges due to no clear direction on its establishment and controlling. As has been criticised by Abdullah (2016) there is no established framework available to guide the funders as well as platform operations until now. The situation is different in other countries such as United States, where all fund raisers must register with the state prior to the fund-raising activities (Othman et al. 2021).

There are number of DCFs which have been established in Malaysia, and these numbers are continuously increasing. The DCF market was at USD1.68 million in 2016 (Statista 2022). At the same time, there are no specific bodies that are responsible to monitor these platforms. Without regulatory direction of Islamic DCF, this situation may open the services to the possibility of agency conflicts.

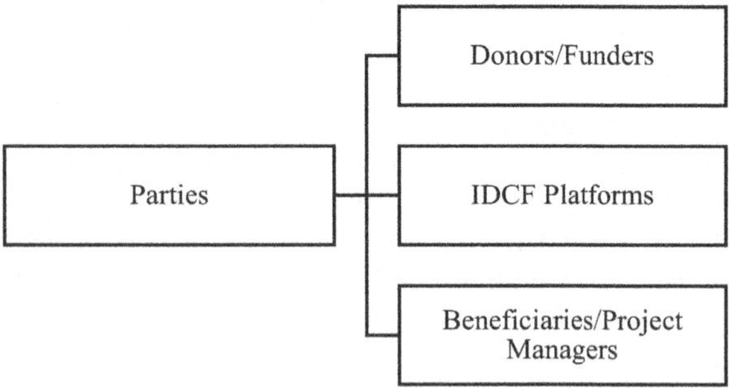

Fig. 18.1 Parties Involved in DCF. *Source* Author's proposed model

Lack of regulations and guidelines on DCF in Malaysia can lead to the disclosure issues (Hong 2018), agency conflicts and transparency issues including on fee structures and the possibility of misconduct.

As most studies have been conducted on equity crowdfunding (Wahjono et al. 2015), not many studies have been focused on the donation-based crowdfunding. There are few studies that attempt to propose the *Shariah* compliance crowdfunding model (Abdullah and Oseni 2017; Wahjono et al. 2015). However, most of these studies focus on ECF by looking at the requirement for *Shariah* screening on investment activities of the beneficiaries.

18.2.2 Theoretical Aspects of DCF

DCF although is based on the concept of altruism and philanthropy has changed the traditional methods of getting financial assistance from funders to fulfil the needs of beneficiaries (Othman et al. 2021). In general, Islamic DCF should be based on Islamic principles of the prohibitions of *riba'*, *gharar*, *maysir*, and other prohibited goods (Achsien and Purnamasari 2016; Faudzi et al. 2021).

In general, all Islamic DCF structures basically involve three main parties; mainly the donors/funders, Islamic DCF platforms, and beneficiaries/project managers. There are some differences between the beneficiary and project manager. Beneficiaries consist of individuals/organisation who apply the funds through the Islamic DCF platforms for their own needs. While, project managers refer to those who apply the funds on behalf of the beneficiaries as end users.

The establishment of Islamic DCF is undisputedly consistent with *Maqasid Shariah* (objective of Islamic law) that aims to preserve and ensure the *maslahah* (goodness) and eliminate the *mafsadah* (harm) in the society. There are three main domains of the *maqasid* to achieve those objectives; (i) *dharuriyyat* (necessities), (ii) *hajiyyat* (wants) and (iii) *tahsinniyyat* (embellishment). The goals of eliminating hunger and poverty (such as Social Development Goals: Goals 1, 2 and 3) are located under the domain of *dharuriyyah*, in which to ensure that all communities can fulfil their needs. *Maqasid Shariah* should be used as the main foundation for product offering (Dusuki and Bouheraoua 2011; Ghazanfar and Islahi 1997). Islamic DCF supports the needs of the needy with no compensation or return from those who are in needs.

Understanding and proposing the possible contracts that are reflected in the relationship between the involved parties, and the responsibility of each of them are the main aspects that should be emphasised in Islamic financial transactions. Nobody should be forced to enter the contracts. Once the contracts are binding, all parties are bound to the terms and condition of those contracts. This will ensure that the elements of fairness are upholding in the contracts. Islam promotes the concepts of justice and the preservation of rights of the participants in any business transactions. Islam promotes the principles of risk sharing, individuals' rights and duties, property right, and sanctity of business contracts (Iqbal and Mirakhor 2011). In this case, four pillars of the contracts (parties, subject matter, offer and acceptance and *majlis aqad*) should be followed in order to ensure that rights of each party are protected.

Parties should enter the business transaction voluntarily, and as such elements of force are discouraged. These aspects should guide the relationship of the platforms, funders and beneficiaries. Lutfi and Ismail (2016) propose the *wakalah* and *qard* contracts in health funds crowdfunding platforms. The issue is whether the process of delivery of the services is following the principles intended.

Governance is the best mechanism to control the entities. *Shariah* governance is a mechanism of corporate governance to ensure the compliance of business entities with *Shariah*. *Shariah* governance thus is needed to support the existing corporate governance model adopted in Islamic-based entities. *Shariah* governance is a set of the institutional and organisational arrangements through which Islamic-based entities including Islamic DCF ensure that there is an effective independent oversight of *Shariah* compliance over the issuance of relevant *Shariah* pronouncements, dissemination of information and an internal *Shariah* compliance review (IFSB 2009).

The internal governance of the crowdfunding platform is very important as assurance towards crowdfunding compliance with *Shariah* principles. As there is no guideline issued by the government on this matter, thus no standard governance structures exist in these platforms. In the event of no strict regulations evolved and non-standard practices, the roles of standard governance and *Shariah* governance are important for Islamic-based entities including Islamic DCF.

18.3 Methodology

This study aims to examine current *Shariah* governance practices and proposes a *Shariah* governance framework for Islamic donation-based crowdfunding (DCF) platforms in Malaysia. This include the status of the establishment, contract used and controlling mechanism related to *Shariah* aspects. Data were collected through a series of interview sessions with six Islamic DCF platforms in Malaysia. These six selected Islamic DCF platforms were chosen based on their similar characteristics which are:

1. Legally registered either as a company under the Companies Commission of Malaysia or as a non-profit organisation under the Registry of Society;
2. Located in Malaysia and operated by Muslim;
3. Adopting the DCF model for its crowdfunding activities;
4. Involve with Islamic social finance funds like *zakat*, *waqf*, *sadaqah*, *hibah* and *infaq*;
5. Claim as Shariah compliant DCF (free from prohibited activities like *riba'*, *maysir*, *gharar* and other *haram* activities).

This study employs the case study method in order to explore *Shariah* governance practices by six selected Islamic DCF platforms. This is because the case study method is able to explore a specific system practiced in an organisation (Yin 2018). Specifically, this study adopted an exploratory case study approach for data collection process and a single-case study approach in reporting the findings.

For data collection process, a series of interviews focused on *Shariah* governance practices were conducted to the representative of each selected Islamic DCF platforms and lasted around 1–2 h for each session. The series of interview sessions were conducted and completed at the end of August 2022 via an online approach through Google Meet. All the findings were analysed and combined into a single case study writing.

18.4 Findings

In general, the sample of the Islamic DCF is taken due to the platform assertion on '*Shariah*-based' in their websites or during the interview. The platform relatively has a common basic structure. The main parties are the beneficiaries/project managers, platforms and donors/funders (Fig. 18.2).

In this case, the beneficiaries/project managers will choose the platforms, enter into the contracts with platforms with certain terms and conditions. The platform will do the due diligence process prior the campaign listing in their websites and promotion activities to reach the donors. The donors who are interested to contribute will get some information on the payments, charges and the current collections through the websites of the platforms. Contributions can be made directly through the payment system.

The platforms offer varies area of involvement including humanitarian needs, reducing poverty programs, natural disaster aids, medical and health aids and many others. This financial assistance according to the specific target is consistent with the *dharuriyyah* dimension under *Maqasid Shariah.*

Findings on legal establishment of the platforms are varying, which support the previous studies such as Othman et al. (2021) and Abdullah (2016). From these six platforms, one platform registered as a foundation under Department of Law (BHEUU), Prime Minister's Department of Malaysia; one platform registered as an enterprise (and later as private limited company) under the Companies Commission of Malaysia (CCM); one platform registered as social club under the Registry of Society (ROS). Lastly, there is one platform established as university centre,

Note: *the sequence of the activity started from 1 to 3; fund flow involved between activity 2 and 3.*

Fig. 18.2 Flow of Fund by Islamic DCIF. *Source* Author's research output

where the university itself was registered under the Universities and Universities Colleges Act 1971).

Interviews also found that there are no regular financial reports that are required to be submitted to the regulators under which they were registered. In this case, the submission of the reports is depending on the platforms' voluntary willingness, rather than due to regulatory requirements. As such, there is some room for misconducts, especially if there is no proper internal governance established in the platforms.

While the assertion of the *Shariah* compliance is maintained in the websites of the platforms, Ttere are inconsistencies in the practices of the platforms. Majority of the platforms did not declare the contracts used in the platforms as well as during the interview process. One of the respondents highlighted the platform uses the *wakalah* contracts, however that contract is not mentioned in any terms and conditions of the documents. From these six platforms, two of them highlighted that they have some *Shariah* governance practices. One of the platforms has the *Shariah* advisors and another one appoints a *Shariah* consultant as reference in the events that the management is not sure on certain practices related to *Shariah.* In terms of the *Shariah* governance functions, the majority of the platforms do not have specific functions for *Shariah.* There is no *Shariah* risk, *Shariah* audit, *Shariah* review and compliance process conducted by the platforms.

18.5 A Proposed *Shariah* Governance Framework for Islamic DCF

Islamic DCF collects the funds from many charitable sources such as *waqf*, *zakat*, *sadaqah* and *hibah.* Fees will be charged either from the funders or collection of funds. While the usage of *sadaqah* and *hibah* may have tolerable practices, other types of sources such as *zakat* and *waqf* have some unique characteristics and thus are very restricted in terms of usage. This study puts some stance on *zakat* and *waqf* contributions. *Zakat* as obligatory *infaq* has some portion for *amil* due to collection services. As such, the fee from *zakat* cannot be taken except from the *amil* portion. In relation to *waqf* contributions, the fee should be taken only from the benefits (*manfaah*). In this case, there should be separate dedicated fee paid to Islamic DCF on their services, not from the same funds. This will preserve the original intention of the funders' contributions.

This study underlines several suggestions to enhance the controlling process of the platform's activities, specifically related to *Shariah* compliance matters. The proposal takes into consideration several aspects such as types of contracts used, roles of the parties involved, and *Shariah* governance structure and roles. This study proposes *wakalah* contract (agency contract) as a suitable underlying contract between the involved parties. *Wakalah* refers to a contract in which a party, as principal (*muwakkil*) authorizes another party as his agent (*wakil*) to perform a particular task in matters that may be delegated, with or without imposition of a fee (BNM 2015).

In specific, the platform involves with dual agency contracts (dual *wakalah* contracts); the first with the founders and the second with the funders. The *wakalah bi ujrah* is proposed on whom the fees are charged by the platform whether the donors of the beneficiaries/project managers. While there is some room for the discussion on this approach, some conditions perhaps should be followed (Fig. 18.3).

In the case of *wakalah,* the contract is adopted by the platforms, where there are two *wakalah* that should be entered by the platform. The first *wakalah* contract is between the platform and beneficiaries/project manager. While the second *wakalah* contract is between the platform and the donors/funders. The platform thus is proposed to establish the *Shariah* governance in its entities. These include the establishment of the *Shariah* Committee and other controlling functions since the platform become the *wakil* (agent) for the *muwakkil* (principal).

There are two main reasons on the need for proper *Shariah* governance. The first reasons are related to the roles of DICF as agents to the beneficiaries/project manager, and the second reason due to the non-existing strict regulatory framework on the Islamic DCF platform. Both of these issues entailed the need for efficient control. *Shariah* committee roles is important to reach efficient *Shariah* decision making and ensure the compliance. The same proposal on Shariah governance establishment has been submitted by Ramli and Ishak (2022), however, on the ECF platform.

Looking at the current practices, only two platforms under observation have been established by the *Shariah* committee, however with no dedicated control functions. The establishment of the controlling functions should include establishing the *Shariah* risk, audit and review (BNM 2019), which are summarized as follows:

- *Shariah* risk functions refer to identifying, assessing, measuring and monitoring risks in the entity. This includes the *Shariah* risk as one type of operational risks in the entity.

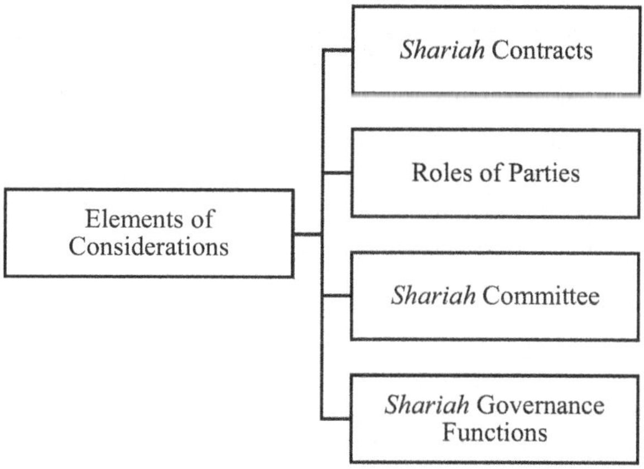

Fig. 18.3 Elements of consideration for Islamic DCF Shariah governance framework. *Source* Author's proposed framework

- *Shariah* review functions refer to continuous assessment on the compliance of entities in relation to their business activities.
- *Shariah* audit refers to independent assessment on the quality and effectiveness of the entities' internal control, risk management systems, governance processes as well as the overall compliance of the Islamic DCFs operations, business, affairs and activities with *Shariah*.

18.6 Conclusion

The sustainability of the crowdfunding platforms can help the sustainable services to the beneficiaries. Trust of beneficiaries can be attained if the Islamic DCF activities are well controlled and enforced. This study argues that that internal controlling functions are important especially when there are no proper regulations developed by the government. While the study highlights the importance of the law and regulation to control the crowdfunding activities, and ensure the transparency, this study also assumed that the *Shariah* governance is important to ensure the entities to be in line with *Shariah* principles.

The aim of this paper is to examine current *Shariah* governance practices and propose a *Shariah* governance framework for Islamic donation-based crowdfunding (DCF) platforms in Malaysia. This includes on the status of the establishment, contract used and controlling mechanism related to *Shariah* aspects. However, the findings show that there is no standard controlling approach that has been practiced by the Islamic DCF. Therefore, this study suggests the structured *Shariah* governance and proper internal control functions which should be adhered by Islamic DCF. This including the existing of *Shariah* contracts, roles of parties, establishment of the *Shariah* committee as well as *Shariah* governance functions.

Based on the findings, several implications and recommendations are proposed. Apart from the issuance of the *Shariah* governance framework for Islamic DCFs by the authority, there is also a need to have a specific criterion in recognising Islamic DCFs. Besides, a proper *Shariah* screening methodology for crowdfunding projects is also needed to ensure crowdfunding projects are in line with *Shariah* principles.

However limitations of the study also need to be recognised, including the scope of the study which is limited to the Islamic DCF model. Hence, future studies are recommended to extend to other Islamic crowdfunding such as LCF and ECF under the SC, as well as RCF. In addition, it is also suggested to focus on other aspects of operations of Islamic DCF in order to examine the Islamic DCF compliance.

Acknowledgements This research was funded by the Ministry of Higher Education (MOHE) under the Fundamental Research Grant Scheme (FRGS) entitled *Developing Shariah Framework for Donation-Based Crowdfunding: Towards Safeguarding Rights of Funders and Beneficiaries* (FRGS/1/2021/SS01/USIM/02/7).

References

Abdullah A (2016) Crowdfunding as an emerging fundraising tool: with special reference to the Malaysian regulatory framework. Islam Civilis Renew 7(1):98–119

Abdullah S, Oseni UA (2017) Towards a Shari'ah compliant equity-based crowdfunding for the halal industry in Malaysia. Int J Bus Soc 18(S1):223–240

Achsien IH, Purnamasari DL (2016) Islamic crowd-funding as the next financial innovation in islamic finance: potential and anticipated regulation in Indonesia. Eur J Islam Financ 5:1–11. https://doi.org/10.13135/2421-2172/1771

BNM (2019) Shariah governance policy document. Bank Negara Malaysia (BNM), Kuala Lumpur

BNM (2015) Wakalah concept paper. Bank Negara Malaysia (BNM), Kuala Lumpur

Chen Y, Dai R, Wang L, Yang S, Li Y, Wei J (2021) Exploring Donor's intention in charitable crowdfunding: intrinsic and extrinsic motivations. Ind Manag Data Syst 121(7):1664–1683. https://doi.org/10.1108/IMDS-11-2020-0631

Dushnitsky G, Guerini M, Piva E, Rossi-Lamastra C (2016) Crowdfunding in Europe: determinants of platform creation across countries. Calif Manag Rev 58(2):44–71. https://doi.org/10.1525/cmr.2016.58.2.44

Dusuki AW, Bouheraoua S (2011) The framework of Maqasid al-Shari'ah and its implication for Islamic finance. Islam Civilis Renew 2(2):316–336

Ghazanfar SM, Islahi AA (1997) Economic thought of Al-Ghazali. Islamic Economics Research Center, Jeddah

Faudzi MSM, Bakar JA, Ahmad S (2021) The relevance of conventional and Islamic crowdfunding as financing instruments for SMEs. J Islam Financ 10(1):59–64

Figueroa-Armijos M, Berns JP (2021) Vulnerable populations and individual social responsibility in prosocial crowdfunding: does the framing matter for female and rural entrepreneurs? J Bus Ethics 177:377–394. https://doi.org/10.1007/s10551-020-04712-0

Hong LC (2018) Crowdfunding: issues pertaining to financial reporting and assurance in Malaysia. J Wealth Manag Financ Plan 5:17–24

Hossain M, Oparaocha GO (2017) Crowdfunding: motives, definitions, typology and ethical challenges. Entrep Res J 7(2):20150045. https://doi.org/10.1515/erj-2015-0045

IFSB (2009) Guiding principles on Shariah governance systems. Islamic Financial Services Board (IFSB), Kuala Lumpur

Iqbal Z, Mirakhor A (2011) An introduction to Islamic finance: theory and practice, 2nd edn. Wiley, Singapore

Kamaruddin MIH, Masruki R, Hanefah MM (2018) Waqf management practices: case study in a Malaysian Waqf institution. World J Soc Sci 8(3):1–12

Lutfi MA, Ismail MA (2016) Sadaqah-based crowdfunding model for microfinancing and health care. J Muamalat Islam Financ Res 13(2):31–51

Othman R, Ali Z, Ilias II, Khalid A (2021) Overview of community-based crowdfunding in Malaysia: legal issues and the way forward. Int J Acad Res Bus Soc Sci 11(11):1213–1224. https://doi.org/10.6007/IJARBSS/v11-i11/11306

Piliyanti I (2019) Fintech achieving sustainable development: the side perspective of crowdfunding platform. *Shirkah* J Econ Bus 3(2):223–242. https://doi.org/10.22515/shirkah.v3i2.207

Ramli HS, Ishak MSI (2022) Equity-crowdfunding in Malaysia: the needs for a comprehensive Shariah governance. J Manag Theory Pract 3(1):48–55. https://doi.org/10.37231/jmtp.2022.3.1.204

Rijanto A (2018) Donation-based crowdfunding as corporate social responsibility activities and financing. J Gen Manag 43(2):79–88. https://doi.org/10.1177/0306307017748125

Salido-Andres N, Rey-Garcia M, Alvarez-Gonzalez LI, Vazquez-Casielles R (2021) Mapping the field of donation-based crowdfunding for charitable causes: systematic review and conceptual framework. VOLUNTAS: Int J Volunt Nonprofit Org 32(2):288–302. https://doi.org/10.1007/s11266-020-00213-w

Statista (2022) Market size of the donation-based crowdfunding industry in Malaysia from 2013 to 2017. https://www.statista.com/statistics/883470/malaysia-donation-based-crowdfunding-market-size/. Accessed on 6 March 2023

Sulaeman S (2020) A conceptual and empirical study on the development of the Islamic donation-based crowdfunding platform model for micro small and medium-sized enterprises (MSMEs) in times of Covid-19 Pandemic in Indonesia. Asian J Islam Manag 2(2):107–122. https://doi.org/10.1108/AJIM.vol2.iss2.art4

Wahjono SI, Marina A, Widayat W (2015) Islamic crowdfunding: alternative funding solution. In: 1st world Islamic social science congress 2015, 1–2 December, Putrajaya, Malaysia, pp 30–38

Yin RK (2018) Case study research and applications: design and methods, 6th ed. Sage Publications, Thousand Oaks, USA

Younkin P, Kashkooli K (2016) What problems does crowdfunding solve? Calif Manag Rev 58(2):20–43. https://doi.org/10.1525/cmr.2016.58.2.20

The manufacturer's authorised representative in the EU is Springer Nature Customer Service Centre GmbH, Europaplatz 3, 69115 Heidelberg, Germany. If you have any concerns regarding our products, please contact ProductSafety@springernature.com

Printed and bound by CPI Group (UK) Ltd, Croydon, CR0 4YY

30/09/2026

02225319-0009